Be Prepared

for the

Calculus Exam

Mark Howell
Gonzaga High School, Washington, D.C.

Martha Montgomery
Fremont City Schools, Fremont, Ohio

Practice exam contributors:

Benita Albert
Oak Ridge High School, Oak Ridge, Tennessee

Thomas Dick
Oregon State University

Joe Milliet
St. Mark's School of Texas, Dallas, Texas

Skylight Publishing
Andover, Massachusetts

Library of Congress Control Number: 2004112724

ISBN 978-0-9727055-5-4

Skylight Publishing
9 Bartlet Street, Suite 70
Andover, MA 01810

web: http://www.skylit.com
e-mail: sales@skylit.com
            support@skylit.com

3 4 5 6 7 8 9 10   12 11 10 09 08 07

Printed in the United States of America

To Maureen and Ryan
To Jack, Beth, and Kevin

Reviewers

Ray Cannon, Baylor University, served as Chief Reader for AP Calculus from 1992 to 1995, Chief Reader Designate in 1991, Exam Leader from 1998 to 1990, and Table Leader from 1982 to 1987. An AP Exam Reader since 1978, he has also served on the AP Calculus Test Development Committee and as a consultant for AP conferences.

Barbara A. Currier, Chairman of the Department of Mathematics at Greenhill School, Addison, Texas, has served as AP Calculus Exam Table Leader since 2001, AP Exam Reader since 1995, and a College Board Consultant in AP Calculus for the Southwestern Region. An AP Calculus teacher since 1980, she has presented at many conferences and summer institutes and is a recipient of the College Board's AP Special Recognition Award for the Southwestern Region.

Maria Litvin, Phillips Academy, Andover, Massachusetts, is the author of *Be Prepared for the AP Computer Science Exam in Java*. An AP Calculus and Computer Science teacher since 1988, she has served as AP Exam Question Leader and College Board consultant in AP Computer Science and is a recipient of the Siemens Award for Advanced Placement and RadioShack National Teacher Award.

Bernard L. Madison, University of Arkansas, served as a Reader, Table Leader, Exam Leader, and Chief Reader for AP Calculus for a total of 18 years. He was Chief Reader from 1996 to 1999 and served on the AP Calculus Test Development Committee from 1989 to 1995. From 1998 to 2001 he was a member of the Commission on the Future of AP. He has served since 2001 as a member of the College Board's Mathematical Sciences Academic Advisory Committee, which he currently chairs. He also serves on the College Board's Academic Assembly Council.

Steven Olson, Hingham High School and Northeastern University, served on the AP Calculus Test Development Committee from 1993 to 1996. An AP Exam Reader since 1980 and a Table Leader, he has taught AP Calculus since 1970.

Nancy W. Stephenson, William P. Clements High School, Sugar Land, Texas, served on the AP Calculus Test Development Committee from 1999 to 2003. An AP Exam reader and College Board Consultant in AP Calculus since 1996, she has taught numerous workshops and AP summer institutes.

JT Sutcliffe, who holds the Founders Master Teaching Chair at St. Mark's School of Texas in Dallas, served as AP Calculus Exam Leader from 1989 to 1992, AP Calculus Reader and Table Leader from 1977 to 1988, and on the AP Calculus Test Development Committee from 1979 to 1982. A recipient of the Presidential Award for Mathematics Teaching and Siemens and Tandy Technology Scholars awards, she has served as a College Board consultant and led many AP institutes since 1980.

Acknowledgements

The production of this work owes a debt of gratitude to many people. Above all, we wish to express our thanks to the legions of colleagues in the Advanced Placement Calculus Program. In many ways, this guide represents the total accumulation of contributions from them all, including fellow readers at the AP Calculus Exam Reading and the teachers in our own schools. Some of them deserve special mention. These include:

The Chairs of the Development Committee we have served with, whose guiding hands were a comfort and whose commitment to the program was exemplary: Anita Solow and Thomas Dick.

The members of the AP Calculus Development Committee we served with: Sally Fischbek, Art Grainger, Joe Milliett, Stella Ashford, Mary Ann Connors, Ben Klein, Nancy Gates, Ray Cannon, Nancy Stephenson, and David Bressoud. The time we spent with them increased our knowledge of and passion for teaching Calculus.

The Chief Readers we worked with at the AP Calculus Reading, whose care and expertise were an inspiration: George Rosenstein, Ray Cannon, Bernie Madison, Larry Riddle, and Caren Diefenderfer. We learned a great deal from each of them.

Our friends from the Educational Testing Service and The College Board: Chan Jones, Gloria Dion, Craig Wright, and Judy Broadwin. We were fortunate to work with such committed, knowledgeable, and friendly professionals.

Our editor, Gary Litvin from Skylight Publishing, offered countless suggestions and improvements, kept the project moving, and treated us with the utmost patience and courtesy in the face of stressful deadlines.

Our reviewers, whose care and attention to detail significantly improved this book: Steve Olson, Bernie Madison, Ray Cannon, Barb Currier, Nancy Stephenson, JT Sutcliffe, and Maria Litvin. They caught many mistakes and offered invaluable advice for better presentation of the material.

Finally, we thank our families, Jack, Beth, and Kevin, Maureen and Ryan, without whose love and support this book would not have been possible.

Mark Howell and Martha Montgomery

About the Authors

 **Mark Howell** is a veteran teacher of Advanced Placement mathematics and computer science at his alma mater, Gonzaga College High School in Washington, DC. Mark has served the Advanced Placement community for many years, as a workshop leader, reader of AP exams, Table Leader, and Question Leader. He served on the AP Calculus Test Development Committee from 1996 to 1999. Mark is the author of the 2005 *AP Calculus Teacher's Guide* and a variety of AP Calculus support materials, and a contributor to *AP Central*. His efforts to further technology use to enhance the teaching and learning of mathematics have taken him throughout the continental US, as well as Puerto Rico, Honolulu, Singapore, Australia, and Thailand. He won the Siemens Award for Advanced Placement in 1999, the Tandy Technology Scholars Award the same year, and a state-level Presidential Teacher Award in 1993.

 **Martha Montgomery** has been the Mathematics Specialist at Fremont City Schools in Fremont, Ohio, since 2002. Prior to that, she taught AP Calculus and chaired the mathematics department at Fremont Ross High School for more than 20 years. Martha has participated in the AP Calculus exam grading as a Reader, Table Leader, Question Leader, Alternate Exam Leader and finally BC Exam Leader. She served on the AP Calculus Test Development Committee from 1996 to 1999 and contributed many reviews and teacher materials to the *AP Central* web site. Martha has conducted College Board workshops for teachers in the Midwest and led many workshops for teachers in northwest Ohio.

 **Benita Albert** has taught AP Calculus for 36 years at Oak Ridge High School in Oak Ridge, Tennessee. She has served as an AP Exam Reader, Table Leader, and member of the AP Calculus Test Development Committee. A consultant for The College Board, Benita has taught more than 100 summer institutes for AP Calculus teachers and authored the College Board's 1985 AP Teacher's Guide to Advanced Placement Mathematics. She has also served on planning committees for the Pre-AP initiatives, Math Vertical Teams, and Building Success in Mathematics. At Oak Ridge High School, Benita has designed and taught a second-year sequel to BC Calculus, covering multivariable calculus, ordinary differential equations, and elementary linear algebra. Benita has won numerous awards: the 1975 Distinguished Classroom Teacher for East Tennessee, the 1991 Presidential Award for Secondary Mathematics Teaching, the 1992 Tandy Outstanding Educator Award, the 1999 Siemens Award for Advanced Placement, and, most recently, the 2003 College Board Southeastern Region Distinguished Service Award.

 **Thomas Dick** is a professor of mathematics and the coordinator of collegiate mathematics education as well as the faculty director at Oregon State University's Mathematics Learning Center. He has served on the Research Advisory Council of the National Council of Teachers of Mathematics, and has chaired The College Board's AP Calculus Test Development Committee. Tom has also chaired the joint committee on research in undergraduate mathematics education of the American Mathematical Society (AMS) and Mathematical Association of America (MAA). Currently, Tom is co-chair of the MAA/College Board Committee on Mutual Concerns. He is also an associate editor for the School Science and Mathematics Journal and a member of the editorial panel for the Journal for Research in Mathematics Education. From 1992 to 1995 he led the Calculus Connections Project, a National Science Foundation program that helped 600 AP Calculus teachers across the nation implement technology-intensive calculus curriculum reforms.

 **Joe Milliet** teaches both middle and high school mathematics and coaches the Upper School Math Team at St. Mark's School of Texas. Joe taught high school mathematics at Beaumont Central High School in Beaumont, Texas, for 22 years. During that time, he served the AP Calculus community as a College Board consultant, a workshop leader, exam reader, and member of the AP Calculus Test Development Committee. During the 1998-1999 academic year, Joe served as Associate Director for the AP Program at the College Board Southwest Region Office (SWRO) in Austin, Texas. From June 1999 to July 2000, Joe worked as a consultant to promote a partnership between the College Board SWRO and the University of Texas, while also developing AP course support material in conjunction with the Dana Center for Math and Science Education at the University of Texas. Joe returned to teaching in 2000, joining the faculty at St. Mark's.

Brief Contents

Contents

How to Use This Book

Practice questions in the review chapters are marked by their number in a box:

1

Their solutions are delimited by ☞ and ☜.

Comments that are relevant only to the BC exam are delimited by ⌈ and ⌋. For example:

⌈ The BC exam also includes improper integrals, series, and parametric, vector, and polar functions. ⌋

The companion web site —

    http://www.skylit.com/calculus

— is an integral part of the book. It contains the latest AP Calculus news, annotated solutions to free-response questions from past exams, and relevant links. Check this book's web site for the current information and to be sure you have the latest edition of *Be Prepared Calculus*.

> **Our practice exams may be slightly more difficult than the actual exams, so don't panic if they take more time.**

AB and BC exams usually share some multiple choice and free-response questions. Our AB and BC practice exams do not overlap. All the questions in our AB exams are useful practice for the BC exam as well. Similarly, many questions in our BC exams require only AB-level material.

Introduction

The AP exams in calculus test your understanding of basic concepts and their applications and your fluency with a graphing calculator. There are two levels of the exam. The AB-level exam covers roughly the material of a one-semester introductory college course in calculus. The BC-level exam adds more advanced material: improper integrals, logistic curves, parametric, vector, and polar functions, and series. BC exam takers receive a subscore that represents their knowledge of AB material. Chapter 1 of this book will help you decide which exam to take.

Exam questions are developed by the AP Calculus Test Development Committee of The College Board, and exams are put together by Educational Testing Service, the same organization that administers the SAT and other exams. In 2004, The College Board offered 34 AP exams in 19 subject areas, and 1,101,802 students took 1,887,770 exams (225,228 of them in AP Calculus AB and BC). The most up-to-date information on the AP exams offered and participation statistics can be found on The College Board's *AP Central* web site, `http://apcentral.collegeboard.com`.

The College Board's course description, *Advanced Placement Course Description for Calculus AB and Calculus BC* is available at `http://apcentral.collegeboard.com`. The AB exam includes limits, derivatives and their applications, integrals and their applications, and simple differential equations. ⌈ The BC exam also includes improper integrals, series, and parametric, vector, and polar functions. ⌋

This is a lot of material to cover, and it is certainly not the goal of this book to teach you everything you need to know from scratch. For that you need a complete textbook with exercises. Most students who take the exam are enrolled in an AP calculus course at their school. A determined student can prepare for the exam on his or her own; it may take anywhere between three and twelve months, and a good textbook will be even more important.

The goals of this book are:

- to describe the exam format and requirements;

- to provide an effective review of what you should know, with emphasis on the more difficult topics and on common omissions and mistakes;

- to help you identify and fill the gaps in your knowledge;

- to offer sample exam questions with answers, hints, and solutions to help you practice and analyze your mistakes.

Chapter 1 of this book explains the format, required materials, and the rules for using calculators, provides information about exam grading, and offers exam-taking tips. Chapters 2-7 cover the core material: limits and continuity, derivatives and their applications, integrals and their applications, differential equations and slope fields. Chapters 8 and 9 cover more advanced BC-only material: parametric, vector, and polar functions and series. Chapter 10 is actually on the web at this book's companion web site, `http://www.skylit.com/calculus`. It offers annotated solutions to free-response questions from past exams. Chapters 2-9 contain sample multiple-choice and free-response questions with detailed explanations of the right and wrong answers. At the end of the book are five complete practice exams — three AB and two BC, with no overlap — followed by answers and solutions.

Good luck!

Chapter 1.   Exam Format, Grading, and Tips

1.1.   Exam Format

Figure 1-1 shows the format of the AP Calculus exam. The exam takes 3 hours and 15 minutes of test time (plus breaks and time for instructions). It is divided into two sections. Section I consists of multiple-choice questions with a total allotted time of 1 hour and 45 minutes. Section II consists of free-response questions with a total allotted time of 1 hour and 30 minutes.

Each section consists of two parts: one with no calculator, the other with a calculator. In Section I, Part A contains 28 questions for 55 minutes; a calculator is not allowed on these questions. Part B contains 17 questions for 50 minutes; some of these questions require a graphing calculator. Section II starts with Part A, the calculator part: three questions for 45 minutes. For Part B, you have to put away your calculator again: there are three questions for 45 minutes. During the Part B time period, you are allowed to go back to the three Part A free-response questions and continue to work on them without your calculator.

The AB and BC practice exams in this book will give you a pretty accurate idea of what the questions are like. The multiple-choice questions cover a wide range of topics, including questions on definitions and fundamental concepts related to limits, continuity, derivatives, and integrals, as well as questions on common methods, rules, and formulas applied to finding limits, derivatives and antiderivatives, analysis of function graphs, related rates, linear approximation, distance-velocity-acceleration problems, areas of regions, volumes of solids, and so on. Both the AB and BC exams also include questions on separable differential equations and slope fields. ⌈ The BC exam adds questions on l'Hôpital's Rule, antidifferentiation by parts and by partial fractions, improper integrals, arc length, Euler's method, the logistic model, polar and parametric curves, series and Taylor and Maclaurin polynomials. ⌋

The free-response questions may include some theoretical elements but are mostly application problems. Each question consists of several parts (usually between 2 and 4); later parts may ask you to use the results from the previous parts.

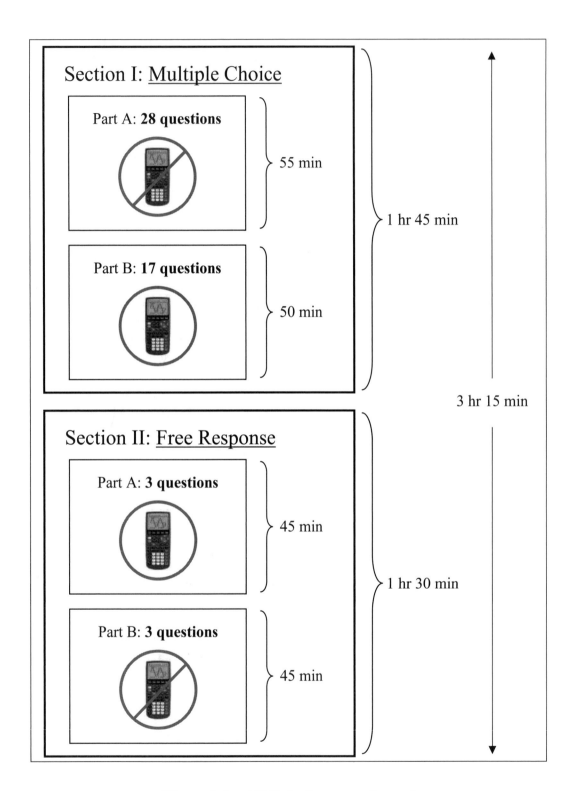

Figure 1-1.   AP Calculus exam format

❚ **It's worth paying attention to recent exam trends.**

Be Prepared "Latest news" link at http://www.skylit.com/calculus takes you to a web page that includes notes on recent changes in content or emphasis. For example, a new policy has been adopted for 2005 with regard to the role of sign charts in solutions to free-response questions.

Chapter 10 at http://www.skylit.com/calculus presents annotated solutions to recent free-response questions (and links to the questions). In recent years, the following types of questions have appeared with some regularity:

1. Questions about data presented in a table and/or a graph.

2. "Application of the integral" questions in which you are given a rate of change and need to find net change.

3. Questions in which you have to interpret the meaning of a derivative or integral in a particular context.

4. Questions that require you to justify the location of an extremum (either global or local) or a point of inflection on a graph. You need to know how to write such a justification.

5. Questions about $g(x) = \int_a^x f(t)\,dt$ for a specified function f, described by a graph or a formula.

1.2.   Grading

The exams are graded on a scale from 1 to 5. Grades of 5 and 4 are called "extremely well qualified" and "well qualified," respectively, and usually will be honored by colleges that give credit or placement for AP exams in calculus. A grade of 3, "qualified," especially on the AB exam, may be denied credit or placement at some colleges. Grades of 2, "possibly qualified," and 1, "no recommendation," are very unlikely to earn you credit or placement.

Table 1-1 presents published statistics and grade distributions on the 2004 AB and BC exams. In 2004, 170,330 candidates nationwide took the AB exam, and 49,332 candidates took the BC exam.

❚ **The multiple-choice and free-response sections weigh equally in the final grade.**

	Calculus AB		Calculus BC	
	Number	%	Number	%
Students	170,330	100.0	49,332	100.0
Grade:				
5	34,226	20.1	19,592	39.7
4	33,831	19.9	9,274	18.8
3	32,484	19.1	10,335	20.9
2	30,202	17.7	3,791	7.7
1	39,587	23.2	6,340	12.9
3 or Higher	100,541	59.1	39,201	79.4

Table 1-1. 2004 grade distributions for AB and BC exams

To determine the grade, the College Board first calculates the total exam score. A weighted combination of the multiple-choice and free-response scores is used to determine the final total score:

$$\text{Total score} = \text{MC coeff} \cdot (\text{correct count} - 0.25 \cdot \text{wrong count}) + \text{FR coeff} \cdot \text{FR score}$$

For multiple-choice questions, one point is given for each correct answer and 1/4 point is subtracted for each wrong answer. Free-response questions are graded by a large invited group of high school teachers and college professors. Scores are assigned based on a *rubric* established by the Chief Reader, the Question Leader, and a group of exam readers. Each free-response question is scored out of 9 points, with partial credit given according to the rubric. The final score is obtained by adding the MC and FR weighted scores. The MC and FR coefficients are chosen in such a way that they give equal weights to the multiple-choice and free-response sections of the exam. For example, if the exam has 45 multiple-choice questions and 6 free-response questions, weights of 1.2 for multiple-choice and 1.0 for free-response will give each section a maximum total of 54, for a maximum possible total score of 108.

AB Max composite score 108 (1.2 * MC + 1.0 * FR)				BC Max composite score 108 (1.2 * MC + 1.0 * FR)			
AP Grade	1997	1998	2003	AP Grade	1997	1998	2003
5	72 - 108	75 - 108	66 - 108	5	66 - 108	65 - 108	64 - 108
4	56 - 71	58 - 74	47 - 65	4	54 - 65	54 - 64	53 - 63
3	39 - 55	40 - 57	29 - 46	3	36 - 53	39 - 53	36 - 52
2	25 - 38	25 - 39	16 - 28	2	29 - 35	30 - 38	26 - 35
1	0 - 24	0 - 24	0 - 15	1	0 - 28	0 - 29	0 - 25

Table 1-2. Score-to-grade conversion

Four cut-off points determine the grade. Table 1-2 shows the maximum composite scores and cut-off points for several years when this information was made public. In 1998, 70% or more correct answers on the AB exam and 60% or more correct answers on the BC exam would get you a 5. The cut-off points are determined by the Chief Reader and vary slightly from year to year based on the score distributions, equalization from year to year, and close examination of a sample of individual exams.

1.3.   AB or BC?

Table 1-1 shows that a larger percentage of BC exam takers got a 5. That's how it should be — if you have a choice and you haven't covered all the BC material, consider taking the AB exam. On the other hand, if you have taken a complete BC course, you should take the BC exam.

Every student who takes the BC exam receives a BC grade and also an AB subscore based on the questions that cover AB topics.

The practice exams in this book will help you make up your mind about which exam to take.

Most colleges will take your AP courses and exam grades, if you take your exams early enough, into account in admissions decisions. But acceptance of AP exam results for credit and/or placement varies widely among colleges. In general, the AB exam is designed to correspond to a one-semester introductory course. The BC exam is designed to correspond to two one-semester courses. Some colleges give one-semester credit for the AB exam and two-semester credit for the BC exam, as intended. But other colleges may only give one semester of credit, regardless of the exam. They may also base their decision on your grade. For example, you may get a full year of credit only if you get a 5 on the BC exam. Some colleges may not give any credit at all. The College Board has collected links to AP acceptance policy statements at many colleges on its website, `www.collegeboard.com/ap/creditpolicy/`.

To do well on the BC exam, you have to be comfortable with l'Hôpital's Rule, antidifferentiation by parts and by partial fractions, improper integrals, Euler's method, the logistic model, arc length, polar and parametric curves, and Taylor polynomials and series.

> **If you know this material, you shouldn't be afraid of the BC exam. Don't assume that the questions are "just harder." The BC exam simply includes more topics. You are not expected to solve problems any faster.**

Once you learn the additional BC topics, the BC exam questions are not necessarily harder than the AB exam questions. In fact, the AB and BC exams usually share three free-response and many multiple-choice questions. A BC course is an extension of an AB course, and BC exam questions have to be more diverse in order to cover all the material in the same number of questions; this may actually make the exam easier for you if you have studied all the BC material. For example, you may understand series very well, but be prone to mistakes in questions that involve analysis of function graphs.

1.4. The Use of Graphing Calculators

> **You must bring a graphing calculator of an allowed model with you when you take the AP Calculus exam.**

Not all calculators are allowed. In particular, non-graphing scientific calculators, computers, calculators with a QWERTY keyboard, tablet computers, and PDAs are not allowed.

> **Consult the list of allowed graphing calculator models in the *AP Calculus Course Description*.**

You may bring two graphing calculators, if you wish, in case one of them stops working or if you have two different models that you commonly use.

You will use your calculator during Part B of Section I (multiple-choice) and Part A of Section II (free-response) only. You will be required to put it away during the remainder of the exam.

> **You are allowed to come in with programs or other information stored in your calculator. You will <u>not</u> be required to clear the memory prior to the exam.**

The capabilities of different calculator models vary widely. Some of them (e.g., TI-89 and several of the HP calculators) have a built-in Computer Algebra System (CAS), which supports symbolic manipulation of expressions, such as symbolic differentiation. To ensure a level playing field, the Test Development Committee has defined four calculator operations that are sufficient to answer all AP exam questions:

1. produce a graph of a function within an arbitrary viewing window;

2. find the zeros of a function (i.e., solve an equation numerically);

3. calculate the derivative of a function at a given value;

4. calculate the value of a definite integral.

> **It is essential to understand the rules of the game clearly. You <u>are allowed</u> to use your calculator in any way you want. At the same time, when writing your solutions to free-response questions and justifying your answers, you may refer only to the above four calculator operations.**

The above rules may sometimes make your life easier and at other times a little harder. On one hand, you do not have to be familiar with all the features of your calculator. If you know how and when to use the above four operations, you can do well on the exam. On the other hand, you have to watch out for inappropriate references to the calculator in your solutions.

> **If you give an answer in decimal approximation, it should be correct to <u>three</u> places after the decimal point, unless the question explicitly specifies other precision.**

Find the relative minimum of $f(x) = x^3 - \sin x$. Justify your answer.

Enter

```
Y₁=X³-sin(X)
```

Plot it and you will see that there is a minimum at around $x = .5$. So far so good. Now you might use the `minimum` command and find $x \approx .535$. Unfortunately, you won't receive credit if you simply write "using the minimum command we find $x \approx .535$," because "find a minimum" is not one of the four allowed calculator operations. Second attempt: you might enter

```
Y₁=X³-sin(X)
Y₂=nDeriv(Y₁(X),X,X)
```

plot Y_2, and find its zero at $x \approx .535$. Again, this solution won't get full credit because plotting a derivative is not one of the four allowed operations. You first need to honestly differentiate $f(x) = x^3 - \sin x$. A correct solution might read like this:

> $f(x) = x^3 - \sin x \implies f'(x) = 3x^2 - \cos x$. Using a calculator graph of $f'(x)$, we find that $f'(x) = 0$ at $x \approx .535$ and $f'(x)$ changes sign from negative to positive there, so f has a relative minimum at $x \approx .535$.

The above example is slightly exaggerated. The people who write AP exam questions do not try to catch you on inappropriate calculator use — they make sure the question calls for a meaningful and unambiguous use of a calculator. Here you would have an advantage if you had a CAS calculator, because it can give you a formula for the derivative. So in reality a question like this would not appear on the open calculator part of the exam. Still, you should learn to play the calculator game by the rules.

To summarize, your familiarity with calculator features can be rather limited, but an excellent command of the four basic operations is essential, and you must understand the rules. The necessary calculator skills and examples are reviewed in this book's Calculator Skills appendix.

1.5.   Exam-Taking Tips

The very first step to doing well on the exam is to make a personal commitment to do the very best you can. You've already taken the next step: you've picked up this book to prepare. Now give yourself ample review time. Hopefully, you haven't waited until the last week to start your studying. What you are seeking is <u>comfort</u> with the material; you want to feel secure that you <u>know calculus</u> as you enter that exam room. This book, besides breaking up the material into manageable chunks and organizing it for you, should help you focus on main ideas, eliminating the need for rote memorization.

The following are some tips for the multiple-choice portion of the exam:

1. If you took the time to read a question and all the answer choices but decided to skip it, take an extra ten seconds and guess. If you can eliminate one or two wrong answers, then it is to your advantage to guess among the remaining three or four choices.

2. If a common paragraph, graph, or table refers to a group of questions and you took the time to read it, try each question in the group.

3. Don't go back and change an answer unless you have found an error in your work. Your first impulse is more likely to be correct.

4. Some multiple choice questions can be answered by working backwards. Plug the answer choices into the problem and see which answer works out.

5. Don't get stuck. Skip a hard question and come back to it if you have time after finishing the rest. But be careful marking your answer sheet when you skip questions.

6. On the calculator portion, you won't need your calculator on all the questions. If all or some of the answers have three digits after the decimal point (e.g., 98.765), it's time to pick up the calculator (but, of course, there may be other times to use it as well).

On the free-response section, try to write clearly, be neat, and keep your exam reader in mind. State the answer clearly, perhaps circle it.

> **Remember that all six free-response questions have equal weight. Don't assume that the first question is the easiest and the last is the hardest. Be sure to try all the questions, including Question 6.**

Other things to remember about free-response questions:

1. Do read the question before jumping to the equations, formulas, or graphs included in the question.

2. Don't waste your time erasing large portions of work. Instead, cross out your work with one neat line, but only after you have something better to replace it with.

3. On the calculator portion, read through the three questions and begin with the ones that require the use of a calculator. After 45 minutes, you'll have to put your calculator away, but you'll be able to keep working on the questions. So save for last those questions where you don't need the calculator.

4. Show your work. Explain your answer in a sentence or two if the question requires it.

5. Store intermediate results on your calculator to be used later in the problem (see Appendix: Calculator Skills). Don't do any rounding until the end of the problem.

6. Write down in the exam book any integral or derivative you evaluate or any equation you solve on your calculator.

7. Try every part of every question. Even if you cannot complete Part (a) of a question, you may still be able to do the remaining parts.

8. Don't give a "recipe" for your answer. Just do the work. Don't waste time announcing how you'll proceed before doing the problem.

9. Be sure to answer the question, but don't do excessive writing. For example, if a question asks when a particle is moving to the right, be sure to answer it, but don't also tell when it is moving to the left.

10. Don't waste time simplifying answers, especially with derivatives and linear equations. You will lose a point if you simplify incorrectly. An answer like $\sin\left(\dfrac{\pi}{3}\right) - 2\sin\left(\dfrac{\pi}{6}\right)$ is just as good as $\dfrac{\sqrt{3}}{2} - 1$. A tangent line written in the form $y - 3 = \dfrac{5}{3}(x - 6)$ is just as good as $y = \dfrac{5}{3}x - 7$.

11. Watch for the phrases like "Justify your answer" or "Explain your reasoning." They mean you must have your work shown and write a sentence or two explaining your answer. A picture or a graph alone is never a justification.

12. Try not to use the pronoun "it." The phrase "it is increasing" could refer back to a given function, its derivative, its antiderivative, or other things. If a reader doesn't know to which function the word "it" refers, the reader will not be able to award credit. Be specific in your explanations.

13. Don't quit until the time is up. Use all the time you have and keep trying. The test will be over before you know it.

Chapter 2.   Limits and Continuity

2.1.   The Concept of Limit

If you graph $f(x) = x^3$, you can see that its graph becomes almost horizontal near the origin (Figure 2-1-a). Can we confirm that analytically? Let us draw a line from the origin to an arbitrary point P with coordinates (x, y) on the graph in the first quadrant. The slope of line OP is $\dfrac{y-0}{x-0} = \dfrac{y}{x}$. We cannot directly evaluate the slope if point P is at the origin because both x and y are equal to 0 there, so $\dfrac{y}{x}$ is undefined. But for $x \neq 0$,

$\dfrac{y}{x} = \dfrac{x^3}{x} = x^2$, so the slope <u>approaches</u> 0 as x gets closer and closer to 0.

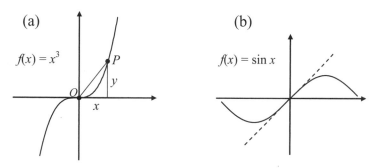

Figure 2-1. Tangent lines at $x = 0$ for $f(x) = x^3$ and $f(x) = \sin x$

Figure 2-1-b illustrates a similar question for $f(x) = \sin x$. Here the slope $m = \dfrac{y}{x} = \dfrac{\sin x}{x}$. When you evaluate $\dfrac{\sin x}{x}$ on your calculator for small x (say, $x = 0.1$ radians, $x = 0.01$ radians, $x = 0.001$ radians, etc.) you see that the corresponding values of $\dfrac{\sin x}{x}$ get close to 1. $\dfrac{\sin x}{x}$ is undefined, of course, at $x = 0$, but it <u>approaches</u> 1 when x approaches 0 from either the left or right side. So $m \to 1$ (we say, "m approaches 1") when $x \to 0$; therefore, the slope of the graph of $f(x) = \sin x$ at $x = 0$ is 1.

The concept of the limit of a function captures the intuitive notion of the function values approaching a certain value when x approaches a given value.

> **We say that L is the limit of f, as x approaches a, and we write $\lim\limits_{x \to a} f(x) = L$, if for all x, sufficiently close to a (but not equal to a), $f(x)$ is close to L.**

Figure 2-2 illustrates this definition.

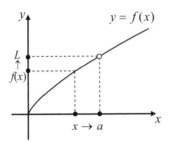

Figure 2-2. $\lim\limits_{x \to a} f(x) = L$

"Is close to L" means as close as we want, based on any criterion of closeness we choose; "x sufficiently close to a" means sufficiently close for the particular criterion chosen earlier. There is a more formal "epsilon-delta" definition of the limit, but it is not required for the AP exam.[*]

> **Note that for $\lim\limits_{x \to a} f(x)$ to exist, $f(a)$ does not have to be defined. If it is defined, $\lim\limits_{x \to a} f(x)$ may or may not be equal to $f(a)$. $f(a)$ is not considered in the definition of $\lim\limits_{x \to a} f(x)$ at all.**

For example, $\lim\limits_{x \to 0} \dfrac{\sin x}{x} = 1$, even though $\dfrac{\sin x}{x}$ is not defined at $x = 0$ (Figure 2-3).

In differential calculus, the limit of the ratio $\lim\limits_{h \to 0} \dfrac{f(a+h) - f(a)}{h}$ is used to define the derivative of the function f at $x = a$ (which represents the instantaneous rate of change of the function and the slope of the graph of the function at $x = a$, as in the above examples). In integral calculus, the limit of sums is used to define the definite integral of a function (which represents the area between the curve and the x-axis for non-negative functions). So the concept of limit is the cornerstone of calculus.

---

[*] The epsilon-delta definition is actually less complicated that it might seem. L is the limit of $f(x)$ when x approaches a if and only if for any $\varepsilon > 0$ [i.e., for any criterion of closeness of $f(x)$ to L that we might choose] there exists a $\delta > 0$, such that $|f(x) - L| < \varepsilon$ [i.e., $f(x)$ is close to L, according to our chosen criterion] whenever $0 < |x - a| < \delta$ [i.e., whenever x is sufficiently close to a (excluding a itself)].

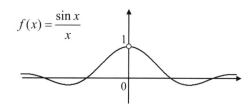

$$f(x) = \frac{\sin x}{x}$$

Figure 2-3. $\displaystyle\lim_{x \to 0} \frac{\sin x}{x} = 1$

1

$\displaystyle\lim_{x \to 1} \frac{\ln x}{x}$ is

    (A) 1   (B) 0   (C) e (D) $-e$ (E) Nonexistent

$\ln 1 = 0$ and $\ln x \to 0$ (approaches 0) when $x \to 1$. Of course, the denominator approaches 1 as $x \to 1$. In fact both the top and the bottom in the fraction are defined at $x = 1$. Therefore, we can use **direct substitution**: we can just substitute $x = 1$ into the function to obtain the answer. The answer is B.

Direct substitution can be used to find $\displaystyle\lim_{x \to a} f(x)$ only if f is **continuous** at $x = a$.

Continuity is discussed in Section 2.4.

2

$\displaystyle\lim_{x \to -2} \frac{x+2}{x^2 - 4}$ is

    (A) $-\dfrac{1}{4}$ (B) $-\dfrac{1}{2}$ (C) 0   (D) 1   (E) Does not exist

Here direct substitution doesn't work: you cannot just substitute –2 for x into $\dfrac{x+2}{x^2-4}$,

because that would give $\dfrac{0}{0}$. Be careful: $\dfrac{0}{0}$ is not equal to 0, or 1, or ∞. When both the numerator and denominator of a fraction approach 0, the limit of the fraction might have any value. For example, $\lim\limits_{x\to 0}\dfrac{kx}{x}=k$, $\lim\limits_{x\to 0}\dfrac{x^2}{x}=0$, and $\lim\limits_{x\to 0}\dfrac{x}{x^3}=\infty$. To find the correct answer, we need to do more work. First, let's factor the denominator:

$\lim\limits_{x\to -2}\dfrac{x+2}{x^2-4}=\lim\limits_{x\to -2}\dfrac{x+2}{(x+2)(x-2)}$. Now we can cancel $(x+2)$ from both numerator and

denominator: $\lim\limits_{x\to -2}\dfrac{x+2}{(x+2)(x-2)}=\lim\limits_{x\to -2}\dfrac{1}{x-2}$. We can do that, because when finding a

limit we only consider the values of the function for $x\neq a$, in this case for $x\neq -2$. Now

we see that $\dfrac{1}{x-2}$ approaches $-\dfrac{1}{4}$ as x approaches -2. The answer is A.

One-Sided Limits

For some functions, as x moves close to a, the corresponding $f(x)$ values approach different limits, depending on whether x approaches a from the left or from the right. In that case, $\lim\limits_{x\to a} f(x)$ does not exist, but it is useful to consider what are called **one-sided limits**. $\lim\limits_{x\to a^+} f(x)=L$ means $f(x)$ approaches L as x approaches a from the right $(x>a)$; $\lim\limits_{x\to a^-} f(x)=L$ means $f(x)$ approaches L as x approaches a from the left $(x<a)$ (Figure 2-4).

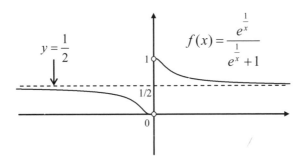

Figure 2-4. A function with different left-hand and right-hand limits at $x=0$:
$$\lim\limits_{x\to 0^-} f(x)=0 \ \text{ and } \ \lim\limits_{x\to 0^+} f(x)=1$$

> For $\lim\limits_{x \to a} f(x)$ to exist, both $\lim\limits_{x \to a^-} f(x)$ and $\lim\limits_{x \to a^+} f(x)$ must exist and they must be equal.

Limits at Infinity

> We say that *L* is the limit of *f*, as *x* approaches infinity, and we write $\lim\limits_{x \to \infty} f(x) = L$ if, for all sufficiently large *x*, $f(x)$ is close to *L*.

A similar definition applies to $x \to -\infty$:

> We say that *L* is the limit of *f*, as *x* approaches negative infinity, and write $\lim\limits_{x \to -\infty} f(x) = L$ if, for all negative *x* with a sufficiently large absolute value, $f(x)$ is close to *L*.

If $\lim\limits_{x \to \infty} f(x) = L$ or $\lim\limits_{x \to -\infty} f(x) = L$ or both, the horizontal line $y = L$ is called a ***horizontal asymptote*** of *f*. The graph of the function approaches its horizontal asymptote on the left ($x \to -\infty$) or on the right ($x \to +\infty$) or both. A function can have a maximum of two horizontal asymptotes: one as $x \to -\infty$ and one as $x \to +\infty$. The graph can approach the asymptote from above or from below (Figure 2-4) or it can cross it or bounce around it with decreasing deviations (Figure 2-3). AP exam questions that involve finding horizontal asymptotes of a function are rather common.

Nonexistent Limits

$\lim\limits_{x \to a} f(x)$ may not exist for one of several reasons:

1. Both right-hand and left-hand limits exist but have different values (Figure 2-4).

2. $\lim\limits_{x \to a^-} f(x)$ and/or $\lim\limits_{x \to a^+} f(x)$ do not exist; the values of *f* jump all over the place and do not all get close to a particular value as $x \to a$ (Figure 2-5-a).

3. The values of *f* go "off the chart" (i.e., their absolute values become larger and larger) as $x \to a$. In other words, $f(x) \to +\infty$ or $f(x) \to -\infty$ as $x \to a$ (Figure 2-5-b).

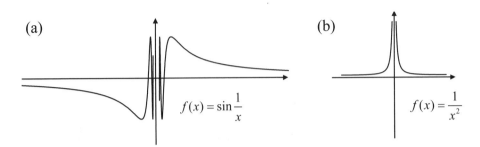

Figure 2-5.   Functions with neither a left nor a right limit at $x = 0$:
    (a) $f(x)$ oscillates from –1 to 1 when $x \to 0$
    (b) $f(x) \to \infty$ when $x \to 0$

Vertical Asymptotes

If $f(x) \to +\infty$ or $f(x) \to -\infty$ as $x \to a^+$ and/or $x \to a^-$, we say that $f(x)$ has a *vertical asymptote* at $x = a$.

The function $f(x) = \dfrac{1}{x^2}$ in Figure 2-5-b above has a vertical asymptote, the line $x = 0$.

The function $f(x) = \dfrac{x}{x^2 - 1}$ in Figure 2-6-a has two vertical asymptotes, $x = -1$ and $x = 1$.

$f(x) = e^{\frac{1}{x-1}}$ in Figure 2-6-b has one vertical asymptote, $x = 1$.

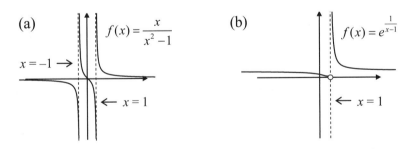

Figure 2-6.   Vertical asymptotes

Sometimes people also consider slanted asymptotes, such as the line $y = x$ in Figure 2-7. But slanted asymptote questions are rarely included in AP Calculus exams.

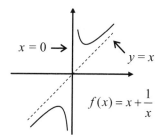

**Figure 2-7. A function with one vertical
and one slanted asymptote**

3

For the function $f(x) = \dfrac{2x-1}{|x|}$, find

(a) $\lim\limits_{x \to \infty} f(x)$ (b) $\lim\limits_{x \to -\infty} f(x)$ (c) $\lim\limits_{x \to 0^+} f(x)$ (d) $\lim\limits_{x \to 0^-} f(x)$

(e)   All horizontal asymptotes

(f)   All vertical asymptotes

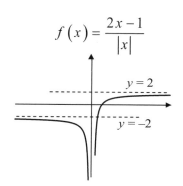

(a)   For $x > 0$, $|x| = x$, so

$$\lim\limits_{x \to \infty} f(x) = \lim\limits_{x \to \infty} \frac{2x-1}{x} = \lim\limits_{x \to \infty} \left(2 - \frac{1}{x}\right) = 2.$$

(Remember the "big-little" principle:

$\dfrac{1}{big} = little$ and $\dfrac{1}{little} = big$).

(b)   For $x < 0$, $|x| = -x$, so

$$\lim\limits_{x \to -\infty} f(x) = \lim\limits_{x \to -\infty} \frac{2x-1}{-x} = \lim\limits_{x \to -\infty} \left(-2 + \frac{1}{x}\right) = -2.$$

(c) $\lim\limits_{x \to 0^+} f(x) = \lim\limits_{x \to 0^+} \dfrac{2x-1}{x} = \lim\limits_{x \to 0^+} \left(2 - \dfrac{1}{x}\right) \Rightarrow f(x) \to -\infty$, so the limit does not exist.

(d) $\lim\limits_{x \to 0^-} f(x) = \lim\limits_{x \to 0^-} \dfrac{2x-1}{-x} = \lim\limits_{x \to 0^-} \left(-2 + \dfrac{1}{x}\right) \Rightarrow f(x) \to -\infty$, so the limit does not exist.

(e)   Horizontal asymptotes are $y = 2$ (from Part (a)) and $y = -2$ (from Part (b)).

(f)   There is only one vertical asymptote, $x = 0$ (from Parts (c) and (d)).

Interpreting Graphs of Functions

Given a graph of a function, you must be able to evaluate limits as x approaches a number. This type of question simply tests your understanding of the concept of limit and your ability to read graphs.

4

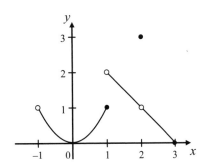

Use the graph of $f(x)$ above to estimate the following limits or explain why they do not exist.

(a) $\lim_{x \to -1^+} f(x)$ (b) $\lim_{x \to 1^-} f(x)$ (c) $\lim_{x \to 1^+} f(x)$

(d) $\lim_{x \to 1} f(x)$ (e) $\lim_{x \to 2} f(x)$ (f) $\lim_{x \to 3^-} f(x)$

Individual open or closed points do not change the value of the limits. Just ignore them.

(a) We approach $x = -1$ from the right, so $\lim_{x \to -1^+} f(x) = 1$.

(b) We approach $x = 1$ from the left, so $\lim_{x \to 1^-} f(x) = 1$.

(c) We approach $x = 1$ from the right, so $\lim_{x \to 1^+} f(x) = 2$.

(d) $\lim_{x \to 1} f(x)$ does not exist because the limit from the left is not equal to the limit from the right at $x = 1$ (see Part (b) and Part (c)).

(e) $\lim_{x \to 2} f(x) = 1$. Don't let the fact that $f(2) = 3$ mislead you. Remember that the definition of the limit at $x \to a$ does not require that $f(a)$ exist. Ignore it if it does.

(f) We approach $x = 3$ from the left, so $\lim_{x \to 3^-} f(x) = 0$.

Note that in the above example, $f(x)$ is not defined for $x < -1$ and $x > 3$, so it wouldn't make sense to ask about $\lim_{x \to -1^-} f(x)$ or $\lim_{x \to 3^+} f(x)$.

2.2.   Properties of Limits

Simple arithmetic properties of limits are summarized in Table 2-1.

1.	$\lim_{x \to a}\left[c \cdot f(x)\right] = c \cdot \lim_{x \to a} f(x)$
2.	$\lim_{x \to a}\left[f(x) + g(x)\right] = \lim_{x \to a} f(x) + \lim_{x \to a} g(x)$
3.	$\lim_{x \to a}\left[f(x) - g(x)\right] = \lim_{x \to a} f(x) - \lim_{x \to a} g(x)$
4.	$\lim_{x \to a}\left[f(x) \cdot g(x)\right] = \lim_{x \to a} f(x) \cdot \lim_{x \to a} g(x)$
5.	$\lim_{x \to a}\dfrac{f(x)}{g(x)} = \dfrac{\lim_{x \to a} f(x)}{\lim_{x \to a} g(x)}$, if $\lim_{x \to a} g(x) \neq 0$

Table 2-1.   Arithmetic properties of limits

Property 1 requires that $\lim_{x \to a} f(x)$ exist; Properties 2 - 5 require that both $\lim_{x \to a} f(x)$ and $\lim_{x \to a} g(x)$ exist.

▌ **One-sided limits and limits at infinity have the same five properties.**

You have to be rather careful with the division property. If $\lim_{x \to a} g(x) \neq 0$, everything works as expected. If, however, $\lim_{x \to a} g(x) = 0$, then things are different. If $\lim_{x \to a} f(x) \neq 0$ then $\left|\dfrac{f(x)}{g(x)}\right| \to \infty$ so $\lim_{x \to a}\dfrac{f(x)}{g(x)}$ does not exist. The tricky case is when both $\lim_{x \to a} f(x) = 0$ and $\lim_{x \to a} g(x) = 0$. The $\dfrac{0}{0}$ situation is called an ***indeterminate form***. In this situation the limit may or may not exist, and if it does exist, it could have any value. We have already seen examples of it: $\lim\limits_{x \to 0} \dfrac{\sin x}{x} = 1$ and $\lim\limits_{x \to -2} \dfrac{x+2}{x^2-4} = -\dfrac{1}{4}$. Another example is

$\lim\limits_{x \to 0} \dfrac{\cos x - 1}{x} = 0$.

You also must be extremely careful when $f(x) \to \pm\infty$ or $g(x) \to \pm\infty$. Remember that ∞ is a symbol that means "arbitrarily large." ∞ and $-\infty$ are not real numbers and we cannot use real number arithmetic with them. For example, $\infty + \infty$ is not $2 \cdot \infty$, $\infty - \infty$ is not 0, and $\dfrac{\infty}{\infty}$ is not 1. In some situations involving infinite limits, properties similar to the ones in Table 2-1 may be useful. For example, if $\lim\limits_{x \to a} f(x) = +\infty$ and $\lim\limits_{x \to a} g(x) = +\infty$, then $\lim\limits_{x \to a} \big[f(x) + g(x) \big] = +\infty$. But if $\lim\limits_{x \to a} f(x) = +\infty$ and $\lim\limits_{x \to a} g(x) = -\infty$, then $\lim\limits_{x \to a} \big[f(x) + g(x) \big]$ may be anything. $\infty - \infty$ is an indeterminate form, as are $0 \cdot \infty$ and $\dfrac{\infty}{\infty}$. We will see how these situations play out through examples in the remaining sections of this chapter.

$\boxed{5}$

$$\lim\limits_{x \to 0} \left(\dfrac{1}{x} + \dfrac{1}{x^2} \right) =$$

(A) 0   (B) $\dfrac{1}{2}$ (C) 1   (D) 2   (E) ∞

Here we cannot just say that $\infty + \infty = \infty$ because $\dfrac{1}{x}$ is negative for negative x. It is best to begin by combining the two fractions over a common denominator:

$\lim\limits_{x \to 0} \left(\dfrac{1}{x} + \dfrac{1}{x^2} \right) = \lim\limits_{x \to 0} \dfrac{x+1}{x^2} = \text{(informally)} \dfrac{1}{0} = \infty$. The answer is E.

6

$$\lim_{x \to \infty}\left[x^2\left(\frac{1}{x-2}-\frac{1}{x-3}\right)\right]=$$

   (A) 0   (B) 1   (C) −1   (D) ∞   (E) −∞

Again, the approach is to combine the two fractions in parentheses over a common denominator. Then

$$\lim_{x \to \infty}\left[x^2\left(\frac{1}{x-2}-\frac{1}{x-3}\right)\right]=\lim_{x \to \infty}\left[x^2\left(\frac{x-3-x+2}{x^2-5x+6}\right)\right]=\lim_{x \to \infty}\left(\frac{-x^2}{x^2-5x+6}\right)$$

Dividing both numerator and denominator by x^2 we get $\lim_{x \to \infty}\dfrac{-1}{1-\dfrac{5}{x}+\dfrac{6}{x^2}}$. As $x \to \infty$, each

of the fractions in the denominator approaches 0, so the limit becomes $\dfrac{-1}{1-0-0}=-1$.

The answer is C.

Compare the above with $\lim_{x \to \infty}\left[x^2\left(\dfrac{2}{x-2}-\dfrac{3}{x-3}\right)\right]$:

$$\lim_{x \to \infty}\left[x^2\left(\frac{2}{x-2}-\frac{3}{x-3}\right)\right]=\lim_{x \to \infty}\left[x^2\left(\frac{2x-6-3x+6}{x^2-5x+6}\right)\right]=\lim_{x \to \infty}\left[\frac{-x^3}{x^2-5x+6}\right]=-\infty$$

2.3.   Techniques for Finding Limits

> **Guessing the limit of a function with your calculator should be done carefully.**

Using a table of values or looking at a graph is a good way to begin solving a limit problem if you don't see another approach. But usually limits can be found faster and more reliably using simple algebraic techniques, calculus facts, ⌈ or l'Hôpital's Rule (Section 2.3.6) ⌋. The calculator may be useful here for doing arithmetic on numbers, if necessary.

2.3.1.   Finding Limits Using Continuity

Many limits can be determined easily if the function is continuous at $x=a$.

> **If f is continuous at $x = a$, we can simply substitute a for x in $f(x)$ to find the limit at a: $\lim_{x \to a} f(x) = f(a)$.**

This is called ***direct substitution***. We will talk more about continuity and discontinuities in Section 2.4. At this point we only need to know the following fact.

> **Every "elementary" function (polynomial, rational, trigonometric, radical, logarithmic, exponential, absolute value) and every combination and composition of elementary functions is <u>continuous</u> at every x that belongs to <u>its natural domain</u> (i.e. where the function has a value).**

This means that if a function f is defined by <u>one</u> "nice" (or ugly) formula, which makes sense for $x = a$, then $\lim_{x \to a} f(x) = f(a)$. For example, $\lim_{x \to a} \dfrac{\ln|\sin x|}{(x-1)\sqrt{x^2 - 4}} = \dfrac{\ln|\sin a|}{(a-1)\sqrt{a^2 - 4}}$

for any a for which this function makes sense, namely, any a such that $\sin a \neq 0$, $a \neq 1$, and $a^2 - 4 > 0$.

At the first glance, this statement may contradict the examples we see in Figure 2-4 and Figure 2-5. But $\sin \dfrac{1}{x}$, $\dfrac{e^{\frac{1}{x}}}{e^{\frac{1}{x}}+1}$, and $\dfrac{1}{x^2}$ are not defined at $x = 0$, so the nonexistent limit at $x = 0$ does not contradict the above rule.

> **Because of the continuity rule, most "interesting" questions on limits involve one of the following: a value of x at which the function is not defined, a limit as $x \to \pm\infty$ or a function described by discontinuous graph or a piecewise function defined by different formulas on different intervals.**

For example, if you have a rational function (a ratio of two polynomials), an exam question is likely to ask for its limit at $x = a$ where the denominator is equal to 0, or at $x \to \pm\infty$.

7

$$\lim_{x\to 2}\frac{x^2-4}{x^3-8}=$$

(A) 0   (B) $\dfrac{1}{3}$ (C) $\dfrac{1}{2}$ (D) $\dfrac{2}{3}$ (E) Does not exist

This rational function is not defined at $x = 2$, but the limit may still exist there. Factoring the numerator and denominator and simplifying the fraction produces a function which is equivalent to the original function at every point except at $x = 2$. This simplified function can then be used to determine the limit:

$$\lim_{x\to 2}\frac{x^2-4}{x^3-8}=\lim_{x\to 2}\frac{(x+2)(x-2)}{(x-2)(x^2+2x+4)}=\lim_{x\to 2}\frac{x+2}{x^2+2x+4}=\frac{1}{3}.$$ The answer is B.

8

$$\lim_{x\to\infty}\frac{x^3-4x+1}{2x^3-5}=$$

(A) $-\dfrac{1}{5}$ (B) $\dfrac{1}{2}$ (C) $\dfrac{2}{3}$ (D) 1   (E) Does not exist

You will see several approaches for solving this problem in the following sections. One method is to divide every term of the numerator and denominator by the largest power of

x, namely x^3: $\lim_{x\to\infty}\dfrac{x^3-4x+1}{2x^3-5}=\lim_{x\to\infty}\dfrac{1-\dfrac{4}{x^2}+\dfrac{1}{x^3}}{2-\dfrac{5}{x^3}}$. Any term with a power of x in its

denominator will approach zero as $x\to\infty$, so the limit equals $\dfrac{1}{2}$ and the answer is B.

2.3.2. Finding Limits of Rational Functions

These questions are concerned with finding $\lim\limits_{x \to a} \dfrac{f(x)}{g(x)}$ or $\lim\limits_{x \to \pm\infty} \dfrac{f(x)}{g(x)}$, where $f(x)$ and $g(x)$ are polynomials.

For $\lim\limits_{x \to a} \dfrac{f(x)}{g(x)}$ there are two possibilities: $g(a) \neq 0$ and $g(a) = 0$. If $g(a) \neq 0$, just substitute a into the function. The interesting case is when $g(a) = 0$.

> **Be careful: $g(a) = 0$ does not automatically mean that the limit does not exist.**

It <u>may</u> exist if $f(a) = 0$, too. In that case, factor both $f(x)$ and $g(x)$ and cancel the highest degree of $(x - a)$ that they share. If one or more $(x - a)$ factors remain at the bottom, the limit does not exist. Otherwise, the limit does exist; to find it just substitute a into the remaining fraction. See Example 7 above for a problem like this.

$$\lim_{x \to 0} \frac{x^2 - x}{x^4 + x^3} =$$

(A) –1   (B) 0   (C) $\dfrac{1}{2}$ (D) 1   (E) Does not exist

Factoring and canceling produces $\lim\limits_{x \to 0} \dfrac{x^2 - x}{x^4 + x^3} = \lim\limits_{x \to 0} \dfrac{x(x-1)}{x^3(x+1)} = \lim\limits_{x \to 0} \dfrac{x-1}{x^2(x+1)}$, but

substituting in zero for x gives a fraction with a denominator of 0. The answer is E.

For what value of k does $\lim\limits_{x \to 4} \dfrac{x^2 - x + k}{x - 4}$ exist?

(A) –12   (B) –4   (C) 3   (D) 7
(E) No such value exists

In order for this limit to exist, the numerator must have a factor of $x - 4$ to cancel out the denominator. The second factor must be $x + 3$ to give a middle term of $-x$. Therefore the numerator is $(x-4)(x+3) = x^2 - x - 12$ and $k = -12$. The answer is A.

For what value(s) of a will $\lim\limits_{x \to a} \dfrac{x^2 - a^2}{x^4 - a^4}$ exist?

(A) No values of a (B) $a = \dfrac{1}{2}$ only   (C) $a > 0$ only

(D) $a \neq 0$ (E) All real numbers a

Factoring the denominator will change the indeterminate form into one where we can see the limit: $\lim\limits_{x \to a} \dfrac{x^2 - a^2}{x^4 - a^4} = \lim\limits_{x \to a} \dfrac{x^2 - a^2}{(x^2 - a^2)(x^2 + a^2)} = \lim\limits_{x \to a} \dfrac{1}{x^2 + a^2} = \dfrac{1}{2a^2}$. For the limit to exist, a must not equal zero. The answer is D.

There is a simple rule for finding the limits of a rational function $\dfrac{f(x)}{g(x)}$

(where f and g are polynomials) as $x \to \infty$ or $x \to -\infty$: you can consider only the leading term (with the highest power of x) in the numerator and in the denominator — the rest do not matter.

We have to distinguish the following three situations:

- If the degree of f is equal to the degree of g, then the limits at ∞ and $-\infty$ are both equal to the ratio of the leading coefficients of f and g.

- If the degree of f is greater than the degree of g, the limit does not exist:
$$\frac{f}{g} \to \infty \text{ or } \frac{f}{g} \to -\infty.$$

- If the degree of f is less than the degree of g, $\displaystyle\lim_{x \to \infty} \frac{f(x)}{g(x)} = \lim_{x \to -\infty} \frac{f(x)}{g(x)} = 0$.

For example, $\displaystyle\lim_{x \to \infty} \frac{2x^3 - 5x + 7}{3x^3 + 1000x - 2500} = \lim_{x \to -\infty} \frac{2x^3 - 5x + 7}{3x^3 + 1000x - 2500} = \frac{2}{3}$. (This is so because

$$\lim_{x \to \pm\infty} \frac{2x^3 - 5x + 7}{3x^3 + 1000x - 2500} = \frac{\displaystyle\lim_{x \to \pm\infty}\left(2 - \frac{5}{x^2} + \frac{7}{x^3}\right)}{\displaystyle\lim_{x \to \pm\infty}\left(3 + \frac{1000}{x^2} - \frac{2500}{x^3}\right)}, \text{ in case you forget the}$$

equal degrees rule.)

$\boxed{12}$

$$\lim_{x \to -\infty} \frac{3x}{\sqrt{3x^2 - 4}} =$$

(A) $\sqrt{3}$ (B) 1   (C) 0   (D) $-\sqrt{3}$ (E) Does not exist

The degree of the numerator is essentially the same as the degree of the denominator, but because of the radical, you cannot just use the ratio of the leading coefficients. You need to multiply the terms in the denominator by $\sqrt{\dfrac{1}{x^2}}$, but when you do the same for the numerator, remember that $\sqrt{x^2} = |x|$, and since $x \to -\infty$, $|x| = -x$. So multiply the

numerator by $-\dfrac{1}{x}$: $\displaystyle\lim_{x \to -\infty} \frac{3x}{\sqrt{3x^2 - 4}} = \lim_{x \to -\infty} \frac{3x\left(-\dfrac{1}{x}\right)}{\sqrt{3x^2 - 4}\left(\sqrt{\dfrac{1}{x^2}}\right)} = \lim_{x \to -\infty} \frac{-3}{\sqrt{3 - \dfrac{4}{x^2}}} = \frac{-3}{\sqrt{3}} = -\sqrt{3}$.

The answer is D. If you get confused with the signs, plug in a large negative number for x to check.

13

For which of the following values of a does $\lim\limits_{x\to a}\dfrac{3x^2-12}{x^3-8}$ exist?

I. $a = 2$ II. a is ∞ III. $a = -2$

(A) I only   (B) II only   (C) III only

(D) I and II only   (E) I, II, and III

When $a = 2$, we have an indeterminate form that can be resolved by factoring:

$\lim\limits_{x\to 2}\dfrac{3x^2-12}{x^3-8} = \lim\limits_{x\to 2}\dfrac{3(x+2)(x-2)}{(x-2)(x^2+2x+4)} = \lim\limits_{x\to 2}\dfrac{3(x+2)}{x^2+2x+4} = 1$. When a is ∞, we can use

the rule for the degree of the numerator being less than the degree of the denominator —
the limit is zero. When $a = -2$, the limit can be found by direct substitution, yielding an
answer of 0. So I, II, and III all produce limits that exist, and the answer is E.

2.3.3.   Finding Limits Using Algebraic Techniques

In many questions you can determine the limit using simple algebra. The usual algebraic
techniques include factoring and canceling factors, dividing or multiplying both the
numerator and denominator of a fraction by the same number or the same power of x,
combining fractions, multiplying by a conjugate expression (sum or difference), and so
on.

14

$\lim\limits_{x\to 0}\dfrac{\tan x}{x} =$

(A) -1 (B) $-\dfrac{1}{2}$ (C) 0 (D) $\dfrac{1}{2}$ (E) 1

We know that $\lim\limits_{x \to 0} \dfrac{\sin x}{x} = 1$ and $\tan x = \dfrac{\sin x}{\cos x}$. Combining the two, we get:

$\lim\limits_{x \to 0} \dfrac{\tan x}{x} = \lim\limits_{x \to 0}\left[\dfrac{\sin x}{x} \cdot \dfrac{1}{\cos x}\right]$. The limits of both fractions exist as $x \to 0$, so, by the

product property, $\lim\limits_{x \to 0} \dfrac{\tan x}{x} = \lim\limits_{x \to 0} \dfrac{\sin x}{x} \cdot \lim\limits_{x \to 0}\dfrac{1}{\cos x} = 1 \cdot 1 = 1$. The answer is E.

15

$\lim\limits_{x \to 0} \dfrac{\cos x - 1}{\sin^2 x} =$

(A) -1 (B) $-\dfrac{1}{2}$ (C) 0 (D) $\dfrac{1}{2}$ (E) 1

Recall that $\sin^2 x = 1 - \cos^2 x$ and substitute it in the denominator:

$$\lim\limits_{x \to 0} \dfrac{\cos x - 1}{\sin^2 x} = \lim\limits_{x \to 0} \dfrac{\cos x - 1}{1 - \cos^2 x} = \lim\limits_{x \to 0} \dfrac{\cos x - 1}{\left(1 - \cos x\right)\left(1 + \cos x\right)} = \lim\limits_{x \to 0} \dfrac{-1}{1 + \cos x} = -\dfrac{1}{2}$$

The answer is B.

16

$\lim\limits_{x \to 0} \dfrac{\sin 3x}{\tan 2x} =$

(A) 0 (B) $\dfrac{2}{3}$ (C) 1 (D) $\dfrac{3}{2}$ (E) 3

Since $\lim\limits_{x \to 0} \dfrac{\sin x}{x} = 1$, it is also true that $\lim\limits_{x \to 0} \dfrac{\sin ax}{ax} = 1$ for any real number $a \neq 0$, because if

$x \to 0$ then $ax \to 0$. The same reasoning follows for $\lim\limits_{x \to 0} \dfrac{\tan x}{x}$. Therefore,

$\lim\limits_{x \to 0} \dfrac{\sin 3x}{3x} = 1$ and $\lim\limits_{x \to 0} \dfrac{2x}{\tan 2x} = 1$ (see Question 14). Combining these two limits, we get

$\lim\limits_{x \to 0}\left[\dfrac{\sin 3x}{3x} \cdot \dfrac{2x}{\tan 2x}\right] = \dfrac{2}{3}\lim\limits_{x \to 0} \dfrac{\sin 3x}{\tan 2x} = 1 \Rightarrow \lim\limits_{x \to 0} \dfrac{\sin 3x}{\tan 2x} = \dfrac{3}{2}$. The answer is D.

$\boxed{17}$

$$\lim_{x\to\infty}\frac{1}{\sqrt{x}\left(\sqrt{x+1}-\sqrt{x-1}\right)}=$$

(A) 0   (B) $\frac{1}{2}$ (C) 1   (D) 2   (E) Does not exist

Trying to combine the limits of the two factors here does not help, because $\sqrt{x}\to\infty$ and $\left(\sqrt{x+1}-\sqrt{x-1}\right)\to 0$, so we get an indeterminate form. Whenever you see the sum or difference of two square roots or two exponents at the top or at the bottom of a fraction, consider multiplying both the top and the bottom of the fraction by the conjugate expression. Here we get:

$$\frac{1}{\sqrt{x}\left(\sqrt{x+1}-\sqrt{x-1}\right)}=\frac{1}{\sqrt{x}\left(\sqrt{x+1}-\sqrt{x-1}\right)}\cdot\frac{\left(\sqrt{x+1}+\sqrt{x-1}\right)}{\left(\sqrt{x+1}+\sqrt{x-1}\right)}=\frac{\left(\sqrt{x+1}+\sqrt{x-1}\right)}{\sqrt{x}\left(\left(\sqrt{x+1}\right)^2-\left(\sqrt{x-1}\right)^2\right)}=$$

$$\frac{\left(\sqrt{x+1}+\sqrt{x-1}\right)}{\sqrt{x}\cdot\left((x+1)-(x-1)\right)}=\frac{\left(\sqrt{x+1}+\sqrt{x-1}\right)}{\sqrt{x}\cdot 2}.$$ Now we can see that the limit is equal to 1 as

$x\to\infty$ (because $\frac{\sqrt{x+1}}{\sqrt{x}}=\sqrt{1+\frac{1}{x}}$ and $\frac{\sqrt{x-1}}{\sqrt{x}}=\sqrt{1-\frac{1}{x}}$). The answer is C.

The above example probably contains more algebra than you will see on the AP exam, but you may encounter a situation where you need to multiply by a conjugate.

2.3.4. Finding Limits Using the Squeeze Theorem

The Squeeze Theorem (a.k.a. the "Sandwich Theorem") states that if $u(x)\le f(x)\le v(x)$ for all x near a and $\lim_{x\to a}u(x)=\lim_{x\to a}v(x)=L$, then $\lim_{x\to a}f(x)$ exists and is equal to L (Figure 2-8-a). In other words, if f is sandwiched between two functions, which have the same limit as $x\to a$, then f has the same limit as $x\to a$. The same theorem applies to the limits at infinity (Figure 2-8-b).

(a)   (b)

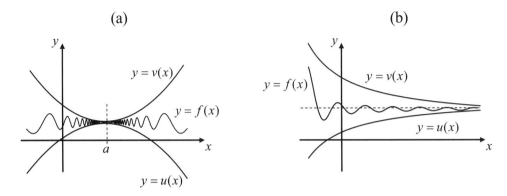

Figure 2-8.   Squeeze (Sandwich) Theorem: (a) $x \to a$; (b) $x \to \infty$

18

Since $-\dfrac{1}{x} \le \dfrac{\sin x}{x} \le \dfrac{1}{x}$ for all $x \ne 0$, the Sandwich Theorem, applied to these three functions, allows us to conclude that

(A) $\lim\limits_{x \to 0} \dfrac{\sin x}{x} = 1$ (B) $\lim\limits_{x \to \infty} \dfrac{\sin x}{x} = 1$ (C) $\lim\limits_{x \to \infty} \dfrac{\sin x}{x} = 0$

(D) $\lim\limits_{x \to 0} \dfrac{\sin x}{x} = \infty$ (E) $\lim\limits_{x \to 0} \dfrac{\sin x}{x} = 0$

First of all, let's consider what happens near zero. The functions $y = \dfrac{1}{x}$ and $y = -\dfrac{1}{x}$ both have no limit at zero. $\lim\limits_{x \to 0^+} \dfrac{1}{x} = \infty$ and $\lim\limits_{x \to 0^-} \dfrac{1}{x} = -\infty$. The left and right-hand limits for $y = -\dfrac{1}{x}$ are the opposite of those for $y = \dfrac{1}{x}$. Since our two "pieces of bread" do not have equal limits, we cannot apply the Sandwich Theorem at zero. Now consider the limits at infinity. $\lim\limits_{x \to \infty} \dfrac{1}{x} = 0$ and $\lim\limits_{x \to \infty}\left(-\dfrac{1}{x} \right) = 0$, so we can apply the Sandwich Theorem and

$\lim\limits_{x \to \infty} \dfrac{\sin x}{x} = 0$. The answer is C.

A similar comparison principle helps us to establish that a limit does not exist: if $u(x) \to \infty$ as $x \to a$ and $f(x) \ge u(x)$ for all x near a, then $f(x) \to \infty$, too.

2.3.5.   Finding Limits that are Derivatives in Disguise

We have not talked about derivatives yet — they are discussed in the next chapter. If you have not studied derivatives, postpone reading this section until you have read the derivatives chapter.

By definition, $f'(c) = \lim\limits_{x \to c} \dfrac{f(x) - f(c)}{x - c}$. If we substitute a number in for c we can construct a limit question. For example, suppose $f(x) = e^x$ and $c = 0$. We get

$$\frac{d}{dx}\left(e^x\right)\bigg|_{x=0} = \lim\limits_{x \to 0}\frac{e^x - e^0}{x - 0} = \lim\limits_{x \to 0}\frac{e^x - 1}{x}. \text{ However, we know that } \frac{d}{dx}\left(e^x\right) = e^x. \text{ So}$$

$$\lim\limits_{x \to 0}\frac{e^x - 1}{x} = e^0 = 1.$$

Similarly, $\lim\limits_{x \to 0}\dfrac{\cos x - 1}{x} = \lim\limits_{x \to 0}\dfrac{\cos x - \cos 0}{x - 0} = \dfrac{d}{dx}\left(\cos x\right)\bigg|_{x=0} = -\sin 0 = 0$.

To solve such a problem you need to recognize that the given limit is a derivative in disguise. This may be difficult when c is chosen in such a way that $f(c) = 0$. For example, the famous limit $\lim\limits_{x \to 0}\dfrac{\sin x}{x}$ is equivalent to

$$\lim\limits_{x \to 0}\frac{\sin x - \sin 0}{x - 0} = \frac{d}{dx}\sin x\bigg|_{x=0} = \cos 0 = 1.$$

A second definition of the derivative, $f'(x) = \lim\limits_{h \to 0}\dfrac{f(x+h) - f(x)}{h}$, is also found in limit questions in a similar manner. For example,

$$\lim\limits_{h \to 0}\frac{\tan(\pi + h)}{h} = \lim\limits_{h \to 0}\frac{\tan(\pi + h) - \tan\pi}{h} = \frac{d}{dx}\left(\tan x\right)\bigg|_{x=\pi} = \sec^2\pi = 1.$$

| 19 |

Find $\lim\limits_{x \to e}\dfrac{\ln x - 1}{x - e}$.

$$\lim\limits_{x \to e}\frac{\ln x - 1}{x - e} = \lim\limits_{x \to e}\frac{\ln x - \ln e}{x - e} = \frac{d}{dx}\left(\ln x\right)\bigg|_{x=e} = \frac{1}{x}\bigg|_{x=e} = \frac{1}{e}.$$

20

If $\lim\limits_{h \to 0} \dfrac{f(4+h) - f(4)}{h} = 6$, which of the following must be true?

(A) $f(4) = 6$ (B) $f(h) = 2$ (C) $f'(4) = 6$

(D) $\lim\limits_{h \to 0} \dfrac{f(h)}{h} = 6$ (E) $\lim\limits_{h \to 0} \dfrac{f(4-h) + f(4)}{h} = 6$

The answer is C. The other choices represent algebraic mistakes or common misconceptions.

2.3.6. L'Hôpital's Rule

Strictly speaking, it is a BC-only topic, but L'Hôpital's Rule is not very complicated, and it provides a general and powerful method for calculating some limits. You won't need to answer specific questions about l'Hôpital's Rule on an AB exam, but you can still use it for answering limit questions. We recommend that you learn it.

To use l'Hôpital's Rule, you must be familiar with derivatives, the subject of the next chapter. If you are not, postpone reading this section until later.

L'Hôpital's Rule helps us resolve indeterminate situations $\lim\limits_{x \to a} \dfrac{f(x)}{g(x)}$ where $f(x) \to 0$ and $g(x) \to 0$ or $f(x)$ and $g(x)$ both approach infinity as $x \to a$ (i.e., we have an indeterminate form $\dfrac{0}{0}$ or $\dfrac{\infty}{\infty}$). These are the most interesting cases, of course. You can also use l'Hôpital's for $\lim f(x) \cdot g(x)$ where $f(x) \to 0$ and $g(x) \to \infty$ (i.e., the indeterminate form $0 \cdot \infty$) by rewriting it first as $\lim \dfrac{f(x)}{\left[\dfrac{1}{g(x)}\right]}$ or $\lim \dfrac{g(x)}{\left[\dfrac{1}{f(x)}\right]}$.

> **l'Hôpital's Rule: if both $f(x) \to 0$ and $g(x) \to 0$ or both $f(x) \to \infty$ and $g(x) \to \infty$, as $x \to a$, then $\lim\limits_{x \to a} \dfrac{f(x)}{g(x)} = \lim\limits_{x \to a} \dfrac{f'(x)}{g'(x)}$, provided the latter limit exists.**

The same is true for $x \to \pm\infty$.

> **Be careful: l'Hôpital's Rule applies only to <u>indeterminate</u> forms $\dfrac{0}{0}$ and $\dfrac{\infty}{\infty}$.**
>
> **If you use l'Hôpital's Rule when the numerator and the denominator of the fraction behave differently (e.g., $f \to 0$, but $g \to L \neq 0$ or $f \to \infty$, but $g \to 0$), you will get an incorrect result.**

It is a very common error to misuse l'Hôpital's Rule where it does not apply. For example, if we apply l'Hôpital's to $\lim\limits_{x \to 0} \dfrac{1 - \cos x}{x + x^2}$, the limit becomes $\lim\limits_{x \to 0} \dfrac{\sin x}{1 + 2x}$, which is equal to 0. You are done — stop here! If you were to apply l'Hôpital's to the second limit, $\lim\limits_{x \to 0} \dfrac{\sin x}{1 + 2x}$, you would have $\lim\limits_{x \to 0} \dfrac{\cos x}{2}$, which is equal to $\dfrac{1}{2}$. Don't be too quick to continue applying l'Hôpital's Rule in succeeding steps in a problem without first checking to see if you still have a $\dfrac{0}{0}$ or $\dfrac{\infty}{\infty}$ situation.

21

$$\lim_{x \to 0^+} x \ln x =$$

(A) -1 (B) 0 (C) 1 (D) ∞ (E) $-\infty$

As $x \to 0^+$, $x \to 0$ and $\ln x \to -\infty$. This limit does not lead to a $\dfrac{0}{0}$ or $\dfrac{\infty}{\infty}$ situation; but we can change it into $\dfrac{\ln x}{\dfrac{1}{x}}$. Now we have $\dfrac{-\infty}{\infty}$ and can apply l'Hôpital's:

$$\lim_{x \to 0^+} x \ln x = \lim_{x \to 0^+} \frac{\ln x}{\dfrac{1}{x}} = \lim_{x \to 0^+} \frac{\dfrac{1}{x}}{\dfrac{-1}{x^2}} = \lim_{x \to 0^+} \frac{-x^2}{x} = \lim_{x \to 0^+} (-x) = 0. \text{ The answer is B.}$$

22

$$\lim_{x \to \infty} \frac{\ln(x+1)^2}{\ln x^3} =$$

(A) $\dfrac{1}{3}$ (B) $\dfrac{2}{3}$ (C) 1   (D) 2   (E) Does not exist

This fraction yields $\dfrac{\infty}{\infty}$, as $x \to \infty$, so we can use l'Hôpital's Rule:

$$\lim_{x \to \infty} \frac{\ln(x+1)^2}{\ln x^3} = \lim_{x \to \infty} \frac{2\ln(x+1)}{3\ln x} = \text{(l'Hôpital's)} \ \lim_{x \to \infty} \frac{\dfrac{2}{x+1}}{\dfrac{3}{x}} = \lim_{x \to \infty} \frac{2x}{3(x+1)} = \frac{2}{3}. \text{ The answer}$$

is B.

23

$$\lim_{x \to 1} \left(\frac{1}{\ln x} - \frac{1}{x-1} \right) =$$

(A) 0   (B) $\dfrac{1}{e}$ (C) $\dfrac{1}{2}$ (D) 1   (E) ∞

This limit has an indeterminate form of $\infty - \infty$. To use l'Hôpital's we can combine the

fractions: $\lim_{x \to 1} \left(\dfrac{1}{\ln x} - \dfrac{1}{x-1} \right) = \lim_{x \to 1} \dfrac{x-1-\ln x}{(\ln x)(x-1)}$. Now we have $\dfrac{0}{0}$ and l'Hôpital's gives us

$$\lim_{x \to 1} \frac{x-1-\ln x}{(\ln x)(x-1)} = \lim_{x \to 1} \frac{1 - \dfrac{1}{x}}{\dfrac{1}{x}(x-1) + \ln x} = \lim_{x \to 1} \frac{x-1}{x-1 + x\ln x}. \text{ Since we still have an}$$

indeterminate form $\dfrac{0}{0}$, we can apply l'Hôpital's a second time. We get

$$\lim_{x \to 1} \frac{1}{1 + \ln x + x\left(\dfrac{1}{x}\right)} = \frac{1}{2}. \text{ The answer is C.}$$

24

$$\lim_{x \to 0} (2x+1)^{\frac{1}{x}} =$$

(A) 0   (B) 1   (C) e (D) e^2 (E) ∞

It is a common error to assume that if $f(x) \to 1$ and $g(x) \to \infty$, then $\lim f(x)^{g(x)} = 1$. In fact, 1^∞ is another indeterminate form. We can resolve it by first taking the natural log of both sides, then applying l'Hôpital's Rule. Here let $L = \lim_{x \to 0} (2x+1)^{\frac{1}{x}}$. Then

$$\ln L = \ln\left[\lim_{x \to 0} (2x+1)^{\frac{1}{x}}\right] = \lim_{x \to 0}\left[\ln\left((2x+1)^{\frac{1}{x}}\right)\right] = \lim_{x \to 0}\left[\frac{1}{x}\ln(2x+1)\right] = \lim_{x \to 0}\frac{\ln(2x+1)}{x}. \text{ Now}$$

we can apply l'Hôpital's: $\lim_{x \to 0}\dfrac{\ln(2x+1)}{x} = \lim_{x \to 0}\dfrac{\dfrac{2}{2x+1}}{1} = 2$. We get $\ln L = 2 \implies L = e^2$.

The answer is D.

2.4. Continuity

> **A function *f* is called *continuous* at *x* = *c* if <u>*c* is in its domain</u> and**
> $$\lim_{x \to c} f(x) = f(c).$$
>
> **We say that *f* is continuous on an open interval (*a*, *b*) if it is defined and continuous at every *x* on (*a*, *b*).**
>
> ***f* is called continuous on a closed interval [*a*, *b*] if it is continuous on the open interval (*a*, *b*), defined at *a* and *b*, and $\lim_{x \to a^+} f(x) = f(a)$ and**
> $$\lim_{x \to b^-} f(x) = f(b).$$
>
> **The definition of continuity does not apply to values outside the function's domain.**

All the common functions that you are familiar with (polynomials, rational, $|x|$, trig, ln, a^x, $\sqrt{x}$, etc.) are continuous on their natural domain (i.e., for every x where the function makes sense). Moreover, any algebraic combination or composition of these functions, that is <u>every function described by a single algebraic formula</u>, is continuous at every x for which that function is defined.

The above fact should not be confused with continuity on an interval. For example, $f(x) = \dfrac{1}{x}$ is continuous at every $x \neq 0$. However, it is not continuous on $[-1, 1]$ (simply because it is not defined at $x = 0$).

As we have seen in Section 2.3.1, continuity can help us find limits. If a function is described by a single algebraic formula and $f(c)$ is defined, then $\lim\limits_{x \to c} f(x) = f(c)$. AP exam questions sometimes use piecewise-defined functions patched together from different formulas on adjacent intervals. To establish continuity of such functions, we have to examine the left and right limits of the function at the "seams."

$\boxed{25}$

The function $f(x)$ is defined as follows:

$$f(x) = \begin{cases} 4x - 11, & x < 3 \\ kx^2, & x \geq 3 \end{cases}$$

Find a value of k such that f is continuous for all x.

We have to find a value k such that there is no jump at the "seam" point $x = 3$. We must ensure that $\lim\limits_{x \to 3^-} f(x) = \lim\limits_{x \to 3^+} f(x) = f(3)$. Each piece of the function is a continuous function (a polynomial), so the limits can be found simply by substituting in 3 for x:

$$4 \cdot 3 - 11 = k \cdot 3^2 \Rightarrow 1 = 9k \Rightarrow k = \frac{1}{9}.$$

26

Which of the following values for k makes the function $f(x) = \begin{cases} \ln(x+k), & 0 < x < 3 \\ \cos(kx), & x \le 0 \end{cases}$

continuous at $x = 0$?

    (A) 0   (B) 1   (C) $\dfrac{\pi}{2}$ (D) e (E) π

Both of the pieces of this function are continuous on their domains, so the right- and left-hand limits at zero can be found by substitution. Equating the limits, we get $\cos 0 = \ln(0+k) \Rightarrow 1 = \ln k \Rightarrow k = e$. The answer is D.

27

Find a value of c that makes the function $f(x) = \begin{cases} \dfrac{1-\cos 3x}{x^2}, & \text{if } x \ne 0 \\ c, & \text{if } x = 0 \end{cases}$ continuous at

$x = 0$.

For f to be continuous at $x = 0$, we must have $\lim\limits_{x \to 0} f(x) = f(0) = c$.

Applying l'Hôpital's Rule twice, we get $\lim\limits_{x \to 0} \dfrac{1-\cos 3x}{x^2} = \lim\limits_{x \to 0} \dfrac{3\sin 3x}{2x} = \lim\limits_{x \to 0} \dfrac{9\cos 3x}{2} = \dfrac{9}{2}$.

The answer is $c = \dfrac{9}{2}$.

Discontinuities

It is important to be able to quickly locate and describe different types of discontinuities of functions. Some AP questions may, among other things, explicitly ask you to find and classify discontinuities of a function. In other questions, you might need to split the domain of the function into intervals where the function is continuous. For example, the Fundamental Theorem of Calculus (Chapter 5) can be used to evaluate a definite integral of a function, but only on an interval where the function is continuous. So, if the interval contains a point of discontinuity, we need to split the interval at that point and work with each subinterval separately.

We have already seen different types of discontinuities in previous examples in this chapter. Let's look at them again — they are summarized in Figure 2-9. All five functions in the figure have a discontinuity at $x = 0$. Discontinuities in (a) and (b) are called ***removable*** because it is easy to repair the function and make it continuous by appropriately setting its value at $x = 0$.

> **A function f has a removable discontinuity at $x = a$ when**
> $$\lim_{x \to a^-} f(x) = \lim_{x \to a^+} f(x), \text{ that is when } \lim_{x \to a} f(x) \text{ exists, but } f(a) \text{ does not exist}$$
> **or $f(a) \neq \lim\limits_{x \to a} f(x)$. To remove this type of discontinuity, we can simply**
> **define (or redefine) $f(a) = \lim\limits_{x \to a} f(x)$.**

The other three functions in Figure 2-9, (c), (d), and (e), have a ***non-removable*** discontinuity at $x = 0$. The (c)-type is called a jump and the (d)-type is called an infinite discontinuity.

> **Exploring discontinuities with calculator graphs is <u>error-prone</u>. The calculator may show jumps and infinite discontinuities as connected by a nearly vertical line if your calculator is in the "connected" mode and the discontinuity happens to fall between pixels. A removable discontinuity, if it falls between pixels, won't show. Oscillating discontinuities may look confusing due to low resolution.**

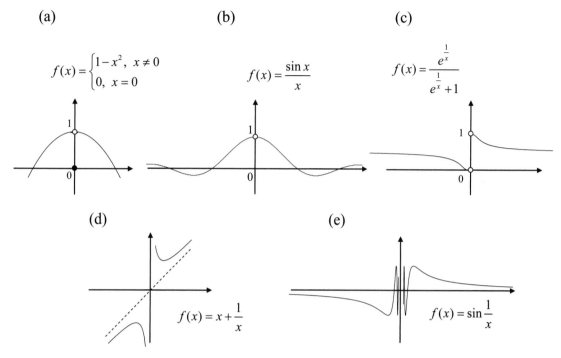

Figure 2-9.   Different types of discontinuities:
(a) and (b) — removable; (c) — jump; (d) — infinite; (e) — oscillation

28

The five functions below are discontinuous at $x = 1$. For each function determine whether the discontinuity is removable or non-removable, and find $\lim\limits_{x \to 1} f(x)$, if it exists.

(a) $f(x) = \dfrac{x^2 + x - 2}{x - 1}$

(b) $f(x) = \begin{cases} \dfrac{x^2 - 1}{x - 1}, & x \neq 1 \\ 3, & x = 1 \end{cases}$

(c) $f(x) = \dfrac{|x - 1|}{x - 1}$

(d) $f(x) = \dfrac{1}{x - 1}$

(e) $f(x) = \cos\left(\dfrac{x}{x - 1}\right)$

(a) $f(x) = \dfrac{x^2 + x - 2}{x - 1}$

Both the numerator and the denominator of this rational function are
equal to 0 at $x = 1$. Let's factor the numerator:
$x^2 + x - 2 = (x-1)(x+2)$. As we can see,

$f(x) = \dfrac{(x-1)(x+2)}{x-1} = x+2$ when $x \neq 1$. $\lim\limits_{x\to1} f(x) = 3$. The

discontinuity at $x = 1$ is removable: if we set $f(1) = 3$, the function

becomes continuous.

(b) $f(x) = \begin{cases} \dfrac{x^2 - 1}{x - 1}, & x \neq 1 \\ 3, & x = 1 \end{cases}$

This function's discontinuity is removable, too. $\lim\limits_{x\to1} f(x) = 2$ (with

work similar to Part (a)). Setting $f(1) = 2$ instead of $f(1) = 3$ would

remove the "hiccup" point to fill the hole, and make $f(x)$ continuous.

(c) $f(x) = \dfrac{|x-1|}{x-1}$

This function has a jump at $x = 1$. This is a non-removable
discontinuity, because the left and right limits at $x = 1$ have different
values: $\lim\limits_{x\to1^-} f(x) = -1$ and $\lim\limits_{x\to1^+} f(x) = 1$. Therefore, $\lim\limits_{x\to1} f(x)$ does
not exist.

(d) $f(x) = \dfrac{1}{x - 1}$

This function has a vertical asymptote at $x = 1$. It is a non-removable
infinite discontinuity; $\lim\limits_{x\to1} f(x)$ does not exist.

(e) $f(x) = \cos\left(\dfrac{x}{x-1}\right)$

This function oscillates between –1 and 1 when $x \to 1$. It is a non-removable discontinuity; $\lim\limits_{x \to 1} f(x)$ does not exist.

The notion of a discontinuity at a given point is somewhat artificial: it implies that the function is continuous in the neighborhood of the point of discontinuity, except that point itself. This is not always the case. For example, the Dirichlet Function —

$$D(x) = \begin{cases} 0, \text{ if } x \text{ is a rational number} \\ 1, \text{ if } x \text{ is an irrational number} \end{cases}$$

— is discontinuous everywhere!

The Intermediate Value Theorem(IVT)

A function *f* that is continuous on a closed interval [*a, b*] takes on every value between $f(a)$ and $f(b)$.

In other words, for any y between $f(a)$ and $f(b)$, there is a c in $[a, b]$ such that $y = f(c)$.

The IVT allows us to ascertain that an equation has a solution on a given interval. For example, let's show that an equation $\sin x = x + 1$ has a solution on $(-\pi, 0)$. The function $f(x) = \sin x - (x+1)$ is continuous over all real numbers since it is a difference of two continuous functions. $f(-\pi) = 0 - (-\pi + 1) = \pi - 1 > 0$ and $f(0) = -1 < 0$, so there is a number c in the interval $[-\pi, 0]$ such that $f(c) = 0$. You can use your graphing calculator to find that solution and be sure that it indeed exists.

29

t (sec)	0	5	7	9	14
$v(t)$ (meters/sec)	3.35	4.25	2.75	2.55	4.70

The table above gives several measurements of the velocity of a particle moving along a straight line. What is the smallest possible number of times where $v(t)$ is exactly 4 meters/sec $(0 \leq t \leq 14)$?

(A) 0   (B) 1   (C) 2   (D) 3   (E) 4

We have to assume that $v(t)$ is a continuous function on [0, 14]. There are at least three values of t for which $v(t) = 4$, one on each of the intervals $(0, 5)$, $(5, 7)$, and $(9, 14)$, because the IVT applies to each of these three intervals. On the other hand, we can construct a function $v(t)$ that has no more than three such values of t. For example, we can simply connect the given points with straight line segments. The answer is D.

2.5.   Relative Rates of Growth

In many applications, we need to compare the relative magnitudes of functions in a somewhat abstract way, disregarding a constant factor. Computer scientists, for example, may be interested in analyzing the running time of an algorithm as a function of n, where n is the size of the computational task, such as computing 2^n. If one algorithm takes $\ln n$ "steps" and another algorithm takes n "steps," the first algorithm is considered much better, even if one "step" in the first algorithm takes longer than one "step" in the second algorithm. For large enough n, $A \ln n < Bn$, even if A is huge and B is small.

> If $f(x)$ and $g(x)$ are positive for sufficiently large values of x and
>
> $$\lim_{x \to \infty} \frac{f(x)}{g(x)} = L \neq 0, \text{ we say that } f \text{ and } g \text{ grow at the same rate as } x \to \infty.$$

If $\lim\limits_{x\to\infty}\dfrac{f(x)}{g(x)}=0$ (or, equivalently, if $\lim\limits_{x\to\infty}\dfrac{g(x)}{f(x)}=\infty$), we say that g grows faster than f.

The rate of growth of a polynomial is determined by its degree. Any two polynomials of the same degree have the same growth rate. $\ln x$ grows much more slowly than $f(x)=x$: as we know, a tenfold increase in x adds only about 2.3 to $\ln x$. Exponential growth, a^x $(a>1)$, is much faster than polynomial growth. Indeed, for any integer $k>0$ and any real $a>1$, $\lim\limits_{x\to\infty}\dfrac{x^k}{a^x}=0$. To show that, apply l'Hôpital's Rule k times. In general,

l'Hôpital's Rule is the tool of choice for comparing rates of growth.

30

Which of the following functions grows at the same rate as e^x?

    (A) x^{1000} (B) 2^x (C) 4^x (D) e^{x+1} (E) xe^x

x^{1000}, a polynomial, grows more slowly than the exponent. To show that, apply l'Hôpital's Rule 1000 times!

$$\lim_{x\to\infty}\frac{2^x}{e^x}=\lim_{x\to\infty}\left(\frac{2}{e}\right)^x=0,\ \text{while}\ \lim_{x\to\infty}\frac{4^x}{e^x}=\lim_{x\to\infty}\left(\frac{4}{e}\right)^x=\infty.\ \lim_{x\to\infty}\frac{xe^x}{e^x}=\lim_{x\to\infty}x=\infty$$

Only $\lim\limits_{x\to\infty}\dfrac{e^{x+1}}{e^x}=\lim\limits_{x\to\infty}\dfrac{e\cdot e^x}{e^x}=e$ is finite and not equal to 0. The answer is D.

To summarize: exponential functions with a base greater than 1 grow faster than polynomials and polynomials grow faster than logarithmic functions (Figure 2-10). When you have two different exponential functions, the one with the greater base grows faster. A polynomial with a greater degree grows faster than a polynomial with a smaller degree.

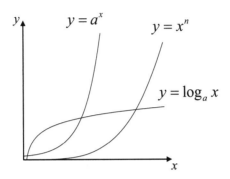

Figure 2-10. Exponential, polynomial, and logarithmic rates of growth

Limits and Continuity Worksheet

1. $\lim\limits_{x \to 1} \dfrac{1 - \sqrt{x}}{x - 1} =$

2. $\lim\limits_{x \to 0} \left(\dfrac{1}{\sin x} - \dfrac{1}{x} \right) =$

3. $\lim\limits_{x \to 0^+} \left(\dfrac{1}{x} - \dfrac{1}{\sqrt{x}} \right) =$

4. $\lim\limits_{x \to \frac{\pi}{2}^-} \dfrac{\tan x + 3}{\tan^2 x} =$

5. $\lim\limits_{x \to -\infty} \dfrac{\sqrt{4x^2 - 1}}{7 - |x|} =$

6. $\lim\limits_{x \to a} \dfrac{\cos x - \cos a}{x - a} =$

7. $\lim\limits_{x \to \infty} x \sin \dfrac{1}{x} =$

8. $\lim\limits_{x \to -4} \dfrac{x^3 + 2x^2 - 8x}{x^2 + 4x} =$

9. $\lim\limits_{x \to 0^+} \dfrac{\ln x}{\ln x - 1} =$

10. $\lim\limits_{x \to 0} \dfrac{(2 + x)^3 - 8}{x} =$

11. $\lim\limits_{x \to \frac{\pi}{2}} \dfrac{\ln(\sin x)}{\dfrac{\pi}{2} - x} =$

12. $\lim\limits_{x \to 0} \dfrac{x \csc x + 1}{x \csc x} =$

13. $\lim\limits_{t \to \infty} \dfrac{1000}{1 + 5e^{-0.2t}} =$

14. Which two functions grow at the same rate: $5x^2$, $2x^5$, $\ln x^2$, $\ln x$, e^{5x}, e^{2x}, as $x \to \infty$?

15. Find a value of k that makes $f(x) = \begin{cases} x^2, & x \le 1 \\ \sin(kx), & x > 1 \end{cases}$ continuous at $x = 1$.

16. If a function f is discontinuous at $x = 4$, which of the following must be true?

    I. $\lim\limits_{x \to 4} f(x)$ does not exist

    II. $f(4)$ does not exist

    III. $\lim\limits_{x \to 4^-} f(x) \ne \lim\limits_{x \to 4^+} f(x)$

    IV. $\lim\limits_{x \to 4} f(x) \ne f(4)$

17. Find all vertical and horizontal asymptotes for the graph of $y = \dfrac{\ln x}{\ln x - 1}$.

18. $\lim\limits_{x \to 0} \dfrac{2\cos x - 2}{x}$ is equal to the derivative of what function at $x = 0$?

19. If $f(x) = \begin{cases} x^2 + 4, & x > 1 \\ 6 - x, & x \le 1 \end{cases}$, then $\lim\limits_{x \to 1} f(x) =$

20. If $\lim\limits_{x \to 3} f(x) = 7$ and $\lim\limits_{x \to 3} g(x) = 5$, then $\lim\limits_{x \to 3} \dfrac{2\left(g(x)\right)^2}{f(x) - 5} =$

21. For what values of k will $\lim\limits_{x \to 3} \dfrac{x - 3}{x^2 - 6x + k}$ exist?

22. For each part draw an example of a function that satisfies the conditions:

    (a) $f(3)$ exists, but $\lim\limits_{x \to 3} f(x)$ does not exist.

    (b) $\lim\limits_{x \to 3} f(x)$ exists, but $f(3)$ does not exist.

23. Find all horizontal and vertical asymptotes for $f(x) = \dfrac{e^{-x}}{x}$.

24. Find the points of discontinuity and identify the type of discontinuity for each function:

    (a) $y = e^{\frac{1}{x}}$ (b) $y = \dfrac{x}{|x|}$ (c) $y = \dfrac{x^2 - 5}{x - \sqrt{5}}$

Worksheet Answers and Solutions

1. $\lim\limits_{x \to 1} \dfrac{1-\sqrt{x}}{x-1} = \lim\limits_{x \to 1} \dfrac{1-\sqrt{x}}{\left(\sqrt{x}-1\right)\left(\sqrt{x}+1\right)} = -\dfrac{1}{2}$.

2. $\lim\limits_{x \to 0}\left(\dfrac{1}{\sin x} - \dfrac{1}{x}\right) = \lim\limits_{x \to 0} \dfrac{x-\sin x}{x\sin x} =$ (l'Hôpital's) $\lim\limits_{x \to 0} \dfrac{1-\cos x}{\sin x + x\cos x} =$ (l'Hôpital's)

    $\lim\limits_{x \to 0} \dfrac{\sin x}{2\cos x - x\sin x} = 0$.

3. $\lim\limits_{x \to 0^+}\left(\dfrac{1}{x} - \dfrac{1}{\sqrt{x}}\right) = \lim\limits_{x \to 0^+} \dfrac{1-\sqrt{x}}{x}$ does not exist.

4. $\lim\limits_{x \to \frac{\pi}{2}^-} \dfrac{\tan x + 3}{\tan^2 x} = \lim\limits_{x \to \frac{\pi}{2}^-} \dfrac{1}{\tan x} + \lim\limits_{x \to \frac{\pi}{2}^-} \dfrac{3}{\tan^2 x} = 0 + 0 = 0$.

5. $\lim\limits_{x \to -\infty} \dfrac{\sqrt{4x^2-1}}{7-|x|} = \lim\limits_{x \to -\infty} \dfrac{\sqrt{4-\dfrac{1}{x^2}}}{\dfrac{7}{|x|}-1} = \dfrac{\sqrt{4}}{-1} = -2$.

6. $\lim\limits_{x \to a} \dfrac{\cos x - \cos a}{x-a} = \dfrac{d}{dx}\cos x\Big|_{x=a} = -\sin a$.

7. Substitute in $\dfrac{1}{x} = y$. Then $\lim\limits_{x \to \infty} x\sin\dfrac{1}{x} = \lim\limits_{y \to 0^+}\dfrac{1}{y}\sin y = 1$.

8. $\lim\limits_{x \to -4} \dfrac{x^3 + 2x^2 - 8x}{x^2 + 4x} = \lim\limits_{x \to -4} \dfrac{x(x+4)(x-2)}{x(x+4)} = -4 - 2 = -6$.

9. $\lim\limits_{x \to 0^+} \dfrac{\ln x}{\ln x - 1} = \lim\limits_{x \to 0^+}\left(1 + \dfrac{1}{\ln x - 1}\right) = 1$.

10. $\lim\limits_{x \to 0} \dfrac{(2+x)^3 - 8}{x} = \dfrac{d}{dx}x^3\Big|_{x=2} = 3x^2\Big|_{x=2} = 12$.

11. $\lim\limits_{x \to \frac{\pi}{2}} \dfrac{\ln(\sin x)}{\dfrac{\pi}{2} - x} =$ (l'Hôpital's) $\lim\limits_{x \to \frac{\pi}{2}} \dfrac{\dfrac{1}{\sin x}\cdot\cos x}{-1} = \lim\limits_{x \to \frac{\pi}{2}} -\cot x = 0$.

12. $\displaystyle\lim_{x\to 0}\frac{x\csc x+1}{x\csc x}=\lim_{x\to 0}\left(\frac{\dfrac{\sin x}{x}+1}{\dfrac{\sin x}{x}}\right)=\frac{1+1}{1}=2$.

13. $\displaystyle\lim_{t\to\infty}\frac{1000}{1+5e^{-0.2t}}=\frac{1000}{1+0}=1000$.

14. $\ln x^2=2\ln|x|$ and $\ln x$.

15. $\sin(kx)=1\Rightarrow k=\dfrac{\pi}{2}+2n\pi$, where n is any integer.

16. IV.

17. $x=e$ and $y=1$.

18. $f(x)=2\cos x+C$.

19. 5

20. $\displaystyle\lim_{x\to 3}\frac{2\big(g(x)\big)^2}{f(x)-5}=\frac{2\cdot 5^2}{7-5}=25$.

21. When $x^2-6x+k\neq(x-3)^2\Rightarrow k\neq 9$.

22.

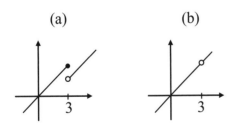

23. $x=0$ and $y=0$.

24. (a) Asymptote (non-removable discontinuity) at $x=0^+$;
    (b) Jump (non-removable discontinuity) at $x=0$;
    (c) Hole (removable discontinuity) at $x=\sqrt{5}$.

Chapter 3.   Derivatives

3.1.   Concepts and Notation

If a person tosses a ball up in the air from the initial position 6 feet above the ground and with an initial velocity of 18 ft/sec, the height of the ball above the ground after t seconds can be found by the formula $s(t) = -16t^2 + 18t + 6$ feet. The velocity of the ball decreases from 18 ft/sec until it reaches zero at the highest point, then changes sign. As the ball falls back to the ground, its velocity is negative and the speed (the absolute value of the velocity) increases. Suppose we want to approximate how fast the ball is moving after 1 second. The average velocity between two points in time, t_1 and t_2, is given by $\dfrac{s(t_2) - s(t_1)}{t_2 - t_1}$. Using the times of $t_1 = 1$ second and $t_2 = 1.1$ seconds, we get

$$\frac{s(1.1) - s(1)}{1.1 - 1} = \frac{6.44 - 8}{0.1} = \frac{-1.56}{0.1} = -15.6 \text{ ft/sec.}$$ The negative number means the ball is falling at an average rate of 15.6 ft/sec. If we use the times of 1 second and 1.01 seconds, we get $\dfrac{s(1.01) - s(1)}{1.01 - 1} = \dfrac{7.8584 - 8}{0.01} = -14.16$ ft/sec. This is a better approximation for the velocity at 1 second. Repeating this process with $t_1 = 1$ second and t_2 closer and closer to 1 second, we can get better and better approximations. Finally, taking the limit $\displaystyle\lim_{t \to 1} \frac{s(t) - s(1)}{t - 1}$, we get a value for the <u>instantaneous</u> velocity of the ball at $t = 1$ second. Substituting in the formula for our $s(t)$, we get

$$\lim_{t \to 1} \frac{(-16t^2 + 18t + 6) - (8)}{t - 1} = \lim_{t \to 1} \frac{-2(8t - 1)(t - 1)}{t - 1} = \lim_{t \to 1}(-2(8t - 1)) = -14.$$ We conclude that at <u>exactly</u> 1 second, the ball is falling at a rate of 14 ft/sec.

Finding the instantaneous rate of change of a function is an essential computation in calculus and its applications. Luckily, we do not have to compute a limit "manually" each time. Differentiation techniques, reviewed later in this chapter, provide powerful tools for finding the **derivative** (i.e. instantaneous rate of change) of a function.

> **Suppose f is defined on the interval $[a, b]$, and $a < c < b$. The derivative of f at $x = c$ is defined as $f'(c) = \displaystyle\lim_{x \to c} \frac{f(x) - f(c)}{x - c}$ provided this limit exists. If the derivative exists, we say that f is *differentiable* at $x = c$.**

In the previous example, we have found that $s'(1) = -14$ ft/sec.

Derivative as the Slope of the Tangent Line

$\dfrac{f(x)-f(c)}{x-c}$ represents the slope of the secant line going through the points $(c, f(c))$ and $(x, f(x))$ on a graph of $f(x)$ (Figure 3-1). Imagine that the point $(x, f(x))$ moves along the curve and approaches the point $(c, f(c))$. If the limit $\displaystyle\lim_{x\to c}\dfrac{f(x)-f(c)}{x-c}$ exists, then the curve has a tangent line at the point $(c, f(c))$ and the limit is equal to the slope of the tangent line at $x = c$.

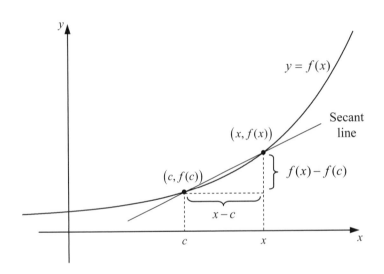

Figure 3-1. Derivative as the slope of the tangent line

The graph of a linear function $f(x) = mx + b$ is a straight line and its derivative $f'(c)$ is equal to m for any c.

Derivative as a Function

The derivative of a function f at a given value of $x = c$ is a number $f'(c) = \lim_{x \to c} \dfrac{f(x) - f(c)}{x - c}$. We can also rewrite it as $f'(c) = \lim_{\Delta x \to 0} \dfrac{f(c + \Delta x) - f(c)}{\Delta x}$ (where $\Delta x = x - c$). If we calculate the derivative for every value of x, we get a function of x:

$$f'(x) = \lim_{\Delta x \to 0} \frac{f(x + \Delta x) - f(x)}{\Delta x}$$

The domain of $f'(x)$ is all values of x where $f(x)$ is defined and differentiable.

Sometimes Δx is replaced with h: $f'(x) = \lim_{h \to 0} \dfrac{f(x + h) - f(x)}{h}$. Writing h rather than Δx makes the formula look a little less cluttered.

1

If $f(x) = -x^2 + x$, which of the following expressions represents $f'(x)$?

(A) $\displaystyle\lim_{h \to 0} \frac{\left(-x^2 + x + h\right) - \left(-x^2 + x\right)}{h}$

(B) $\displaystyle\lim_{h \to x} \frac{\left(-x^2 + x + h\right) - \left(-x^2 + x\right)}{h}$

(C) $\displaystyle \frac{\left[-(x+h)^2 + (x+h)\right] - \left(-x^2 + x\right)}{h}$

(D) $\displaystyle\lim_{h \to 0} \frac{\left[-(x+h)^2 + (x+h)\right] - \left(-x^2 + x\right)}{x}$

(E) None of the above

The correct expression for $f'(x)$ would be

$$\lim_{h \to 0} \frac{f(x+h) - f(h)}{h} = \lim_{h \to 0} \frac{\left[-(x+h)^2 + (x+h)\right] - \left(-x^2 + x\right)}{h}.$$ In Choices A and B, we have

$\dfrac{f(x) + h - f(x)}{h}$ rather than $\dfrac{f(x+h) - f(h)}{h}$. You must be careful to substitute the

quantity $(x + h)$ for each x in the function: $f(x+h)$ is not the same as $f(x) + h$. In addition, in Choice B $h \to x$ should be $h \to 0$. Choice C offers the correct quotient but $\lim_{h \to 0}$ is missing altogether. In Choice D, the denominator should be h, not x.

The answer is E.

Variations in Terminology and Notation

Several expressions can be used in calculus problems to mean "find the derivative":

- Differentiate

- Find the rate of change

- Find the instantaneous rate of change

- Find the slope of the tangent line to the curve

The following notations can be used to represent the derivative of a function $y = f(x)$:

$$f'(x) \qquad\qquad y'$$

$$\frac{dy}{dx} \qquad\qquad \frac{d}{dx} f(x)$$

$$\frac{df}{dx}$$

For example, if $y = \ln x$, then $y' = (\ln x)' = \dfrac{d(\ln x)}{dx} = \dfrac{d}{dx} \ln x$.

If we want to represent the value of $y' = f'(x)$ at a particular $x = c$, we can write $f'(c)$ or

$y'(c)$ or $\dfrac{dy}{dx}\bigg|_{x=c}$.

One-Sided Derivatives

A function is differentiable on an open interval (a, b) if the derivative exists at every point in the interval. At the ends of the interval we can define the right-hand derivative at a as $\lim_{h \to 0^+} \dfrac{f(a+h) - f(a)}{h}$, and the left-hand derivative at b, as $\lim_{h \to 0^-} \dfrac{f(b+h) - f(b)}{h}$. If the function is differentiable on (a, b) and the one-sided derivatives exist at $x = a$ and $x = b$, then we can say that the function is differentiable on the closed interval $[a, b]$.

3.2.   Differentiation Methods

If a function is defined by one formula that combines algebraic operations and "elementary" functions (algebraic, trig, log and exponential), we can use the inventory of derivative formulas and general rules for derivatives to find a formula for its derivative. These techniques are the subject of this section. We can also use a calculator to calculate the derivative of a function at a given x. The use of a calculator is discussed in Section 3.7.

3.2.1.   Derivatives of Algebraic Functions

The formulas for the derivatives of algebraic functions are summarized in Table 3-1. You need to be able to use these formulas very fluently.

$\dfrac{d}{dx}(C) = 0$	The derivative of a constant is 0.
$\dfrac{d}{dx}(mx + b) = m$	The derivative of a linear function is a constant equal to its slope.
$\dfrac{d}{dx}(x^n) = nx^{n-1}$	The *Power Rule*
$\dfrac{d}{dx}\left(\dfrac{1}{x}\right) = -\dfrac{1}{x^2}$	Special case of the Power Rule for $n = -1$
$\dfrac{d}{dx}(\sqrt{x}) = \dfrac{1}{2\sqrt{x}}$	Special case of the Power Rule for $n = \dfrac{1}{2}$

Table 3-1.   Derivatives of algebraic functions

The derivatives of $\dfrac{1}{x}$ and $\sqrt{x}$ are special cases of the Power Rule. For example,

$$\frac{1}{x} = x^{-1} \Rightarrow \frac{d}{dx}\left(\frac{1}{x}\right) = \frac{d}{dx}\left(x^{-1}\right) = (-1)\left(x^{-1-1}\right) = (-1)\left(x^{-2}\right) = -\frac{1}{x^2}.$$

> **These two functions occur often on the AP exam, so it is well worth your time to memorize these special cases of the Power Rule, since time at the exam is at a premium.**

The formulas in Table 3-1 can be combined with general rules for derivatives of a sum, product, and quotient, listed in Table 3-2.

$(C \cdot f)' = C \cdot f'$	A constant factor "slides out" of a derivative
$(f + g)' = f' + g'$	The derivative of a sum
$(f - g)' = f' - g'$	The derivative of a difference
$(f \cdot g)' = fg' + gf'$	The *Product Rule*
$\left(\dfrac{f}{g}\right)' = \dfrac{gf' - fg'}{g^2}$	The *Quotient Rule*

Table 3-2. Algebraic rules for derivatives

> **Be especially careful with derivatives of products and quotients — do not make false assumptions.**

Differentiation questions often require you to combine some of these formulas, so you must be able to recognize which formulas to use and in which order.

2

If $y = \dfrac{6-x}{x^2+1}$ then $\dfrac{dy}{dx} =$

(A) $\dfrac{3x^2-12x+1}{\left(x^2+1\right)^2}$

(B) $\dfrac{-3x^2+12x-1}{\left(x^2+1\right)^2}$

(C) $\dfrac{x^2-12x-1}{\left(x^2+1\right)^2}$

(D) $\dfrac{-x^2+12x+1}{\left(x^2+1\right)^2}$

(E) $\dfrac{-x^2-7}{\left(x^2+1\right)^2}$

Using the Quotient Rule, we get:

$$\frac{dy}{dx} = \frac{\left(x^2+1\right)(-1)-(6-x)(2x)}{\left(x^2+1\right)^2} = \frac{-x^2-1-12x+2x^2}{\left(x^2+1\right)^2} = \frac{x^2-12x-1}{\left(x^2+1\right)^2}.$$

The answer is C. Be careful! The most common mistakes are reversing the two terms in the numerator (choice D) and adding the two terms in the numerator instead of subtracting them (choice B).

3

If $y = \dfrac{f}{gh}$, where f, g, and h are differentiable functions of x, $\dfrac{dy}{dx} =$

(A) $\dfrac{f'gh - fg'h'}{g^2h^2}$

(B) $\dfrac{fg'h' - f'gh}{g^2h^2}$

(C) $\dfrac{f'gh - fgh' + fg'h}{g^2h^2}$

(D) $\dfrac{f'gh - fgh' - fg'h}{g^2h^2}$

(E) $\dfrac{f'gh + fgh' + fg'h}{g^2h^2}$

This problem requires both the Quotient Rule and the Product Rule:

$$\left(\frac{f}{gh}\right)' = \frac{(gh)f' - f(gh)'}{(gh)^2} = \frac{(gh)f' - f(gh' + hg')}{g^2h^2} = \frac{f'gh - fgh' - fg'h}{g^2h^2}. \text{ The answer}$$

is D. Be careful distributing the negative sign or you may mistakenly choose C.

3.2.2.   Derivatives of Trig, Exponential, and Logarithmic Functions

The formulas for the derivatives of trig, exponential and logarithmic functions are summarized in Table 3-3. Many of them can be derived from each other and from more general rules, but there is no time for that during an AP exam, so

you need to memorize all of them.

$\dfrac{d}{dx}(\sin x) = \cos x$	$\dfrac{d}{dx}(\cos x) = -\sin x$
$\dfrac{d}{dx}(\tan x) = \sec^2 x$	$\dfrac{d}{dx}(\cot x) = -\csc^2 x$
$\dfrac{d}{dx}(\sec x) = \sec x \tan x$	$\dfrac{d}{dx}(\csc x) = -\csc x \cot x$
$\dfrac{d}{dx}(\sin^{-1} x) = \dfrac{1}{\sqrt{1-x^2}}$	$\dfrac{d}{dx}(\cos^{-1} x) = -\dfrac{1}{\sqrt{1-x^2}}$
$\dfrac{d}{dx}(\tan^{-1} x) = \dfrac{1}{1+x^2}$	$\dfrac{d}{dx}(\cot^{-1} x) = -\dfrac{1}{1+x^2}$
$\dfrac{d}{dx}(e^x) = e^x$	$\dfrac{d}{dx}(a^x) = a^x(\ln a)$
$\dfrac{d}{dx}(\ln x) = \dfrac{1}{x}$	$\dfrac{d}{dx}(\log_a x) = \dfrac{1}{x(\ln a)}$

Table 3-3.   Derivatives of trig, exponential, and logarithmic functions

The derivatives of the six trig functions and the four inverse trig functions are used in many AP problems, both multiple-choice and free-response. You should memorize these formulas in pairs, sin and cos, tan and cot, sec and csc, because of their similarities. It will also help a lot if you can remember four basic trig identities —

$$\sin^2\theta + \cos^2\theta = 1$$

$$\sin(2\theta) = 2\sin\theta\cos\theta$$

$$\cos(2\theta) = \cos^2\theta - \sin^2\theta$$

$$\tan\theta = \frac{\sin\theta}{\cos\theta}; \quad \cot\theta = \frac{\cos\theta}{\sin\theta} = \frac{1}{\tan\theta}$$

— and be able to recall quickly the sin, cos, and tan values for

$$0, \frac{\pi}{6}, \frac{\pi}{4}, \frac{\pi}{3}, \frac{\pi}{2}, \pi, \frac{3\pi}{2}, 2\pi.$$

Review the simplification rules for exponents and logs. You are expected to know, of course, that $e^0 = 1$, $\ln 1 = 0$, and $\ln e = 1$.

Which of the following defines the rate of change of the function $f(t) = \dfrac{\sin t}{\sin t - \cos t}$?

(A) $\dfrac{1}{\left(\sin t - \cos t\right)^2}$
(B) $\dfrac{-1}{\left(\sin t - \cos t\right)^2}$
(C) $\dfrac{1 + \sin 2t}{\left(\sin t - \cos t\right)^2}$

(D) $\dfrac{-\cos 2t}{\left(\sin t - \cos t\right)^2}$
(E) $\dfrac{\sin 2t - \cos 2t}{\left(\sin t - \cos t\right)^2}$

Apply the Quotient Rule and the derivative formulas for $\sin t$ and $\cos t$:

$$f'(t) = \frac{\left(\sin t - \cos t\right)\left(\cos t\right) - \left(\sin t\right)\left(\cos t + \sin t\right)}{\left(\sin t - \cos t\right)^2} = \frac{\sin t \cos t - \cos^2 t - \sin t \cos t - \sin^2 t}{\left(\sin t - \cos t\right)^2} =$$

$$\frac{-\left(\cos^2 t + \sin^2 t\right)}{\left(\sin t - \cos t\right)^2} = \frac{-1}{\left(\sin t - \cos t\right)^2}. \text{ The answer is B.}$$

5

If $s(t) = \dfrac{\ln t}{e^t}$ then $\dfrac{ds}{dt} =$

(A) $\dfrac{1 - \ln t}{e^t}$ (B) $\dfrac{1 - t \ln t}{e^t}$ (C) $\dfrac{t \ln t - 1}{te^t}$ (D) $\dfrac{1 - t \ln t}{te^t}$ (E) $\dfrac{1 - e^t \ln t}{e^{2t}}$

This is one more problem that requires the Quotient Rule: $s'(t) = \dfrac{e^t\left(\dfrac{1}{t}\right) - \ln t\left(e^t\right)}{\left(e^t\right)^2}$.

Divide the numerator and denominator by e^t and multiply by t to get $s'(t) = \dfrac{1 - t \ln t}{te^t}$.

The answer is D.

3.2.3.   The Chain Rule

The Chain Rule allows us to find the derivative of a function that is a composition of two or more functions. In combination with the rules and formulas discussed in the previous sections, the Chain Rule allows us to find the derivative of almost any function defined by one reasonable formula.

You may encounter several equivalent forms of the Chain Rule:

Form 1: If $y = f(u)$ where u is a function of x, then $\dfrac{dy}{dx} = \dfrac{dy}{du} \cdot \dfrac{du}{dx}$.

Form 2: If f and g are differentiable functions of x, then $\dfrac{d}{dx}\big(f(g(x))\big) = f'(g(x)) \cdot g'(x)$.

> **Forgetting to apply the Chain Rule when differentiating a composite function is a common mistake.**

One special case of the Chain Rule involves finding the derivative of $f(mx + b)$. If $f'(x) = g(x)$ then $f'(mx + b) = g(mx + b) \cdot m = m \cdot g(mx + b)$. For example, if $y = \sin\left(2x + \dfrac{\pi}{4}\right)$, then $y' = 2\cos\left(2x + \dfrac{\pi}{4}\right)$. Don't forget the factor m.

6

If $y = \cos\left(4x^3\right)$, which of the following is $\dfrac{dy}{dx}$?

(A) $\sin\left(4x^3\right)$ (B) $-\sin\left(4x^3\right)$ (C) $\sin\left(12x^2\right)$

(D) $-\sin\left(12x^2\right)$ (E) $-12x^2\sin\left(4x^3\right)$

First think of y as cos(*something*) and recall its derivative, –sin(*something*). But do not stop there! Remember that *something* is not just x but $4x^3$, a function of x. To get it right, you have to multiply –sin(*something*) by the derivative of *something*. Thus you get $-\sin(4x^3)\left(4 \cdot 3x^2\right) = -12x^2\sin(4x^3)$. The answer is E.

7

$\dfrac{d}{dx}\left(\cos^3 x\right) =$

(A) $\sin^3 x$ (B) $-\sin^3 x$ (C) $-3\cos^2 x\sin x$

(D) $3\cos^2 x\sin^2 x$ (E) $-\cos^3 x\sin x$

Here you can think of $\left(\cos x\right)^3$ as $(...)^3$. Using the Chain Rule, you get

$\dfrac{d}{dx}\left(\cos^3 x\right) = 3\left(\cos x\right)^2\left(-\sin x\right)$. The answer is C.

Sometimes it helps to think of this as the "outside-inside" method. The cubic function is the "outside" and $\cos x$ is the "inside" function. First think of the inside function as "u" and find the derivative of the outside function with respect to u; then multiply it by the derivative of the inside function.

8

What is the derivative of $y = \ln(\cos t)$?

(A) $\dfrac{1}{\cos t}$ (B) $\dfrac{-1}{\sin t}$ (C) $\cot t$ (D) $-\tan t$ (E) $-\sec t$

The derivative of $\ln u$ is $\dfrac{1}{u} \cdot \dfrac{du}{dt}$. In this case, $u = \cos t$, so we get $y' = \dfrac{-\sin t}{\cos t} = -\tan t$.

The answer is D.

9

$\dfrac{d}{dx} \ln\left(\sin\sqrt{x}\right) =$

(A) $2\sqrt{x}\cos x$ (B) $\dfrac{\cot x}{2}$ (C) $\dfrac{\cos\sqrt{x}}{2\sqrt{x}}$

(D) $\dfrac{\cot\sqrt{x}}{2\sqrt{x}}$ (E) $\dfrac{1}{2\sqrt{x}\sin\sqrt{x}}$

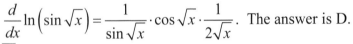

In the previous examples, the "chain" had only two links. In this example, the "chain" has three links — $f\big(g(h(x))\big)$ — unless, of course, you remember the derivative of $\ln(\sin u)$. Applying the Chain Rule carefully, starting from the outside, we get

$\dfrac{d}{dx} \ln\left(\sin\sqrt{x}\right) = \dfrac{1}{\sin\sqrt{x}} \cdot \cos\sqrt{x} \cdot \dfrac{1}{2\sqrt{x}}$. The answer is D.

10

Suppose that the functions f and g have the following values:
$f(-1) = 4$, $g(-1) = 2$, $f'(-1) = 7$, $f'(2) = 5$ and $g'(-1) = 3$.
What is the value of the derivative of $f\big(g(x)\big)$ at $x = -1$?

(A) 5   (B) 14   (C) 15   (D) 21   (E) 32

Using the Chain Rule we get,

$$\frac{d}{dx}f\big(g(x)\big)\Big|_{x=-1} = f'\big(g(-1)\big)\cdot g'(-1) = f'(2)\cdot g'(-1) = 5\cdot 3 = 15. \quad \text{The answer is C.}$$

Find the rate of change of the function $f(x) = x\tan^{-1}x - \ln\sqrt{1+x^2}$ when $x = 1$.

It is easier to compute the derivative if we rewrite the function as

$f(x) = x\tan^{-1}x - \dfrac{1}{2}\ln\big(1+x^2\big)$. Use the Product Rule on the first term and the Chain

Rule on the second: $f'(x) = \left[x\left(\dfrac{1}{1+x^2}\right) + \tan^{-1}x \right] - \dfrac{1}{2}\left(\dfrac{1}{1+x^2}\right)(2x) = \tan^{-1}x$. Then

$f'(1) = \dfrac{\pi}{4}$.

3.3. Higher Order Derivatives

In many applications, it is necessary to find the derivative of the derivative of *f*. This function is called the second derivative of *f*. The second derivative is often used in curve sketching and in distance-velocity-acceleration problems.

The following notations are used for the second derivative of $y = f(x)$:

$$f''(x) = \frac{d^2y}{dx^2} = \frac{d}{dx}\big[f'(x)\big] = y''$$

There are occasions (once in a while) when it is necessary to find a third or even a fourth derivative. In general, the *n*-th derivative of a function is the derivative of its $(n-1)$-th derivative:

$$f^{(n)}(x) = \frac{d}{dx}\big[f^{(n-1)}(x)\big]$$

12

If $f(x) = x^3 - x^2 + \dfrac{1}{x}$, which of the following is $f''(2)$?

(A) $\dfrac{31}{4}$ (B) $\dfrac{39}{4}$ (C) $\dfrac{79}{8}$ (D) $\dfrac{81}{8}$ (E) $\dfrac{41}{4}$

$f'(x) = 3x^2 - 2x - \dfrac{1}{x^2}$ and $f''(x) = 6x - 2 + \dfrac{2}{x^3}$, so $f''(2) = 12 - 2 + \dfrac{2}{8} = \dfrac{41}{4}$. The answer is E.

13

The fourth derivative of $f(x)$ equals $f(x)$ for all of the following functions except

(A) $\sin x$ (B) $\cos x$ (C) $-5e^x$ (D) e^{2x} (E) e^{-x}

If you simply examine the functions quickly, you might notice that differentiating e^{2x} multiplies it by 2 each time. A more formal solution comes from the following table:

	$f(x)$	$f'(x)$	$f''(x)$	$f'''(x)$	$f^{(4)}(x)$
(A)	$\sin x$	$\cos x$	$-\sin x$	$-\cos x$	$\sin x$
(B)	$\cos x$	$-\sin x$	$-\cos x$	$\sin x$	$\cos x$
(C)	$-5e^x$	$-5e^x$	$-5e^x$	$-5e^x$	$-5e^x$
(D)	e^{2x}	$2e^{2x}$	$4e^{2x}$	$8e^{2x}$	$16e^{2x}$
(E)	e^{-x}	$-e^{-x}$	e^{-x}	$-e^{-x}$	e^{-x}

The answer is D.

3.4.   Implicit Differentiation

In most AP calculus problems, the relationship between y and x is given as an <u>explicit</u> function $y = f(x)$. But a function also can be defined implicitly through an equation in x and y. The equation defines a curve in the xy-plane. For example, $x^2 + y^2 = 9$ is the equation of a circle. An equation like this may represent a relationship that is not actually a function because more than one y corresponds to a given x. But it is often possible to split the curve into two or more curves that represent functions. For example, the circle defined by the equation $x^2 + y^2 = 9$ can be viewed as the union of two functions:

$y = \sqrt{9 - x^2}$ (upper semicircle) and $y = -\sqrt{9 - x^2}$ (lower semicircle). In problems that we encounter in AP calculus exams, we can assume that the implicit relationship represents a function or the union of several functions.

> **When you need to find the derivative of a function given implicitly, it is <u>not</u> necessary to solve the equation for y in terms of x.**

In fact, many problems are very difficult or impossible to solve for y. Try, for example, $\tan(xy) + 3xy^3 - 10 = e^{4y} - x$.

Instead, we use the method called ***implicit differentiation***. It consists of three steps:

1. Differentiate both sides of the equation with respect to x. When you take the derivative of a term that includes y, think of y as $y(x)$ and apply the Chain Rule (multiply by y') when appropriate. We do not know the derivative of y at this point in the process, so just trust that it exists and write y' for it.

2. Collect the terms with y' on one side of the equation and the terms without it on the other side.

3. Solve for y'.

The formula that you get for y' is an expression that may be in terms of x or y, or both. When you use it to find the slope at a point on the curve, you may need to know both coordinates of the point.

Many problems involving implicit differentiation are no more complicated than finding derivatives of regular functions, but we often cannot refer to a graph to check the result, even if calculator is allowed, since not all graphing calculators can graph implicit relations.

14

What is the slope of the tangent line to the ellipse with equation $x^2 - xy + y^2 = 21$ at the point $(1, 5)$?

(A) $-\dfrac{7}{9}$ (B) $-\dfrac{2}{9}$ (C) $\dfrac{1}{3}$ (D) $\dfrac{24}{9}$ (E) 7

Use implicit differentiation.

Step 1: $2x - (xy' + y) + 2yy' = 0$

Step 2: Substitute in $x = 1$, $y = 5$: $2 \cdot 1 - (1 \cdot y' + 5) + 2 \cdot 5 \cdot y' = 0 \Rightarrow 9y' = 3 \Rightarrow y' = \dfrac{1}{3}$. Or

solve first: $y' = \dfrac{2x - y}{x - 2y}$, then substitute in the values for x and y. The answer is C.

A very common error would be to forget to use the Product Rule for the xy term or the Chain Rule for the y^2 term. Another common error would be to neglect distributing the negative over both terms in the derivative of $-xy$. Still another error would be to write 21 instead of 0 for the right side of the derivative in Step 1.

15

For what values of x in the interval $[0, 2]$ is the tangent line to the curve given by $y = x^2 - \cos y$ parallel to the y-axis?

(A) 0 (B) $\sqrt{\dfrac{\pi}{2}}$ (C) $\sqrt{\dfrac{3\pi}{2}}$ (D) $\dfrac{\pi}{2}$ (E) No such values

A line parallel to the y-axis has an undefined slope, so we are looking for the values of x where the derivative is undefined. Using implicit differentiation we get:
$\dfrac{dy}{dx} = 2x + \sin y \dfrac{dy}{dx} \Rightarrow \dfrac{dy}{dx}(1 - \sin y) = 2x \Rightarrow \dfrac{dy}{dx} = \dfrac{2x}{1 - \sin y}$. It is undefined when

$1 - \sin y = 0 \Rightarrow y = \dfrac{\pi}{2} + 2\pi k$ (where k is any integer). Since the question asked for

values of x, we need to solve for x: $x^2 - \cos\dfrac{\pi}{2} = \dfrac{\pi}{2} + 2\pi k \Rightarrow x = \pm\sqrt{\dfrac{\pi}{2} + 2\pi k}$. Of these

values, only $x = \sqrt{\dfrac{\pi}{2}}$ lies in the interval $[0, 2]$. The answer is B.

Second Derivative of an Implicit Function

To find the second derivative for an implicit function, find the first derivative first. In AP questions, the first derivative is often a fraction, in which both the numerator and the denominator may include x and y. For example, for $x^2 + y^2 = 9$ we get

$2x + 2yy' = 0 \Rightarrow y' = \dfrac{-2x}{2y} = -\dfrac{x}{y}$. Apply $\dfrac{d}{dx}$ again using the Quotient Rule:

$\dfrac{d^2y}{dx^2} = -\dfrac{y \cdot 1 - xy'}{y^2}$. Now substitute for y' the expression that you found earlier:

$\dfrac{d^2y}{dx^2} = -\dfrac{y \cdot 1 - xy'}{y^2} = -\dfrac{y - x\left(-\dfrac{x}{y}\right)}{y^2} = -\dfrac{y^2 + x^2}{y^3}$. In this case you can simplify the result

further: $x^2 + y^2 = 9 \Rightarrow \dfrac{d^2y}{dx^2} = -\dfrac{9}{y^3}$. In other cases you may end up with a pretty

complicated expression.

16

Find $\dfrac{d^2y}{dx^2}$ if y and x are related by the formula $y^3 = 4 - \ln x$.

We could solve the formula for y as a function of x, but then taking the derivatives would require fractional exponents and more complicated expressions. Using implicit

differentiation on $y^3 = 4 - \ln x$, we get $3y^2 y' = -\dfrac{1}{x} \Rightarrow y' = -\dfrac{1}{3xy^2}$. Differentiate again:

$\dfrac{d^2y}{dx^2} = -\left[-\dfrac{3x(2y)y' + 3y^2}{\left(3xy^2\right)^2}\right] = \dfrac{6xyy' + 3y^2}{9x^2y^4}$. Now substitute in the y' found above:

$\dfrac{d^2y}{dx^2} = \dfrac{6xy\left(\dfrac{-1}{3xy^2}\right) + 3y^2}{9x^2y^4} = \dfrac{-\dfrac{2}{y} + 3y^2}{9x^2y^4} = \dfrac{-2 + 3y^3}{9x^2y^5}.$

> **In problems where you need to find $\dfrac{d^2y}{dx^2}$ of an implicit function <u>at a given point</u> on the curve, you can simplify algebra by first finding the value of y' at that point, then substituting the <u>values</u> of $x, y,$ and y' into the expression for $\dfrac{d^2y}{dx^2}$.**

Suppose in the previous example we only had to find the value of $\dfrac{d^2y}{dx^2}$ at the point

$\left(e^3, 1\right)$. After finding $\dfrac{d^2y}{dx^2} = ... = \dfrac{6xyy' + 3y^2}{9x^2 y^4}$, we can substitute in the value for

$y'\big|_{\left(e^3, 1\right)} = -\dfrac{1}{3xy^2}\bigg|_{x=e^3,\ y=1} = -\dfrac{1}{3e^3}$ and then simplify:

$$\dfrac{d^2y}{dx^2}\bigg|_{\left(e^3, 1\right)} = ... = \dfrac{6xyy' + 3y^2}{9x^2 y^4}\bigg|_{x=e^3,\ y=1} = \dfrac{6 \cdot e^3 \cdot 1 \cdot \left(-\dfrac{1}{3e^3}\right) + 3 \cdot 1^2}{9 \cdot \left(e^3\right)^2 1^4} = \dfrac{1}{9e^6}.$$

3.5.   The Derivative of the Inverse Function

Functions f and g are inverse to each other if and only if $f\left(g(x)\right) = g\left(f(x)\right) = x$ for all x such that both sides are defined. For example, $f(x) = e^x$ and $g(x) = \ln x$ are mutually inverse. If a point with coordinates (a, b) is on the graph of $f(x)$, then the point (b, a) is on the graph of $g(x)$.

> **If functions f and g are inverse to each other and the point (a, b) is on the graph of f, then $g'(b) = \dfrac{1}{f'(a)}$.**

This can be rewritten as $g'(x) = \dfrac{1}{f'\left(g(x)\right)}.$ [*]

For example, if $f(x) = \sin x$ and $g(x) = \sin^{-1} x$, then $\dfrac{d}{dx}\left(\sin^{-1} x\right) = \dfrac{1}{\cos\left(\sin^{-1} x\right)}$. The

denominator can be simplified further. Recall that $\cos\theta = \sqrt{1 - \sin^2\theta}$, for $\dfrac{-\pi}{2} < \theta < \dfrac{\pi}{2}$.

So we get $\dfrac{d}{dx}\left(\sin^{-1} x\right) = \dfrac{1}{\sqrt{1 - \left[\sin\left(\sin^{-1} x\right)\right]^2}} = \dfrac{1}{\sqrt{1 - x^2}}$, the formula from Table 3-3.

---

[*] You can obtain the formula for the derivative of the inverse function through implicit differentiation of the equation $f\left(g(x)\right) = x$. Taking derivatives of both sides and applying the Chain Rule to the left side,

we get: $\dfrac{d}{dx}\left[f\left(g(x)\right)\right] = f'\left(g(x)\right) \cdot g'(x) = 1 \Rightarrow g'(x) = \dfrac{1}{f'\left(g(x)\right)}.$

The slopes of the tangent lines to the graph of f at (a, b) and to the graph of its inverse function g at (b, a) are reciprocal to each other (Figure 3-2).

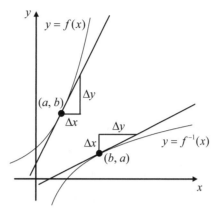

Figure 3-2.   Slope at (a, b) on the graph of f is reciprocal to slope at (b, a) on the graph of f^{-1}

17

If $(4, 5)$ is a point on the graph of $y = f(x)$, a one-to-one function, and $f'(4) = 2$, then which of the following must be true?

(A) $\dfrac{d}{dx}\left[f^{-1}(x)\right]\bigg|_{x=4} = \dfrac{1}{2}$ (B) $\dfrac{d}{dx}\left[f^{-1}(x)\right]\bigg|_{x=5} = \dfrac{1}{2}$

(C) $\dfrac{d}{dx}\left[f^{-1}(x)\right]\bigg|_{x=4} = -\dfrac{1}{2}$ (D) $\dfrac{d}{dx}\left[f^{-1}(x)\right]\bigg|_{x=5} = -\dfrac{1}{2}$

(E) $\dfrac{d}{dx}\left[f^{-1}(x)\right]\bigg|_{x=4} = -2$

Remember that the point on the inverse graph is $(5, 4)$. The answer is B.

18

If $f(x) = \dfrac{4}{x} + 2$ and g is the inverse of f, then $g'(10) =$

(A) -16 (B) $-\dfrac{1}{2}$ (C) $-\dfrac{1}{16}$ (D) $\dfrac{1}{2}$ (E) 16

There is an x such that $f(x) = 10$: $\dfrac{4}{x} + 2 = 10 \Rightarrow x = \dfrac{1}{2}$. Since $f'(x) = -\dfrac{4}{x^2}$,

$f'\left(\dfrac{1}{2}\right) = -\dfrac{4}{\left(\dfrac{1}{2}\right)^2} = -16$. Then $g'(10) = -\dfrac{1}{16}$. The answer is C.

3.6.   Differentiability and Continuity

> **If a function has a derivative at $x = c$, it must be continuous at that point. In other words, differentiability implies continuity.**

Indeed, if $\lim\limits_{x \to c} \dfrac{f(x) - f(c)}{x - c}$ exists, then $\lim\limits_{x \to c}[f(x) - f(c)] = 0 \Rightarrow \lim\limits_{x \to c} f(x) = f(c)$.

> **By definition, a function with a discontinuity at $x = c$ does not have a derivative at $x = c$, even if it is a removable discontinuity.**

A continuous function does not have to be differentiable. In particular, the derivative of a continuous function does not exist if the function has a "corner", a "cusp," or a vertical tangent. Figure 3-3 illustrates these situations.

If the discontinuity were removed in the function in Figure 3-3-a, it would become differentiable. For the function in Figure 3-3-b, both one-sided derivatives exist at $x = 3$, but they have different values. At the cusp in Figure 3-3-c, the slopes of the secant lines approach ∞ from one side and $-\infty$ from the other side. Thus a vertical tangent line goes through a cusp. You can think of a cusp as an extreme case of a corner. The vertical tangent in Figure 3-3-d is where the slopes of the secant lines approach ∞ from both sides.

You may encounter the above situations on the AP exam.[*]

---

[*] There are also more interesting examples. Consider, for instance, $f(x) = \begin{cases} x\sin\dfrac{1}{x}, & \text{if } x \neq 0 \\ 0, & \text{if } x = 0 \end{cases}$. This function is continuous everywhere. But if you try to find its derivative at $x = 0$, by definition you get

$\lim\limits_{x \to 0} \dfrac{x\sin\dfrac{1}{x} - 0}{x - 0} = \lim\limits_{x \to 0}\left[\sin\dfrac{1}{x}\right]$, which does not exist.

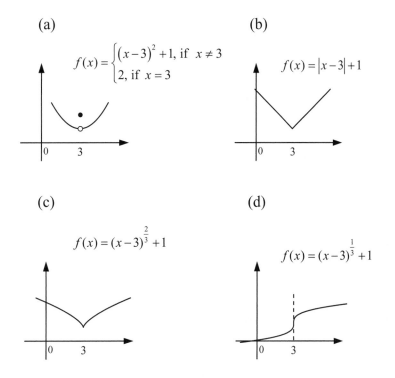

Figure 3-3.   Functions with no derivative at $x = 3$:
(a) discontinuity (b) "corner" (c) "cusp" (d) vertical tangent

---

19

Let $f(x) = \begin{cases} x^2 - 4x + 3, & x \le 4 \\ e^{x+b} + a, & x > 4 \end{cases}$. Find the values of a and b such that $f(x)$ is

differentiable at $x = 4$.

First of all, $f(x)$ must be continuous, so $\lim_{x \to 4^-} f(x) = \lim_{x \to 4^+} f(x) \Rightarrow 3 = e^{4+b} + a$. A

piecewise formula for f' is $f'(x) = \begin{cases} 2x - 4, & x < 4 \\ e^{x+b}, & x > 4 \end{cases}$. If f is differentiable, the left-hand

and the right-hand derivatives must be equal at $x = 4$:

$4 = e^{4+b} \Rightarrow \ln 4 = 4 + b \Rightarrow b = \ln 4 - 4$. Substituting this into the first equation, we get

$3 = e^{4+\ln 4 - 4} + a \Rightarrow 3 = 4 + a \Rightarrow a = -1$. Thus, the values for a and b are

$a = -1$ and $b = \ln 4 - 4$.

20

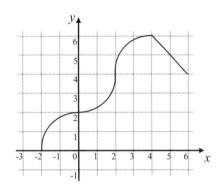

The graph of $y = f(x)$ above consists of three quarter circles and a line segment. For which values of x in the open interval $(-2, 6)$ is f not differentiable?

(A) 0 only   (B) 2 only   C)   0 and 2 only

(D) 2 and 4 only   (E) 0, 2, and 4

f has a vertical tangent at $x = 2$ and a corner at $x = 4$. At $x = 0$, the function has a horizontal tangent; so $f'(x) = 0$. The answer is D.

21

For what values of x is $f(x) = \dfrac{(x-1)^{\frac{2}{3}}}{x+2}$ not differentiable?

(A) 1 only   (B) −2 only   (C) −2 and 1 only

(D) −2, 1, and 7 only   (E) ±1 and ±2 only

$f(x)$ is not continuous and, therefore, not differentiable at $x = -2$.

$$f'(x) = \frac{(x+2)\left(\dfrac{2}{3}\right)(x-1)^{-\frac{1}{3}} - (x-1)^{\frac{2}{3}}}{(x+2)^2} = \frac{\dfrac{(x-1)^{-\frac{1}{3}}}{3}\left[2(x+2) - 3(x-1)\right]}{(x+2)^2} =$$

$\dfrac{2(x+2) - 3(x-1)}{3(x-1)^{\frac{1}{3}}(x+2)^2}$, so the derivative does not exist at $x = 1$ and at $x = -2$. The answer is C.

3.7.   Finding Derivatives with a Calculator

On the calculator portion of the AP exam, it may be appropriate to use your graphing calculator to find derivatives. You must be familiar with the procedure for finding the value of the derivative at a point. It may be also helpful to know how to produce the graph of the derivative of a function.

Your graphing calculator has a command for finding a numerical estimate for the derivative of a function at a given value of x. However, this procedure is not foolproof. You have to understand how your calculator produces the numerical estimate for the derivative in order to know when you can trust the result.

Most calculators evaluate the derivative by using a symmetric difference quotient $\dfrac{f(x+h)-f(x-h)}{2h}$ where h is a small number. TI-83 and other TI models set $h = 0.001$ by default, but you can enter your own value for h as a parameter in the `nDeriv` command.

> **Be careful: your calculator always gives you some result, even if the function is non-differentiable.**

For example, if $f(x)=|x|$, your calculator may tell you that $f'(0)=0$. If $y=\sqrt[3]{x}$, your calculator may say that $\left.\dfrac{dy}{dx}\right|_{x=0}=100$. Actually, in both of these cases the derivative does not exist. For functions that are differentiable, the symmetric difference quotient gives results that are quite accurate for most derivatives. Usually you can trust the accuracy to at least the three places after the decimal point required in AP exam questions. A good procedure for assuring the required accuracy is to decrease the value of h to see if that changes the value of f' in the first three decimal places. If not, you have the desired accuracy.

| 22 |

What is the rate of change for $f(x)=\left(x^2-8.5\right)^x$ at $x=3.04$?

You could try to find a formula for the derivative, but even if you did it correctly, you'd still have to use your calculator to evaluate it at $x = 3.04$. It is much easier and faster to use your calculator to get the numerical derivative. The syntax for entering this problem into the calculator varies with the model. With a TI-83 calculator, for example, you enter `nDeriv(x²-8.5)^x,x,3.04)`. The calculator gives ▮ ≈ 9.924 (rounded to the third decimal place).

Your calculator can find the derivative of a function at a given point, but, strictly speaking, AP questions will not assume that your calculator can also graph the derivative of a function. Still, most calculators provide this operation and occasionally it may be useful to speed up your work. Consider the following example.

23

Given $f(x) = \sin(16x^2 - e^{2x})\cos x$, what is the smallest positive value of x at which the tangent line to the graph of f has a slope of $\frac{1}{2}$?

(A) 0.068   (B) 0.103   (C) 0.357   (D) 0.429   (E) 2.183

You could, of course find the derivative of f analytically, but it would be complicated. Most graphing calculators are capable of graphing the derivative of a function. This will allow you to visualize the derivative, even though you don't know its formula. With TI-83, for example, enter $f(x)$ in as Y₁. Then, define Y₂ as `nDeriv(Y₁,x,x)`. Define Y₃=0.5. Graph only the functions Y₂ and Y₃, since it is not necessary to see the graph of f to answer this question. Use your calculator to find the first intersection of Y₂ and Y₃ to the right of the y-axis. The x-coordinate of the intersection is 0.103, answer B.

Working with CAS Calculators

If you have a calculator with a built-in Computer Algebra System (CAS) and are comfortable using it, you should plan to use it on the AP exam. The TI-89 and several of the HP calculators have a built-in CAS. These calculators give results that you can trust for most problems that you will encounter on the exam. For example, if you try to find the derivative of $y = \sqrt[3]{x}$ at $x = 0$, a non-CAS calculator gives you an incorrect result (e.g., 100), while a CAS calculator will correctly give you `undef`.

CAS calculators will sometimes give you results that are not what you expect, and you have to know some mathematics to be able to interpret them. If you find the derivative of $y = \tan x$, the result is $(\tan(x))^2 + 1$, which equals $\sec^2 x$, but you must remember the trig identity to know that.

Derivatives Worksheet

In problems 1 – 14, find the derivative:

1. $f(x) = \dfrac{x^2 - 3}{2x - 5}$

2. $f(x) = e^x \cos x$

3. $y = \tan\left(\ln(2x+1)\right)$

4. $y = \cos^4\left(x^2\right)$

5. $y = \dfrac{\cos x}{1 + \sin x}$

6. $y = \tan^{-1}\left(3x^2\right)$

7. $f(x) = \ln\left(1 + e^x\right)$

8. $f(x) = e^{-\frac{x^2}{2}}$

9. $y = \left(\dfrac{1 + \sin x}{1 - \cos x}\right)^2$

10. $y = \sin\left(\dfrac{2}{x}\right)$

11. $y = x\sqrt{2x+1}$

12. $y = 2x^{3.5} + x^{-3.5} + \tan(3.5x) + 3.5^2$

13. $y = \sec x \tan x$

14. $y = \sqrt{1 + \cos x}$

15. Find the derivative of $y = \dfrac{fg}{h}$, in terms of f, g, h, f', g', and h', where $f, g,$ and h are functions of x.

16. Find the rate of change of $y = \left(4x^3 + 7x^2 + 1\right)^2$ at $x = -1$.

17. If $f(x) = \sin^3 x$, find $f''(x)$.

18. Find the slope of the line tangent to the curve $y = \cos(2x)$ at the point where $x = \dfrac{\pi}{6}$.

19. Let $f(x) = \begin{cases} \sin(\pi x), & 0 \le x \le 1 \\ ax + b, & 1 < x \le 2 \end{cases}$. Find the values of a and b such that $f(x)$ is differentiable at $x = 1$.

20. Which of the following functions are NOT differentiable at $x = 0$?

    I. $y = \sqrt{4 - x^2}$ II. $y = x^{2/3}$ III. $y = x^{4/3}$ IV. $y = x^{-2}$ V. $y = \left|\sin(2x)\right|$

21. Consider the function given by $f(x) = \begin{cases} 2-x \ , \ x \le 1 \\ x^2 - x + 1 \ , \ x > 1 \end{cases}$. Is the function continuous, differentiable, neither, or both at $x = 1$?

22. If $y^2 = 1 + \dfrac{2}{x}$, find $\dfrac{d^2 y}{dx^2}$ in terms of y and x.

23. What is the n-th derivative of $y = e^{nx}$?

24. What is the slope of the tangent line to the curve with equation $x^3 + 2xy + y^3 = 3$ at the point $(-1, \ 2)$?

25. If $f(x) = 4x - \dfrac{1}{x}$ $(x > 0)$ and g is the inverse of f, then what is the value of $g'(0)$?

26. What is the derivative of $y = x + \sin(xy)$, in terms of x and y, at any point on the curve?

27. $f(g(x)) = x$, $f(2) = 7$, and $f'(2) = 6$. What is $g'(7)$?

Worksheet Answers and Solutions

1. $\dfrac{(2x-5)2x-(x^2-3)2}{(2x-5)^2}=\dfrac{2x^2-10x+6}{(2x-5)^2}$

2. $e^x\left(\cos x-\sin x\right)$

3. $\dfrac{2\sec^2\left(\ln\left(2x+1\right)\right)}{2x+1}$

4. $4\cos^3\left(x^2\right)\cdot\left(-\sin\left(x^2\right)\right)\cdot 2x=-8x\cos^3\left(x^2\right)\sin\left(x^2\right)$

5. $\dfrac{\left(1+\sin x\right)\left(-\sin x\right)-\cos x\cdot\cos x}{\left(1+\sin x\right)^2}=-\dfrac{1+\sin x}{\left(1+\sin x\right)^2}=-\dfrac{1}{1+\sin x}$

6. $\dfrac{6x}{1+9x^4}$

7. $\dfrac{e^x}{1+e^x}$

8. $-xe^{-\frac{x^2}{2}}$

9. $\dfrac{2\left(1+\sin x\right)}{1-\cos x}\cdot\dfrac{\left(1-\cos x\right)\cos x-\left(1+\sin x\right)\sin x}{\left(1-\cos x\right)^2}=\dfrac{2\left(1+\sin x\right)\left(\cos x-\sin x-1\right)}{\left(1-\cos x\right)^3}$

10. $-\dfrac{2}{x^2}\cos\left(\dfrac{2}{x}\right)$

11. $\sqrt{2x+1}+\dfrac{x}{\sqrt{2x+1}}=\dfrac{3x+1}{\sqrt{2x+1}}$

12. $7x^{2.5}-3.5x^{-4.5}+3.5\sec^2\left(3.5x\right)$

13. $\sec x\left(\sec^2 x+\tan^2 x\right)$

14. $\dfrac{-\sin x}{2\sqrt{1+\cos x}}$

15. $\dfrac{f'gh+fg'h-fgh'}{h^2}$

16. $2\left(4x^3+7x^2+1\right)\left(12x^2+14x\right)\Big|_{x=-1}=-16$

17. $\dfrac{d}{dx}\left[3\sin^2 x\cos x\right]=3\left[-\sin^3 x+2\sin x\cos^2 x\right]=3\sin x\left(2\cos^2 x-\sin^2 x\right)$

18. $-2\sin\left(2x\right)\big|_{x=\frac{\pi}{6}}=-\sqrt{3}$

19. $\sin\pi=a+b;\ \pi\cos\pi=a\ \Rightarrow\ a=-\pi,\ b=\pi$

20.   II, IV, and V

21.   Continuous but not differentiable

22. $2yy'=\dfrac{-2}{x^2}\ \Rightarrow\ y'=-\dfrac{1}{yx^2}\ \Rightarrow\ y''=\dfrac{x^2y'+2yx}{y^2x^4}=\dfrac{-\dfrac{x^2}{yx^2}+2yx}{y^2x^4}=\dfrac{2y^2x-1}{y^3x^4}$

23. $n^n e^{nx}$

24. $3x^2+2y+2xy'+3y^2y'=0\ \Rightarrow\ y'=-\dfrac{3x^2+2y}{2x+3y^2}\ \Rightarrow\ y'\big|_{x=-1,y=2}=-\dfrac{7}{10}$

25. $f(c)=0\ \Rightarrow\ \dfrac{1}{c^2}=4\ \Rightarrow\ f'(c)=4+\dfrac{1}{c^2}=8\ \Rightarrow\ g'(0)=\dfrac{1}{f'(c)}=\dfrac{1}{8}$

26. $y'=1+\cos\left(xy\right)\left(y+xy'\right)\ \Rightarrow\ y'=\dfrac{1+y\cos\left(xy\right)}{1-x\cos\left(xy\right)}$

27. $g'(7)=\dfrac{1}{f'(2)}=\dfrac{1}{6}$

Chapter 4.   Applications of Derivatives

4.1.   Tangent and Normal Lines and Linear Approximation

Writing the equation of the tangent line to a curve is the most basic application of the derivative and one of the most often seen in AP problems. Recall that

> **the slope of the tangent line to the graph of a function at a point on the graph is equal to the derivative of the function at that point.**

1

What is the slope of the tangent line to the graph of $y = \sin^3(2x)$ at $x = \dfrac{\pi}{6}$?

(A) $\dfrac{1}{8}$ (B) $\dfrac{3}{4}$ (C) $\dfrac{9}{8}$ (D) $\dfrac{9}{4}$ (E) $\dfrac{9}{2}$

This problem requires application of the Chain Rule with three "links." Working from the outside in, we get $y' = 3\sin^2(2x)\cos(2x)(2) = 6\sin^2(2x)\cos(2x)$. The slope of the tangent line at $x = \dfrac{\pi}{6}$ is $y'\left(\dfrac{\pi}{6}\right) = 6\sin^2\left(\dfrac{\pi}{3}\right)\cos\left(\dfrac{\pi}{3}\right) = 6\left(\dfrac{\sqrt{3}}{2}\right)^2\left(\dfrac{1}{2}\right) = \dfrac{9}{4}$. The answer is D.

2

Which of the following is an equation for the tangent line to the graph of $f(x) = \sin(\cos x)$ at $x = \dfrac{\pi}{2}$?

(A) $y = x - \dfrac{\pi}{2}$ (B) $y = -x + \dfrac{\pi}{2}$ (C) $y = \dfrac{\pi}{2}$

(D) $y = \sin(1)\left(x - \dfrac{\pi}{2}\right)$ (E) $y = \cos(1)\left(x - \dfrac{\pi}{2}\right)$

$f'(x) = -\cos(\cos x) \cdot \sin x$ so the slope of the tangent line is

$f'\left(\dfrac{\pi}{2}\right) = -\cos\left(\cos\dfrac{\pi}{2}\right)\sin\dfrac{\pi}{2} = -1$. The point on the curve has coordinates $\left(\dfrac{\pi}{2}, 0\right)$. The

equation of the tangent line is $y - 0 = -1\left(x - \dfrac{\pi}{2}\right)$. The answer is B.

| 3 |

For what values of x is the tangent line to the graph of $f(x) = e^x(x-1)$ parallel to the x-axis?

   (A) 0 only   (B) 1 only   (C) 0 and 1   (D) 2 only

   (E) There are no such values of x

If the tangent line is parallel to the x-axis, the derivative of f must be equal to zero.
$f'(x) = e^x + e^x(x-1) = xe^x \Rightarrow x = 0$. The answer is A.

The tangent line to a curve can be used to approximate values for the function near the point of tangency. Approximating a function by its tangent line is called ***linear approximation*** of a function.

In general terms,

> **If f is differentiable at $x = a$, then the linear approximation of f at a is the linear function $L(x) = f(a) + f'(a)(x-a)$.**

(If f were a straight line, we would get exactly $f(x) = L(x)$.)

| 4 |

Use the linear approximation to $f(x) = \sqrt[3]{x}$ at $x = 64$ to estimate $\sqrt[3]{63}$.

The linear approximation of *f* at $x = 64$ is $L(x) = f(64) + f'(64)(x - 64)$. $f(64) = 4$.

Find the derivative at $x = 64$: $f(x) = x^{\frac{1}{3}} \Rightarrow f'(x) = \frac{1}{3}x^{-\frac{2}{3}} \Rightarrow f'(64) = \frac{1}{3}(64)^{-\frac{2}{3}} = \frac{1}{48}$.

The linear approximation gives us $f(63) = 4 + \frac{1}{48}(63 - 64) = 4 - \frac{1}{48} \approx 3.979166$. This is

a pretty good approximation: $\sqrt[3]{63} \approx 3.9790572$, so the linear approximation is correct
to the thousandths place.

$\boxed{5}$

Use linear approximation for the area of a circle to estimate the increase in that area when
the radius changes from 8 cm to 8.1 cm.

$L(r) = A(8) + A'(8)(r - 8) \Rightarrow L(r) - A(8) = A'(8)(r - 8)$. The area of a circle is

$A = \pi r^2$, so $A'(r) = 2\pi r$ and $A'(8) = 16\pi$. The increase is approximately

$16\pi(8.1 - 8) = 1.6\pi$. So, as the radius increases from 8 cm to 8.1 cm, the area increases

by approximately 1.6π cm^2.

Normal Lines

The normal line to a curve is "perpendicular" to the curve; more precisely it is
perpendicular to the tangent line at the point of tangency. The term "normal line" does
not appear in the AP calculus course description; AP questions are likely to refer to a
normal line as a line perpendicular to a tangent line.

If the slope of the tangent line is *m*, then the slope of the normal line is $-\dfrac{1}{m}$.

Which of the following is an equation for the line perpendicular to the tangent line to the graph of $f(x) = \sin^{-1} x$ at the point where $x = \dfrac{\sqrt{3}}{2}$?

(A) $y - \dfrac{\pi}{3} = 2\left(x - \dfrac{\sqrt{3}}{2}\right)$ (B) $y - \dfrac{\pi}{3} = -2\left(x - \dfrac{\sqrt{3}}{2}\right)$

(C) $y - \dfrac{\pi}{3} = -\dfrac{1}{2}\left(x - \dfrac{\sqrt{3}}{2}\right)$ (D) $y - \dfrac{\pi}{3} = \dfrac{1}{2}\left(x - \dfrac{\sqrt{3}}{2}\right)$

(E) $y - \dfrac{\pi}{3} = \dfrac{1}{2}\left(x - \sqrt{3}\right)$

$f'(x) = \dfrac{1}{\sqrt{1-x^2}}$, so $f'\left(\dfrac{\sqrt{3}}{2}\right) = 2$. Thus, the slope of the normal line is $-\dfrac{1}{2}$.

$f\left(\dfrac{\sqrt{3}}{2}\right) = \dfrac{\pi}{3}$, so the normal line contains the point $\left(\dfrac{\sqrt{3}}{2}, \dfrac{\pi}{3}\right)$ and has a slope of $-\dfrac{1}{2}$.

The point-slope equation is $y - \dfrac{\pi}{3} = -\dfrac{1}{2}\left(x - \dfrac{\sqrt{3}}{2}\right)$, answer C.

4.2.   The Mean Value Theorem

The Mean Value Theorem (MVT) for derivatives relates the average rate of change for a function over an interval to the instantaneous rate of change at some point in the interval. Recall that the average rate of change for f on $[a, b]$ is $\dfrac{f(b) - f(a)}{b - a}$.

If f is continuous on a closed interval [*a*, *b*] and differentiable on the open interval (*a*, *b*), then there exists at least one number c in the interval (*a*, *b*), such that $f'(c) = \dfrac{f(b) - f(a)}{b - a}$.

In the graphical interpretation, the MVT means that if you connect the endpoints of the graph on the interval with a chord, there will be at least one point on the graph where the tangent line is parallel to the chord (Figure 4-1).

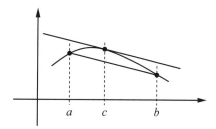

Figure 4-1. The Mean Value Theorem for derivatives

The conditions of the theorem are very important and cannot be ignored. If there is just <u>one point</u> in the interval at which *f* fails to have a derivative, the conclusion of the theorem may not be true (Figure 4-2).

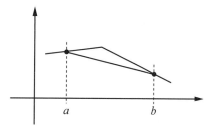

Figure 4-2. The MVT conditions are not met

7

Find all the values of *c* on the closed interval [0, 1] that satisfy the Mean Value Theorem for $f(x) = \sqrt{1 - x^2}$, or explain why no such values exist.

f is continuous on [0, 1]. $f'(x) = \dfrac{1}{2\sqrt{1 - x^2}}(-2x) = \dfrac{-x}{\sqrt{1 - x^2}}$ exists for all *x* in (0, 1). (The derivative does not exist for *x* = 1, but the conditions of the MVT do not include differentiability at the endpoints of the interval.) So the conditions of the MVT are met, and at least one value *c* must exist such that $\dfrac{-c}{\sqrt{1 - c^2}} = \dfrac{f(1) - f(0)}{1 - 0} = \dfrac{0 - 1}{1 - 0} = -1$. Solving for *c* we get: $c = \sqrt{1 - c^2} \Rightarrow c^2 = \dfrac{1}{2}$ and $c = \dfrac{1}{\sqrt{2}}$. (The second solution, $c = -\dfrac{1}{\sqrt{2}}$, is not in our interval.)

Rolle's Theorem is a special case of the Mean Value Theorem in which the values of $f(a) = f(b) = 0$.

> **If f is continuous on a closed interval [a, b], differentiable on the open interval (a, b), and $f(a) = f(b) = 0$, then there exists at least one number c in the interval (a, b) at which $f'(c) = 0$.**

Rolle's Theorem is not mentioned by name in AP questions, but you can use it to justify your answers when appropriate. Or just always refer to the MVT.

8

Suppose a function f is differentiable for all x and $f(0) = 0$. If $g(x)$ is defined as $g(x) = f(x)\cos x$, which of the following statements must be true?

I.   There exists a number c in $\left(0, \dfrac{\pi}{2}\right)$ such that $g'(c) = 0$.

II.   There exists a number c in $\left(\dfrac{\pi}{2}, \pi\right)$ such that $g'(c) = 0$.

III.   There exists a number c in $\left(-\dfrac{\pi}{2}, 0\right)$ such that $g'(c) = 0$.

(A) I only   (B) II only   (C) I and II only

(D) I and III only   (E) I, II, and III

g is differentiable for all real numbers and $g\left(-\dfrac{\pi}{2}\right) = g(0) = g\left(\dfrac{\pi}{2}\right) = 0$. So the intervals $\left(-\dfrac{\pi}{2}, 0\right)$ and $\left(0, \dfrac{\pi}{2}\right)$ satisfy the conditions of Rolle's Theorem (or the MVT with an average slope of 0). This means that statements I and III must both be true. II is not necessarily true. The answer is D.

The proof of Rolle's Theorem is based on the fact that any function g, continuous on a closed interval $[a, b]$, reaches both its maximum and its minimum on that interval. If $g(a) = g(b) = 0$, then either $x_{\min}$ or $x_{\max}$ (or both) must be strictly inside the interval (unless g is simply equal to 0 everywhere on the interval). It is fairly easy to show that the derivative at that point, if it exists, must be equal to 0. (This follows from the definition of the derivative.)

Even though the MVT is more general than Rolle's Theorem, the MVT can be proved using Rolle's Theorem. We can reduce the more general case of any function f to the special case of a function that satisfies the conditions of Rolle's Theorem by subtracting from f the secant line that connects the endpoints $(a, f(a))$ and $(b, f(b))$. In other words, consider $g(x) = f(x) - \left[f(a) + \dfrac{f(b) - f(a)}{b - a} (x - a) \right]$. It satisfies the conditions of Rolle's Theorem on $[a, b]$ and $g'(x) = f'(x) - \dfrac{f(b) - f(a)}{b - a}$, so if $g'(c) = 0$ then

$$f'(c) = \frac{f(b) - f(a)}{b - a}.$$

For the AP exam, you need to know how and when to use the MVT to justify statements about functions.

4.3.   Analysis of Function Graphs

The advent of graphing calculators has changed the way function graph analysis is taught and tested. We now must be able to analyze a function to see whether its calculator graph is complete. There may be extrema that are not in the graphing window, or there may be hidden behavior within the viewing window. We need to use calculus techniques to help us find all the important aspects of the graph of a function. Also, many AP questions deal with properties of function graphs.

The analysis of graphs involves looking at "interesting" points and intervals and at horizontal and vertical asymptotes. Below is a brief catalog of the "interesting" features of graphs. In the following sections we will look at them in more detail and explore some of the calculus tools that allow us to detect these features.

<u>Intervals where the function is increasing or decreasing.</u> f is called ***increasing*** on a given interval (open or closed) if for any x_1 and x_2 in the interval, such that $x_2 > x_1$, $f(x_2) > f(x_1)$. f is called ***decreasing*** on a given interval if for any x_1 and x_2 in the interval, such that $x_2 > x_1$, $f(x_2) < f(x_1)$ (Figure 4-3-a).

<u>Critical numbers.</u> c is called a ***critical number*** for f if $f'(c) = 0$ or $f'(c)$ does not exist (Figure 4-3-b).

Local (relative) maxima and minima (known together as *extreme values* or *extrema*). f has a *local* (or *relative*) *maximum* at $x = c$ if $f(c) \geq f(x)$ for all x in the vicinity of c. f has a *local minimum* at $x = c$ if $f(c) \leq f(x)$ for all x in the vicinity of c (Figure 4-3-c). If f is defined on a closed interval, do we count an endpoint as a local maximum or minimum? Different people hold different opinions on that, and AP questions avoid this ambiguity.

Global (absolute) maxima and minima. f has a *global* (or *absolute*) *maximum* at $x = c$ in a given interval if $f(c) \geq f(x)$ for all x in that interval. f has a *global* (or *absolute*) *minimum* at $x = c$ in a given interval if $f(c) \leq f(x)$ for all x in that interval (Figure 4-3-d). Any continuous function defined on a closed interval $[a, b]$ has both a global maximum and a global minimum on that interval.

Intervals where the function's graph is concave up or down. The graph of f is called *concave up* (or concave upward) on a given interval if for any two points in the interval, the graph of f lies below the chord that connects these points. The graph of f is called *concave down* (or concave downward) on a given interval if for any two points in the interval, the graph of f lies above the chord that connects these points (Figure 4-4-a).

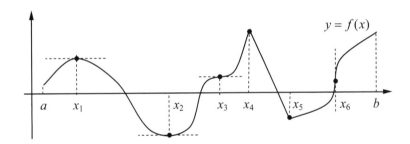

Figure 4-3.

(a) f is increasing on $[a, x_1]$, $[x_2, x_4]$ and $[x_5, b]$; f is decreasing on $[x_1, x_2]$ and $[x_4, x_5]$.

(b)   Critical numbers for f are $x_1, x_2, x_3, x_4, x_5,$ and x_6: $f'(x) = 0$ at $x_1, x_2,$ and x_3 and $f'(x)$ does not exist at x_4, $x_5,$ and x_6.

(c) f has local (relative) maxima at x_1 and x_4 and local (relative) minima at x_2 and x_5.

(d)   The global (absolute) maximum is at $x = x_4$ and the global (absolute) minimum is at $x = x_2$.

> **Remember: a function shaped like a U is concave Up.**

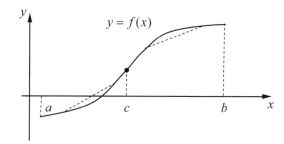

Figure 4-4.   The graph of $y = f(x)$ **is concave up on [a, c] and
concave down on [c, b].** $\left(c, f(c)\right)$ **is an inflection point.**

<u>Points of Inflection.</u> $\left(c, f(c)\right)$ is called a point of inflection on the graph of f if f is
continuous at c and the graph of f changes its concavity at c from up to down or vice
versa.[*]

<u>Horizontal asymptotes.</u> The line $y = L$ is called a horizontal asymptote for f if
$\lim\limits_{x \to +\infty} f(x) = L$ and/or $\lim\limits_{x \to -\infty} f(x) = L$ (see Section 2.1).

<u>Vertical asymptotes.</u> The line $x = c$ is called a vertical asymptote for f at $x = c$ if
$\lim\limits_{x \to c^+} f(x) = \pm\infty$ and/or $\lim\limits_{x \to c^-} f(x) = \pm\infty$ (see Section 2.1).

4.3.1.   Increasing and Decreasing Functions

> **If f is continuous on [a, b] and f' exists and is positive everywhere on (a, b),
> then f is increasing on [a, b]. Likewise, if f' exists and is negative
> everywhere on (a, b), then f is decreasing on [a, b].**

The above theorem follows from the MVT.

The converse is not quite true. It is true that if f is increasing and $f'(x)$ exists
everywhere, then $f'(x) \geq 0$. $f'(x) = 0$ may occur at a high point (local maximum) or at
a low point (local minimum), but it can also happen when the graph crosses the tangent
line at 0 slope. For example, $y = x^3$ is increasing for all x, but $\dfrac{d}{dx}\left(x^3\right)\Big|_{x=0} = 0$.

---

[*] Many textbooks also insist that f must have a tangent line at a point of inflection. It is unlikely, though,
that an AP question would hinge on that distinction.

Some textbooks will say that the function is increasing (decreasing) on an open interval rather than on a closed interval (even if the function is defined on the closed interval). Because of the differences among popular textbooks, the writers of the AP exam do not write multiple-choice questions about increasing/decreasing functions with both [a, b] and (a, b) as choices. In free response questions, either an open or a closed interval will be accepted as a correct answer.

9

If $f(x) = -x^3 - 3x^2 + 5$, on which of the following intervals is f increasing?

(A) $(-\infty, -2)$ and $(0, \infty)$ (B) $(-\infty, 0)$ and $(2, \infty)$ (C) $(-2, 0)$

(D) $(0, 2)$ (E) $(0, \infty)$ only

We need to find where $f'(x) > 0$. $f'(x) = -3x^2 - 6x = -3x(x+2)$. It has two zeros, -2 and 0, which divide the real number line into three intervals: $(-\infty, -2)$, $(-2, 0)$ and $(0, \infty)$. If we test values in these intervals, we find that only the middle one produces a positive result for the derivative. The answer is C.

10

Suppose $f'(x) = (x+1)(x-2)^2 g(x)$ where g is a continuous function and $g(x) < 0$ for all x. On what interval(s) is f decreasing?

(A) $(-\infty, -1)$ (B) $(-\infty, -1) \cup (2, \infty)$ (C) $(-1, 2)$

(D) $(-1, \infty)$ (E) $(-\infty, \infty)$

f' has zeros at $x = -1$ and 2. These zeros partition the x-axis into intervals as follows:

Interval	$x < -1$	$-1 < x < 2$	$x > 2$
Sign of f'	$+$	$-$	$-$
Behavior of f	Increasing	Decreasing	Decreasing

The answer is D.

The same analysis can be represented graphically:

However, a picture alone, without further explanation, would not get full credit for an AP question like this, if it were on the free-response section of the exam. A correct solution would have to say something like this: f is a continuous function, $f'(x) > 0$ for $x < -1$ and $f'(x) < 0$ for $-1 < x$, $x \neq 2$, therefore f is decreasing for $x > -1$.

Note that we do not know the formula for $f(x)$ or for $f'(x)$, so a graphing calculator would not be helpful in this problem.

4.3.2. Extreme Values (Relative and Absolute)

Luckily, we do not need to examine all points on a graph to find its extreme values.

> **To find all <u>local maxima</u> of *f* on [*a*, *b*], find all the critical numbers where *f* is continuous and *f'* changes from positive to negative (as we move from left to right). Similarly, to find all <u>local minima</u> of *f* on [*a*, *b*], choose all the critical numbers where *f'* changes from negative to positive.**

The above rules are known as the ***First Derivative Test***. They assume that $f'(x)$ exists between discrete critical points. You may refer to the First Derivative Test by name when asked to explain why a function has a local minimum or maximum at a given point.

> **If *f* has a local maximum or a local minimum at *c* and $f'(c)$ exists, then $f'(c) = 0$.**

Be careful not to read this theorem backwards:

> **The converse is not necessarily true. If $f'(c) = 0$, there may not necessarily be a local maximum or minimum at c: if $f'(x)$ does not change sign at c, then $f(c)$ is <u>not</u> a local extremum.**

| 11 |

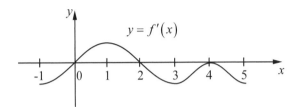

The graph of $y = f'(x)$ is shown above. Which of the following statements correctly describes f on the open interval $(-1, 5)$?

> (A) f has a local maximum at 1 and 4 and a local minimum at 3.
> (B) f has a local maximum at 2 and 4 and a local minimum at 0.
> (C) f has a local maximum at 2 and a local minimum at 0.
> (D) f has a local maximum at 0 and a local minimum at 2.
> (E) f has a local maximum at 1 and a local minimum at 3.

 f' changes from negative to positive at $x = 0$, so f has a local minimum there. f' changes from positive to negative at $x = 2$, so f has a local maximum there. There are no more locations where f' changes sign. The answer is C.

> **In questions like this, it is a common error to look at the graph as if it were the graph of f, even though the question states that the given graph is the graph of f' .**

If you can't immediately visualize what the graph of the derivative is telling you, write the words "negative" and "positive" along the appropriate segments of the graph.

The second derivative, if it exists, is also useful for finding local extrema.

> If $f'(c) = 0$ and $f''(c) < 0$, then $f(x)$ has a local maximum at $x = c$.
>
> If $f'(c) = 0$ and $f''(c) > 0$, then $f(x)$ has a local minimum at $x = c$.

The above is called the ***Second Derivative Test*** for local extrema. There are many problems in which it is easier to find the second derivative at the critical point than to determine whether the first derivative changes sign there. You don't actually have to find the value of the second derivative; you only need to determine whether it is positive or negative at the critical point.

12

Let $y = f(x)$ be a function that contains the point $P(5, 1)$ and has a slope given by $\dfrac{dy}{dx} = \dfrac{x-5}{y+4}$. Which of the following statements is true?

(A) The graph of f has a zero at P.
(B) The graph of f has a vertical tangent line at P.
(C) The slope of f is undefined at P.
(D) f has a local maximum at point P.
(E) f has a local minimum at point P.

$\left.\dfrac{dy}{dx}\right|_{x=5,\,y=1} = 0$ and $f(5) = 1$, so Choices A, B, and C are incorrect. We cannot apply the First Derivative Test to decide between D and E here, because the values of y around the point P are not readily available to us. However, we can use the Second Derivative Test:

$$\frac{d^2y}{dx^2} = \frac{(y+4) - (x-5)\dfrac{dy}{dx}}{(y+4)^2} \Rightarrow \left.\frac{d^2y}{dx^2}\right|_{x=5,\,y=1} = \frac{5-0}{(1+4)^2} > 0.$$ Thus, f has a local minimum at P. The answer is E.

> If f is continuous on $[a, b]$, then f has both an absolute maximum and an absolute minimum on $[a, b]$.

This property is called the ***Extreme Value Theorem***.

The absolute maximum and minimum may occur at the endpoints of the interval or inside the interval (Figure 4-5).

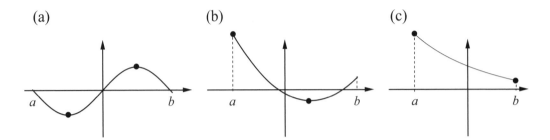

(a)   (b)   (c)

Figure 4-5. Positions of extreme values:

    (a)   Both absolute maximum and minimum occur inside the interval.

    (b)   One occurs inside, another at an endpoint.

    (c)   Both absolute extreme values are reached at the endpoints.

> **To find the <u>absolute</u> (global) maximum of f on $[a, b]$, compare the values of f at all the critical numbers and at the endpoints a and b, or find all local maxima and compare their values with the values at the endpoints. Use similar methods for the absolute minimum.**

13

Determine the absolute extreme values for the function $y = x\sqrt{1-x^2}$ on its domain.

The function is continuous on its domain $[-1, 1]$, so it must have an absolute maximum and minimum. First find the critical numbers:

$y' = \dfrac{-x^2}{\sqrt{1-x^2}} + \sqrt{1-x^2} = \dfrac{1-2x^2}{\sqrt{1-x^2}} = 0 \Rightarrow x = \pm\dfrac{1}{\sqrt{2}}$. The two critical numbers $x = \dfrac{1}{\sqrt{2}}$

and $x = -\dfrac{1}{\sqrt{2}}$ divide the domain into three intervals, and we can analyze the behavior of

y as follows:

Intervals	$-1 < x < -\dfrac{1}{\sqrt{2}}$	$-\dfrac{1}{\sqrt{2}} < x < \dfrac{1}{\sqrt{2}}$	$\dfrac{1}{\sqrt{2}} < x < 1$
Sign of y'	$-$	$+$	$-$
Behavior of y	Decreasing	Increasing	Decreasing

There is a local minimum at $x = -\dfrac{1}{\sqrt{2}}$, since y' changes from negative to positive there.

Similarly, there is a local maximum at $x = \dfrac{1}{\sqrt{2}}$. $y\left(-\dfrac{1}{\sqrt{2}}\right) = -\dfrac{1}{2}$ and $y\left(\dfrac{1}{\sqrt{2}}\right) = \dfrac{1}{2}$.

Compare these with the values at the endpoints: $y(-1) = y(1) = 0$. Therefore, the absolute minimum is $-\dfrac{1}{2}$ and the absolute maximum is $\dfrac{1}{2}$. (Looking at a graph of this function on a calculator can help you to verify these results, but such a problem could very well appear on a non-calculator portion of the exam.)

In the above problem, a solution based solely on a chart indicating the signs of the derivative and the increasing/decreasing behavior of the function will not suffice to get full credit for such a problem. You must justify your answer clearly in words.

> **Some questions ask about the <u>values</u> of relative or absolute extrema, while other questions ask you to find <u>at what point</u> or <u>at what x</u> extrema occur. Read extrema questions carefully.**

In the above example, $x = -\dfrac{1}{\sqrt{2}}$ and $x = \dfrac{1}{\sqrt{2}}$ would be an incorrect answer.

4.3.3. Concavity and Points of Inflection

Let f be a differentiable function on (a, b).

> **If f' is increasing on (a, b), then the graph of f is concave up on (a, b). If f' is decreasing on (a, b), then the graph of f is concave down on (a, b).**

The tangent lines will lie below the graph of f if it is concave up and above the graph if it is concave down.

A *point of inflection* is a point where the concavity of the graph changes from concave up to down or vice-versa. The tangent at the point of inflection, if it exists, crosses the graph of the function.

If f'' exists, it can tell us where the graph of f is concave up or down, too: if $f''(x) > 0$, then f' is increasing near x, and the graph of f is concave up; if $f''(x) < 0$, then f' is decreasing, and the graph of f is concave down. At an inflection point, $f''(x) = 0$, or it does not exist.

If you forget the above rules, you can recall them by comparing your function with $y = x^2$ or $y = x^3$. $y = x^2$ (shaped like a U) is concave up everywhere because $y'' > 0$ (Figure 4-6-a). For $y = x^3$, $y' = 3x^2$ and $y'' = 6x$. For $x < 0$, $y'' < 0$, y' is decreasing, and $y = x^3$ is concave down; for $x > 0$, $y'' > 0$, y' is increasing, and $y = x^3$ is concave up. At $x = 0$ there is a point of inflection (Figure 4-6-b).

> **The fact that $f''(x) = 0$ alone does not mean the graph of f has an inflection point at x (Figure 4-6-c).**

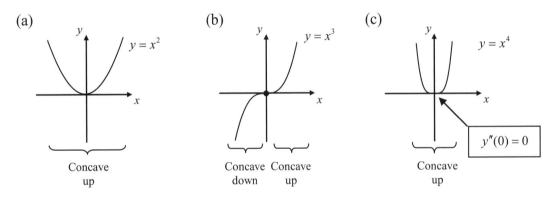

Figure 4-6. $y = x^2$ **and** $y = x^3$ **as mnemonics for concavity**

14

Find the x-coordinates of all points of inflection on the graph of $y = x^{\frac{1}{5}} - \dfrac{1}{10}x^2$.

We must find the points where the concavity changes. The first and second derivatives of y are $y' = \frac{1}{5}x^{-\frac{4}{5}} - \frac{1}{5}x$ and $y'' = \frac{-4}{25}x^{-\frac{9}{5}} - \frac{1}{5}$. The second derivative equals zero at

$x = \left(-\frac{4}{5}\right)^{\frac{5}{9}}$ ≈ -0.833 and does not exist at $x = 0$. These two values partition the x-axis into three intervals:

Interval	$x < -0.883$	$-0.883 < x < 0$	$0 < x$
Sign of y''	$-$	$+$	$-$
Behavior of y	concave down	concave up	concave down

There are two points of inflection: at $x = -0.883$ and at $x = 0$. Common errors include looking at the sign of the first derivative instead of the second derivative or ignoring the point where the first or the second derivative is undefined. Remember that a tangent line may be vertical at a point of inflection.

15

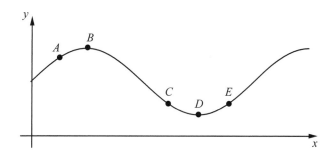

The graph of a function f is shown above. At which of the marked points are both f' and f'' positive?

    (A) A (B) B (C) C (D) D (E) E

If both derivatives are positive, then the graph must be both increasing and concave up. The answer is E.

16

A function is defined by $f(x) = x^3 + ax^2 + bx + c$, where a, b, and c are constants. $f(-1) = 4$ is a local maximum, and the graph of f has a point of inflection at $x = 2$. Find the values of a, b, and c.

$f'(x) = 3x^2 + 2ax + b = 0$ when $x = -1$, so $3 - 2a + b = 0$. $f''(x) = 6x + 2a = 0$ when $x = 2$, so $12 + 2a = 0 \Rightarrow a = -6$. Using the value of a in the first derivative equation yields $15 + b = 0 \Rightarrow b = -15$. Now, since $f(-1) = 4$, $-1 - 6 + 15 + c = 4 \Rightarrow c = -4$. Thus, $a = -6$, $b = -15$, $c = -4$.

4.3.4. Using the Graphs of f, f', and f'' in Combination

Even without any formulas for a function and its derivatives, we can often deduce a lot of information about the function from its derivatives. For example, at a local maximum of f, the graph of f' must change from above the x-axis to below; at a local minimum of f, the graph of f' must change from below the x-axis to above (f' must consequently either be zero or undefined there). A point of inflection on the graph of f must coincide with a local extremum for f'; the graph of the second derivative must change from below the x-axis to above, or vice versa. The second derivative of f must either be zero or undefined there.

17

The graphs of f, f', and f'' are shown in the picture to the right as graphs (i), (ii), and (iii), but not necessarily in that order. Which of these graphs corresponds to which function?

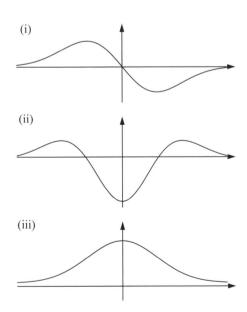

	(i)	(ii)	(iii)
(A)	f	f'	f''
(B)	f''	f	f'
(C)	f	f''	f'
(D)	f''	f'	f
(E)	f'	f''	f

Each of the graphs has at least one local extremum, so the derivative must be zero at the point of the extrema. But the function in graph (iii) has no zeros. Therefore, (iii) must be a graph of f. $f'(0)$ must be equal to zero, so (i) is a graph of f' and (ii) is a graph of f''. The answer is E.

18

	$x = -1$	$-1 < x < 1$	$x = 1$	$1 < x < 2$	$x = 2$	$2 < x < 3$	$x = 3$
$f(x)$	-2		0		-1		-0.5
$f'(x)$		Positive	Does not exist	Negative	Does not exist	Positive	
$f''(x)$		Positive	Does not exist	Negative	Does not exist	0	

A function f, continuous on the closed interval $[-1, 3]$, and its derivatives have the values indicated in the table above.

(a) Find the x-coordinates of all local extrema of f on $(-1, 3)$ and indicate which are maxima and which are minima.
(b) Find the x-coordinates of the global maximum and minimum of f on $[-1, 3]$.
(c) Find all the intervals on which the graph of f is concave up or concave down.
(d) Sketch the graph of a function $y = f(x)$ with all the given features.

(a)

The critical numbers of f on $(-1, 3)$ are $x = 1$ and $x = 2$. At $x = 1$, the first derivative changes from positive to negative, so there is a local maximum there. At $x = 2$, the first derivative changes from negative to positive, so there is a local minimum there.

(b)

Since the value of f at the point of local maximum is $f(1) = 0$, and the values of f at the endpoints are -2 and -0.5, the global maximum is 0 at $x = 1$. Comparing the value at the point of local minimum $f(2) = -1$ with the values of f at the endpoints, we find that the global minimum is at $x = -1$.

(c)

Since $f''(x) > 0$ on $(-1, 1)$, the graph of f is concave up there. Since $f''(x) < 0$ on $(1, 2)$, the graph of f is concave down there. Since $f''(x) = 0$ on $(2, 3)$, the graph of f is a straight line there, so it is neither concave up nor down.

(d)

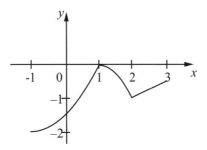

4.4.   Modeling and Optimization

The AP exams may include word problems that rely on calculus techniques for finding maxima and minima. The ones most likely to occur are not too complicated.

19

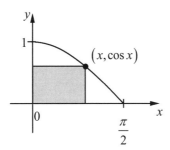

A rectangle with two sides on the x- and y-axes has one vertex at the origin and the opposite vertex on the graph of $y = \cos x$ on $\left[0, \dfrac{\pi}{2}\right]$. Find the dimensions of such a rectangle with the largest area.

Let $(x, \cos x)$ be the coordinates of the vertex of the rectangle that is on the cosine curve. Our task is to maximize the area of the rectangle, $A = L \cdot W = x \cos x$. Note that the domain for this area function is also $\left[0, \dfrac{\pi}{2}\right]$ and the area at both endpoints of the domain is zero. As usual, we can find the maximum of $A(x)$ by looking for x where $A'(x) = 0$: $A'(x) = \cos x - x \sin x = 0 \Rightarrow$ ▪ $x \approx 0.860$ (rounded to three decimal places). Store this value in a named variable in your calculator for future use. When $x = 0.860$, $A = 0.561$. This must be the maximum, since this is the only critical point and the value at this point is higher than at the endpoints of the domain. So the dimensions of the rectangle with the largest area are $L = 0.860$ by $W = \dfrac{A}{L} = 0.652$.

> **Note that in this type of problem you cannot just use your calculator to find the maximum of $A(x)$ because that is not one of the four allowed operations.**
>
> **You have to write the formula for $\dfrac{dA}{dx}$, find its zero, and show that A has a minimum there to receive full credit.**

When solving this type of problem, follow these steps:

1. Read the problem carefully and identify what information is needed to solve the problem. Draw a picture or a diagram, if it would be helpful.

2. Create a mathematical model for the problem. This should be a formula, a function in terms of a single variable that represents the quantity to be optimized.

3. Find the domain of your function and the values that make sense in the real-world situation. Draw a graph, if necessary, to help you understand the function.

4. Find the critical numbers for your function: values where the derivative is zero or does not exist.

5. Find the maximum or minimum for your function.

6. Consider whether your answer makes sense in the real world.

> **Always remember that the maximum or minimum may occur at an endpoint of the domain.**

 20

What are the dimensions of the square-based, open-top box that has a volume of 20 cubic inches and the smallest surface area?

A formula for the surface area of a box with an open top and a square base is $S = x^2 + 4xh$, where x represents the length and width of the base and h represents the height. But we need to have this formula in terms of <u>one</u> variable only, not both x and h. Since the volume of the box is 20, $x^2 h = 20$ or $h = \dfrac{20}{x^2}$. Now $S = x^2 + 4x\left(\dfrac{20}{x^2}\right) = x^2 + \dfrac{80}{x}$, and we have our surface area model in terms of one variable. The domain for this function is all $x \neq 0$, but only $x > 0$ make sense for the real-world situation.

$\dfrac{dS}{dx} = 2x - \dfrac{80}{x^2} = 0 \Rightarrow$ ▪ $x \approx 3.420$. $\dfrac{d^2S}{dx^2} = 2 + \dfrac{160}{x^3} > 0$ for all $x > 0$, so, by the Second Derivative Test, our critical point produces a relative minimum for the surface area. But since there is only one critical point in the domain, it is the point of absolute minimum. The dimensions of the box are 3.420 by 3.420 by 1.710 inches.

4.5.   Related Rates Problems

In this type of problem, you are given the rate, usually constant, at which a quantity is changing with respect to time, and you need to find the rate of change of a second quantity related to the first one. Suppose, for example, you have an expanding circle whose radius is increasing at the rate of 2 cm/min. How fast is the area of the circle increasing when $r = 5$? Here the radius of the circle has a rate of change $\dfrac{dr}{dt} = 2$. The area of the circle is $A(t) = \pi \left[r(t)\right]^2$ and the rate of change of the area is

$\dfrac{dA}{dt} = 2\pi r \dfrac{dr}{dt} = 4\pi r$. When $r = 5$, $\dfrac{dA}{dt} = 20\pi$.

To solve a Related Rates problem, follow these steps:

1. Draw a picture (if helpful) and name the variables. Use the variable t for time. Any rate of change is a derivative with respect to t. Be sure to note which quantities vary and which ones are constant.

2. Write down the given information and what you are asked to find in terms of your variables.

3. Write an equation that relates the variables.

4. Differentiate with respect to t. Assume that all variables are differentiable functions of t. Use the Chain Rule and/or implicit differentiation where appropriate.

5. Substitute in your known values and solve for the unknown value.

6. Verify that your answer is realistic.

If units are specified in the question, be sure to include the correct units in your answer, too.

21

A man is driving north at a rate of 17 m/s. He sees a railroad track 20 m ahead of him that is perpendicular to the road. There is a train going east on the track crossing the road and the man determines with a radar gun that the engine is 35 m from him and the distance between his car and the engine is increasing at the rate of 5 m/s. What is the speed of the train?

Let x be the distance from the railroad crossing to the engine. Let y represent the distance from the man to the crossing and z be the distance from the man to the engine. When $y = 20\,\text{m}$ and $z = 35\,\text{m}$,

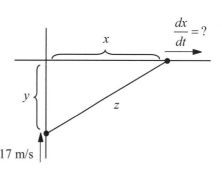

$\dfrac{dy}{dt} = -17\,\text{m/s}$ and $\dfrac{dz}{dt} = 5\,\text{m/s}$. We need to find $\dfrac{dx}{dt}$.

From the Pythagorean Theorem, $x^2 + y^2 = z^2$.

Implicit differentiation of this equation gives us

$2x\dfrac{dx}{dt} + 2y\dfrac{dy}{dt} = 2z\dfrac{dz}{dt} \Rightarrow x\dfrac{dx}{dt} + y\dfrac{dy}{dt} = z\dfrac{dz}{dt}$.

> **Caution: be sure to differentiate your formula with respect to time <u>before</u> you substitute in values for any of the variables.**

We can find x at the given point in time from the Pythagorean Theorem:

$x = \sqrt{35^2 - 20^2} \approx 28.72281323\,\text{m}$. Substituting all the given information into the formula that relates the rates of change, we get $28.72281323\dfrac{dx}{dt} + 20(-17) = 35 \cdot 5$. So

$\dfrac{dx}{dt} = 17.930\,\text{m/s}$. (This is about 65 km/hour, which is a reasonable speed.)

22

The length of a rectangle increases at a rate of 0.5 cm/sec and the width decreases at a rate of 0.5 cm/sec. At the time when the length is 10 cm and the width is 7 cm, what is the rate of change in the area of the rectangle?

(A) $-1.75\ \text{cm}^2/\sec$ (B) $-1.5\ \text{cm}^2/\sec$ (C) $-0.25\ \text{cm}^2/\sec$

(D) $8.5\ \text{cm}^2/\sec$ (E) There is no change in the area

$\dfrac{dl}{dt} = 0.5\,\text{cm/sec}$ and $\dfrac{dw}{dt} = -0.5\,\text{cm/sec}$. The rate of change in the area can be found by

differentiating $A = lw$ with respect to time: $\dfrac{dA}{dt} = l\dfrac{dw}{dt} + w\dfrac{dl}{dt}$. At the time in question,

$\dfrac{dA}{dt} = 10(-0.5) + 7(0.5) = -1.5$ cm^2/sec. The answer is B. Common mistakes are to

omit the negative sign in the value of $\dfrac{dw}{dt}$ or to not use the Product Rule in

differentiating.

23

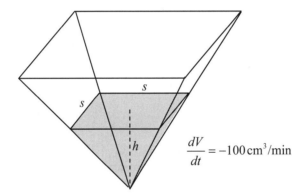

$\dfrac{dV}{dt} = -100\,\text{cm}^3/\text{min}$

Water drains out of a container in the shape of an inverted square pyramid at the rate of
$100\,\text{cm}^3/\text{min}$. The height of the pyramid is equal to one-half the edge of the square
base. How fast is the water level falling when the water is 10 cm deep? (Recall that the
volume of a pyramid is $V = \dfrac{1}{3}Ah$, where A is the area of the base of the pyramid and h is

its height.)

We know that $\dfrac{dV}{dt} = -100 \, \text{cm}^3 / \text{min}$ (use the units given to help you determine which quantity is given: cubic centimeters per minute indicates a volume per unit of time.)

$h = \dfrac{1}{2}s$, where s is the side of the square base and h is the height. The volume of a square pyramid can be found by the formula $V = \dfrac{1}{3}s^2h$, or in our pyramid

$V = \dfrac{1}{3}(2h)^2 \cdot h = \dfrac{4}{3}h^3$. Differentiate with respect to t: $\dfrac{dV}{dt} = 4h^2 \dfrac{dh}{dt}$. Substitute in the

given information: $-100 = 4 \cdot 10^2 \cdot \dfrac{dh}{dt}$, so $\dfrac{dh}{dt} = -0.25 \, \text{cm} / \text{min}$. The water is falling at

the rate of 0.25 cm/min. This rate may seem too low, but in general volume may change rapidly even with a slow change in linear dimensions. Also, the rate at which the water level decreases will get larger as the depth decreases.

24

A particle moves along the curve $y = x^3$ so that its x-coordinate increases at a constant rate of 3 cm/sec. Assuming the same scale on the x- and y-axes, how fast, in radians/sec, is the angle of inclination of the line connecting the particle to the origin changing when $x = 2$?

(A) $\dfrac{3}{17}$ (B) $\dfrac{4}{17}$ (C) $\dfrac{3}{5}$ (D) $\dfrac{12}{17}$ (E) $\dfrac{12}{5}$

Let θ be the angle of inclination of the line connecting the particle to the origin. We need to find the value of $\dfrac{d\theta}{dt}$ when $x = 2$, given that $\dfrac{dx}{dt} = 3 \, \text{cm/sec}$.

$\tan\theta = \dfrac{y}{x} = \dfrac{x^3}{x} = x^2$ so $\theta = \tan^{-1}(x^2)$.

$\dfrac{d\theta}{dt} = \dfrac{1}{1+x^4}(2x)\left(\dfrac{dx}{dt}\right)$. When $x = 2$,

$\dfrac{d\theta}{dt} = \dfrac{1}{1+16}(4)(3) = \dfrac{12}{17}$ radians/sec. Be careful: it is a

common error to omit one or both steps of the Chain Rule when taking the derivative.

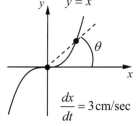

4.6. Distance, Velocity, Acceleration

Particle motion is an important topic in AP Calculus. There are problems concerning particle motion on the AP exam almost every year. Motion along a horizontal or vertical line is an AB topic, and motion along a curve is a BC topic. ⌈ The latter will be discussed in Chapter 8 along with vectors. ⌋ If the particle moves along the x-axis, its position is sometimes represented as $x(t)$, and if the motion is along the y-axis, it may be represented as $y(t)$.

In general, if the position of a particle is given as $s(t)$, then

> the (instantaneous) velocity of the particle is $v(t) = s'(t) = \dfrac{ds}{dt}$;
>
> the acceleration of the particle is $a(t) = v'(t) = s''(t) = \dfrac{d^2 s}{dt^2}$;
>
> the speed of the particle is the absolute value of its velocity: speed $= |v(t)|$.

By convention, motion to the right or up is considered positive ($v(t) > 0$) and motion to the left or down is considered negative ($v(t) < 0$).

25

What is the maximum velocity attained on the interval $0 \le t \le 3$ by a particle whose position is given by $s(t) = t^3 - 6t^2 + 9t - 1$?

(A) –3   (B) 0   (C) 3   (D) 6   (E) 9

For this particle, $v(t) = 3t^2 - 12t + 9$ and $a(t) = 6t - 12$. To find when the maximum velocity occurs, we must find the critical numbers of $a(t) = v'(t)$. There is only one, $t = 2$. But this value of t produces a minimum velocity, since $a(t) < 0$ for $0 < t < 2$ and $a(t) > 0$ for $2 < t < 3$. So the maximum velocity must occur at an endpoint of the domain. Comparing the values of the velocity at the endpoints, $v(0) = 9$ and $v(3) = 0$, we find that the maximum velocity is 9. The answer is E. Actually, since the velocity graph is a parabola that opens upward, we could have stated right away that the maximum must occur at one of the endpoints of the interval instead of finding the acceleration and analyzing its critical numbers.

Consider the following two examples together. The first one analyzes a position graph; the second analyzes a velocity graph.

 26

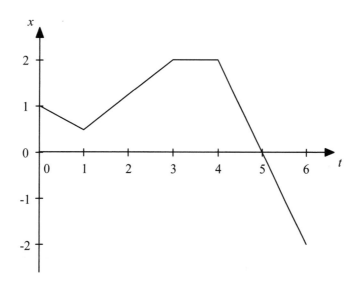

A particle *P* moves along a horizontal line. The graph above shows its position as a function of time *t*.

(a)   When is *P* moving to the left?
(b)   When is *P* standing still?
(c)   When is the first time that *P* reverses direction?
(d)   When does *P* move at its greatest speed?

(a)   *P* moves to the left when the graph has a negative slope: $0 < t < 1$ and $4 < t < 6$.
(b)   *P* is standing still when the graph is horizontal: $3 < t < 4$.
(c)   *P* reverses direction for the first time at $t = 1$, since for $0 < t < 1$, *P* moves to the left and for $1 < t < 2$, *P* moves to the right.
(d)   *P* moves at its greatest speed for $4 < t < 6$, since the slope of that portion of the graph has the largest absolute value.

27

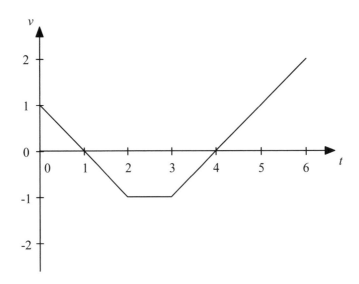

A particle *P* moves along a horizontal line. The graph above shows its velocity as a function of time *t*.

(a)   When is *P* moving to the left?
(b)   When is P standing still?
(c)   When is the first time that *P* reverses direction?
(d)   When does *P* move at its greatest speed?

(a)   *P* moves to the left when the velocity graph has negative values: $1 < t < 4$.
(b)   *P* is standing still when the velocity has a zero value: at $t = 1$ and at $t = 4$.
(c)   *P* changes direction at $t = 1$, since for $0 < t < 1$, the velocity is positive (*P* moves to the right) and for $1 < t < 4$, the velocity is negative (*P* moves to the left).
(d)   *P* moves at its greatest speed at $t = 6$, since that is when the velocity achieves its greatest absolute value of 2.

We will see other particle motion problems in Chapter 6, Integral Applications. When the velocity or the acceleration is given, the position function can be determined by finding an antiderivative of the velocity. Many of these problems require the use of both derivatives and integrals within parts of the same problem.

4.7.   Data-Driven Problems

In recent years, with the use of the graphing calculators on the AP exam, problems that begin with a set of data have become common. Sometimes this data is presented in a table, sometimes as a graph, and sometimes as both. The last two examples in the previous section are problems presented in a graphical format. The following example uses a table. When solving such problems, you must be careful not to make assumptions that may not necessarily be true. For instance, in these problems, you can compute an average rate of change or an approximation for a derivative, but it may not be possible to compute an exact value for a derivative.

28

Age	Height (in inches)
9	46
10	47
11	48
12	50
13	53
14	58
15	62
16	67
17	69

The table above shows Andrew's height on the last day of summer for consecutive years when his age was 9 through 17.

(a)   What is the average rate of Andrew's growth between the ages of 9 and 17?
(b)   Give an approximation for the rate of Andrew's growth when he was 15.
(c)   Find an approximation for Andrew's height on the summer when he was 18.

(a)   The average rate of change is $\dfrac{69-46}{17-9} = \dfrac{23}{8} = 2.875$ inches/year.

(b)   Since Andrew's height was measured at discrete points, we cannot compute the derivative at the age 15 exactly. We can, however, approximate the derivative, using values that are close to 15: $\dfrac{h(15)-h(14)}{15-14} = 4$, $\dfrac{h(16)-h(15)}{16-15} = 5$, and $\dfrac{h(16)-S(14)}{16-14} = 4.5$ inches/year are three reasonable approximations for the rate of Andrew's growth at the age 15. Any one of these answers would be acceptable on an AP exam.

(c)   If we look at the rate of growth from the age of 16 to 17, we get 2 inches per year. Linear extrapolation from the age of 17 to 18 gives $69 + 2\cdot(18-17) = 71$.

Chapter 5.   Integration

5.1.   The Fundamental Theorem of Calculus

In differential calculus, we have learned how to calculate the rate at which one quantity is changing with respect to another quantity. If we have a differentiable function, its derivative at a point represents the instantaneous rate of change of the function at that point. The derivative of a function over an interval represents the rate of change of the function at each point of the interval.

Given a derivative, can we recover the function that has that derivative? Does such an **antiderivative** always exist? Is it unique? What good is it? Integral calculus answers these questions.

Since derivative measures change or difference, it is not surprising that antiderivative is related to addition, accumulation, **integration**. The main idea of integration is straightforward. Suppose the points $x_0, x_1, x_2, \ldots, x_{n-1}, x_n$, where $x_0 = a$ and $x_n = b$, are evenly spaced over the closed interval $[a, b]$, so that $x_i = x_{i-1} + \Delta x$ $(i = 1, 2, \ldots, n)$. We can represent the total ("integral") change of a function F over $[a, b]$ as the sum of smaller, step-wise changes of F:

$$F(b) - F(a) = \left[F(x_1) - F(x_0) \right] + \left[F(x_2) - F(x_1) \right] + \ldots + \left[F(x_{n-1}) - F(x_{n-2}) \right] + \left[F(x_n) - F(x_{n-1}) \right]$$

Let's multiply and divide each term in the above sum by Δx :

$$F(b) - F(a) = \Delta x \left(\frac{F(x_1) - F(x_0)}{\Delta x} + \frac{F(x_2) - F(x_1)}{\Delta x} + \ldots + \frac{F(x_n) - F(x_{n-1})}{\Delta x} \right)$$

Since $x_i = x_{i-1} + \Delta x$, we get

$$F(b) - F(a) = \Delta x \left(\frac{F(x_0 + \Delta x) - F(x_0)}{\Delta x} + \frac{F(x_1 + \Delta x) - F(x_1)}{\Delta x} + \ldots + \frac{F(x_{n-1} + \Delta x) - F(x_{n-1})}{\Delta x} \right)$$

If Δx is small, each fraction in the above expression approximates $F'(x_i)$, assuming that $F'(x)$ exists on $[a, b]$. In fact, due to the Mean Value Theorem each fraction is equal to the derivative of F at some point $\tilde{x}_i \in [x_{i-1},\, x_i]$. So

$$F(b) - F(a) = \Delta x \left[F'(\tilde{x}_1) + F'(\tilde{x}_2) + \ldots + F'(\tilde{x}_n) \right] = \sum_{i=1}^{n} \Delta x\, F'(\tilde{x}_i) = \sum_{i=1}^{n} \Delta x\, f(\tilde{x}_i)$$

where $f(x) = F'(x)$ (i.e., F is an antiderivative of f).

The sum

$$\sum_{i=1}^{n} \Delta x\, f(\tilde{x}_i)$$

where $\Delta x = \dfrac{b-a}{n}$, is called a ***Riemann sum*** for the function f on $[a, b]$. As we make Δx smaller and smaller, the choice of points $\tilde{x}_i$ in Riemann sums matters less and less as all the Riemann sums converge to a number:

$$F(b) - F(a) = \lim_{\Delta x \to 0} \sum_{i=1}^{n} \Delta x\, f(\tilde{x}_i)$$

> **The above formula represents the key relationship between the limit of Riemann sums of a function f and its antiderivative F.**

This essential fact (or one of its variations) is known as the ***Fundamental Theorem of Calculus***.

We have chosen all the distances between consecutive points x_i to be the same. In that case

$$\sum_{i=1}^{n} \Delta x\, f(\tilde{x}_i) = \Delta x \sum_{i=1}^{n} f(\tilde{x}_i)$$

In the more general case, we can use different distances in a Riemann sum:

$$\sum_{i=1}^{n} \Delta x_i\, f(\tilde{x}_i)$$

It doesn't matter for the limit as long as all Δx_i approach 0.

The Fundamental Theorem of Calculus (**FTC**) has profound implications for both theory and applications. On the theoretical side, the FTC assures us that any continuous function *f* has an antiderivative and helps us approximate that antiderivative using Riemann sums. On the applications side, the FTC allows us to calculate a quantity represented by the limit of Riemann sums of some function *f* — such a quantity as the area under a curve or the volume of a solid — using an antiderivative of *f*.

5.2.   Definite and Indefinite Integrals

The quantity

$$\lim_{\Delta x \to 0} \sum_{i=1}^{n} \Delta x \, f(\tilde{x}_i)$$

is called the **definite integral** of *f* on [*a*, *b*] and is written as

$$\int_a^b f(x)\,dx$$

In the above notation, *a* and *b* are called the **limits of integration** (*a* is the **lower limit** and *b* is the **upper limit**) and the function $f(x)$ is called the **integrand**. *dx* indicates which variable we are integrating on (and remains as a vestige of Δx in Riemann sums).

> **The Fundamental Theorem of Calculus states that if *F* is an antiderivative of *f*, then**
> $$F(b) - F(a) = \int_a^b f(x)\,dx$$

In calculations, $F(b) - F(a)$ is often written as $F(x)\Big|_a^b$.

An antiderivative of $f(x)$ is called the **indefinite integral** of *f*, and is written as

$$\int f(x)\,dx$$

The same terms and similar notations are used in definite and indefinite integrals because, according to the FTC, they are closely related. We can rewrite the FTC as

$$\int f(x)\,dx\Big|_a^b = \int_a^b f(x)\,dx$$

Antiderivatives of a Function

A definite integral is a number. If you make the upper limit of integration vary, the integral becomes a function of the upper limit of integration:

$$F(x) = \int_a^x f(t)\, dt$$

Plug in different values for x, and you get different numbers. Such a function is called an **accumulation function**, since it measures the total accumulation of $f(t)$ on $[a, x]$.

> **Another formulation of the FTC is that the accumulation function is an antiderivative of the integrand. In other words,**
>
> $$\frac{d}{dx}\left[\int_a^x f(t)\, dt \right] = f(x)$$

Think about what this says. Remember that the derivative of a function represents its rate of change. The left-hand side of this equation represents the rate at which the function that measures accumulation changes. The right-hand side tells you that the rate of change of the accumulated quantity is <u>exactly</u> the size of the quantity you are accumulating! The bigger the function values you are "adding on", the faster the accumulation function grows.

Any continuous function has an infinite set of antiderivatives, but any two antiderivatives differ from each other only by a constant (because the derivative of their difference is 0). So, $\int f(x)\, dx$ is defined as a family of functions that differ from each other by a constant.

For example, $\int \cos x\, dx = \sin x + C$; $\sin x + 7$, $\sin x - 15$, and $\sin x - \dfrac{\sqrt{e}}{\pi^2}$ are all

antiderivatives of $\cos x$.

Find $\dfrac{d}{dx}\left[\int_{-1}^x \sin t\, dt \right]$.

Due to the FTC, the answer is $\sin x$. The operations of differentiation and integration are inverse to each other.

When you apply the FTC to find the derivative of an integral, pay attention
to the limits of integration.

The rule $\dfrac{d}{dx}\left[\displaystyle\int_a^x f(t)\,dt\right] = f(x)$ applies only when the lower limit is a constant and the

upper limit is x. Occasionally, an AP question may ask you to find $\dfrac{d}{dx}\left[\displaystyle\int_a^{g(x)} f(t)\,dt\right]$.

Then you have to use the Chain Rule, as in the following example.

[2]

What is $\dfrac{d}{dx}\left[\displaystyle\int_1^{x^3} \sqrt{t}\,dt\right]$?

(A) $\sqrt{x^3}$ (B) $3x^2\sqrt{x^3}$ (C) $\dfrac{1}{2\sqrt{x^3}}$ (D) $\dfrac{3x^2}{2\sqrt{x^3}}$ (E) 0

Recall the Chain Rule from Chapter 3: $\dfrac{d}{dx}F\big(g(x)\big) = F'\big(g(x)\big)g'(x)$. Here,

$F(x) = \displaystyle\int_1^x \sqrt{t}\,dt$ and $g(x) = x^3$. From the FTC, $F'(x) = \sqrt{x}$, so

$F'\big(g(x)\big) = F'\big(x^3\big) = \sqrt{x^3}$. $g'(x) = 3x^2$. Therefore, $\dfrac{d}{dx}\left[\displaystyle\int_1^{x^3} \sqrt{t}\,dt\right] = 3x^2\sqrt{x^3}$. The

answer is B.

5.3.   Approximating Definite Integrals with Sums

As we have seen, you can approximate a definite integral by the sum of several products:
$\displaystyle\int_a^b f(x)\,dx \approx \sum_{i=1}^n \Delta x_i\, f(\tilde{x}_i)$. The two factors in each product are an increment of the

independent variable (Δx_i) and a value of the function being integrated at some point in a
small subinterval. If the function is positive, each product represents the area of a
rectangle that approximates the area underneath the graph of the function. If we use more
and more rectangles with smaller and smaller Δx_i, the sum of their areas gives us a more
and more accurate approximation of the area under the curve (Figure 5-1-a). The exact
area under the curve is $\displaystyle\int_a^b f(x)\,dx$.

What happens if f is negative on the interval over which you are integrating? In that case, the value of the "height" of each rectangle will be negative, and when you multiply that negative number by the positive width of the rectangle, you end up with a negative product. Therefore, if you integrate a negative function, each product represents the area of a rectangle, but with a negative sign. Then the area above the curve and below the x-axis is $-\int_a^b f(x)\,dx$ (Figure 5-1-b).

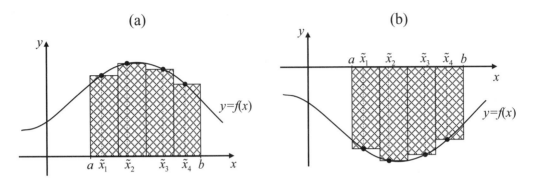

Figure 5-1. Riemann-sum approximation of the area (a) under, and (b) above a curve

If f changes sign within the interval, then $\int_a^b f(x)\,dx$ is a combination of the areas of regions that lie above and below the x-axis. The areas of the regions above the x-axis contribute to the integral with a positive sign and the areas of the regions below the x-axis contribute to the integral with a negative sign (Figure 5-2).

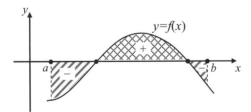

Figure 5-2. An integral with positive and negative contributing areas

Figure 5-3 shows how the sum of the areas of four rectangles approximates $\int_0^2 f(x)\,dx$.

$$\int_0^2 f(x)\,dx \approx 0.5 \cdot f(0) + 0.5 \cdot f(0.5) + 0.5 \cdot f(1) + 0.5 \cdot f(1.5) =$$

$$0.5 \cdot \left[f(0) + f(0.5) + f(1) + f(1.5) \right].$$

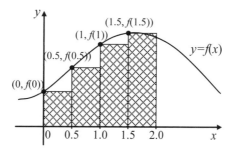

Figure 5-3. Left-endpoint approximation of $\int_0^2 f(x)\,dx$

In this example, the height of each rectangle is the value of f at the <u>left</u> endpoint of each subinterval. This is called a ***left-hand Riemann sum***. You could just as easily have used the right endpoint or the midpoint of each subinterval (Figure 5-4). These sums are called the ***right-hand*** and ***midpoint Riemann sums***, respectively.

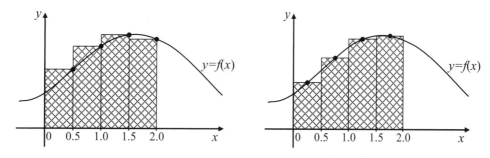

Figure 5-4. Right-hand and midpoint Riemann sums for $\int_0^2 f(x)\,dx$

Note that if the integrand is increasing, then all left-hand sums <u>underestimate</u> the exact value of the integral and all right-hand sums <u>overestimate</u> the integral. The reverse is true when the function is decreasing.

3

L and R are the left-hand and right-hand Riemann sums, respectively, of $f(x) = 3x - x^2$ on [1, 3], divided into 4 subintervals of equal length. Which of the following statements is true?

(A) $R = 0$ (B) $L < R$ (C) $L > R$ (D) $L = R$

(E) We cannot determine whether L is greater than R or less than R from the given information.

Use your calculator to graph $y = 3x - x^2$ in the window [1, 3] by [–3, 3]. Since f is a polynomial, it is continuous on [1, 3]. It has a maximum at $x = 1.5$, so it is neither increasing nor decreasing on [1, 3]. Therefore, we can't tell right away whether L or R underestimate or overestimate $\int_1^3 (3x - x^2) \, dx$ and which one of them is greater. Let us calculate these sums, then. First tabulate the function values at the ends of the subintervals:

x	$y = 3x - x^2$
1.0	2
1.5	2.25
2.0	2
2.5	1.25
3.0	0

We get $L = 0.5 \cdot (2 + 2.25 + 2 + 1.25) = 0.5 \cdot 7.5$ and $R = 0.5 \cdot (2.25 + 2 + 1.25 + 0) = 0.5 \cdot 5.5$. Note that since all the subintervals have the same length, we can factor out $\Delta x = 0.5$ from the sums to make the computations a little easier. Also note that if we know L, we don't really have to compute R from scratch. All the addends in L and R overlap, except the first and the last one, so, if we subtract the first term from L and add the last term, we get R:

$$R = L - 0.5 \cdot 2 + 0.5 \cdot 0 = L - 1$$

So $R - L = -1$ and the answer is C. In fact, we could have gotten the answer without computing the Riemann sums at all. All we had to do was compare the values of f at the ends of the interval, because, for a continuous function, $R - L = \Delta x \cdot [f(b) - f(a)]$.

4

The graph of the function shown to the right consists of a quarter of a circle and two line segments. $\int_{-2}^{4} f(x)\,dx =$

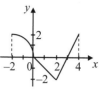

(A) $-\pi-2$ (B) $2-\pi$ (C) $\pi-2$ (D) $\pi+2$ (E) $\pi+4$

The integral consists of three regions: one quarter of a circle and two triangles. The quarter-circle region and the smaller triangle on the right are above the x-axis, so their areas, π and 1, respectively, are added to the integral. The bigger triangle in the middle is below the x-axis, so its area, 3, is subtracted from the integral. The total integral is $\pi - 3 + 1 = \pi - 2$. The answer is C.

This question is contrived, of course, but questions like this appear regularly on AP exams. In such questions, geometry is used to evaluate integrals. We will revisit this topic in Section 5.6.

5

Find the left-hand and right-hand Riemann sums for $\int_{2}^{3} \dfrac{1}{\ln x}\,dx$ using five subintervals of equal length and determine whether these sums under- or overestimate the exact value of the integral.

A question like this could appear in the free-response portion of the exam. Graph $f(x) = \dfrac{1}{\ln x}$ in the window [2, 3] by [0, 2] on your calculator. Note that the integrand is decreasing over the entire interval; therefore, any left-hand sum will overestimate the integral, and any right-hand sum will underestimate the integral. The left-hand sum L is

$$0.2 \cdot \left(\frac{1}{\ln 2} + \frac{1}{\ln 2.2} + \frac{1}{\ln 2.4} + \frac{1}{\ln 2.6} + \frac{1}{\ln 2.8} \right) \blacksquare \approx 1.174.$$ The right-hand sum R is

$$R = L + 0.2 \cdot \left[f(3) - f(2) \right] = L + 0.2 \cdot \left[\frac{1}{\ln 3} - \frac{1}{\ln 2} \right] \blacksquare \approx 1.068.$$ (Using the numerical

integration command on the calculator, you get $\int_{2}^{3} \dfrac{1}{\ln x}\,dx \blacksquare \approx 1.118$.)

6

The table below shows the velocity readings of a car taken every 30 seconds over a five-minute interval.

Time (seconds)	Velocity (miles per hour)
0	60
30	55
60	50
90	45
120	40
150	45
180	50
210	60
240	30
270	40
300	45

What is the approximate distance (in miles) traveled by the car during this five-minute interval, using a midpoint Riemann sum with 60-second subintervals?

(A) 3.583   (B) 3.708   (C) 3.750   (D) 3.833   (E) 4.083

Recall that velocity is the rate of change of distance over time. From the FTC, the distance traveled is the definite integral of velocity over the time interval. For the midpoint sum estimate, consider the velocity at points 30, 90, 150, 210, and 270:

$$S_M = 60 \text{ sec} \cdot (55 + 45 + 45 + 60 + 40) \text{ mi/hr} = \frac{1}{60} \text{ hr} \cdot 245 \text{ mi/hr} = 4.083 \text{ mi}$$

The answer is E.

Choice D is the left-hand sum approximation, S_L, using the velocity at points 0, 60, 120, 180, and 240:

$$S_L = 60 \text{ sec} \cdot (60 + 50 + 40 + 50 + 30) \text{ mi/hr} = \frac{1}{60} \text{ hr} \cdot 230 \text{ mi/hr} = 3.833 \text{ mi}$$

Choice A is the right-hand sum estimate, dropping the value at $t = 0$ and adding the value at $t = 300$:

$$S_R = 60 \sec \cdot (50 + 40 + 50 + 30 + 45)\,\text{mi/hr} = \frac{1}{60}\,\text{hr} \cdot 215\,\text{mi/hr} = 3.583\,\text{mi}$$

Choice C comes from using just the velocity reading at $t = 150$:

$$300\,\sec \cdot 45\,\text{mi/hr} = \frac{1}{12}\,\text{hr} \cdot 45\,\text{mi/hr} = 3.750\,\text{mi}$$

> **Note that in questions that deal with physical quantities you have to be careful with the units: be sure to use consistent units and make conversions as necessary.**

In the above example, the length of the interval is given in minutes, the length of the subintervals Δt is given in seconds, and the velocity is given in miles per hour. We have converted Δt into hours to make the units compatible. We could have converted all the velocity readings into miles per second, but that would have taken a little more work.

7

We know from geometry that $\int_0^1 \sqrt{1 - x^2}\,dx = \dfrac{\pi}{4}$, because it represents the area of one quarter of the unit circle. Find the midpoint Riemann sum approximation for this integral using three subintervals, $\left[0, \dfrac{1}{4}\right]$, $\left[\dfrac{1}{4}, \dfrac{3}{4}\right]$, and $\left[\dfrac{3}{4}, 1\right]$, and use it to estimate π.

In this question, the subintervals do not have the same length. Their lengths are $\dfrac{1}{4}$, $\dfrac{1}{2}$, and $\dfrac{1}{4}$, and the corresponding midpoints are $\dfrac{1}{8}$, $\dfrac{1}{2}$, and $\dfrac{7}{8}$. The midpoint

Riemann sum approximation is $\sqrt{1 - \left(\dfrac{1}{8}\right)^2} \cdot \dfrac{1}{4} + \sqrt{1 - \left(\dfrac{1}{2}\right)^2} \cdot \dfrac{1}{2} + \sqrt{1 - \left(\dfrac{7}{8}\right)^2} \cdot \dfrac{1}{4} \approx$

$0.9921567 \cdot 0.25 + 0.8660254 \cdot 0.5 + 0.4841229 \cdot 0.25 \approx 0.802$. This gives an estimate for $\pi \approx 0.802 \cdot 4 = 3.208$.

Trapezoidal Sums

This method for estimating the definite integral uses a trapezoid instead of a rectangle on each subinterval to approximate the area under (or above) a curve (Figure 5-5).

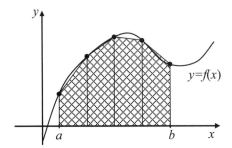

Figure 5-5.   Trapezoidal sum approximation of a definite integral

As you may recall from geometry, the area of a trapezoid is equal to its height multiplied by the average of the lengths of its two bases (Figure 5-6).

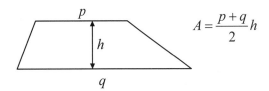

Figure 5-6.   The area of a trapezoid

In a trapezoidal sum, the trapezoids are turned on their sides. Δx is the "height" and $f(x_{i-1})$ and $f(x_i)$ are the "bases" (where $i = 1, 2, ..., n$). The area of the i-th trapezoid is $\left| \dfrac{f(x_{i-1}) + f(x_i)}{2} \right| \cdot \Delta x$. Then

$$\int_a^b f(x)\,dx \approx \Delta x \cdot \left[\frac{f(a)+f(x_1)}{2} + \frac{f(x_1)+f(x_2)}{2} + \ldots + \frac{f(x_{n-1})+f(b)}{2} \right] =$$

$$\frac{\Delta x}{2}\left[f(a)+2f(x_1)+\ldots+2f(x_{n-1})+f(b) \right]$$

Note that

$$\frac{\Delta x}{2}\left[f(a)+2f(x_1)+\ldots+2f(x_{n-1})+f(b) \right] =$$

$$\frac{\Delta x\left[f(a)+f(x_1)+\ldots+f(x_{n-1}) \right] + \Delta x\left[f(x_1)+\ldots+f(x_{n-1})+f(b) \right]}{2}$$

Therefore,

> **the trapezoidal sum is simply the average of the corresponding left-hand and right-hand sums:** $S_T = \dfrac{S_L + S_R}{2}$.

Trapezoidal sums work for estimating the definite integral even if the function changes its sign. There may be no actual trapezoid in that case, but the formulas still work.

What is the approximation of the area under the graph of $f(x)=\sqrt{1+x^3}$ using the trapezoidal sum with all the points in the partition $\left\{1, \dfrac{5}{4}, 2, 3\right\}$.

(A) 4.642   (B) 6.307   (C) 7.971   (D) 8.071   (E) 12.614

In practice, you probably wouldn't use one of the numerical methods to approximate an integral when you knew an explicit formula for the integrand. Nonetheless, some AP exam questions may ask you to do just that.

Note that here the subintervals do not have equal lengths, so you do need to find the area of each individual trapezoid. The area of the first trapezoid is

$$\frac{1}{2}\left(\sqrt{1+1^3} + \sqrt{1+\left(\frac{5}{4}\right)^3}\right)\cdot\left(\frac{5}{4}-1\right) \approx .39158\,;$$ the area of the second trapezoid is

$$\frac{1}{2}\left(\sqrt{1+\left(\frac{5}{4}\right)^3} + \sqrt{1+2^3}\right)\cdot\left(2-\frac{5}{4}\right) \approx 1.76942\,;$$ and the area of the third trapezoid is

$$\frac{1}{2}\left(\sqrt{1+2^3} + \sqrt{1+3^3}\right)\cdot\left(3-2\right) \approx 4.14575\,.$$ The sum of these three trapezoids is about 6.3068. The answer is B.

Another way to get the same result is to find the left-hand sum

$$\left(\sqrt{1+1^3}\right)\cdot\frac{1}{4} + \left(\sqrt{1+\left(\frac{5}{4}\right)^3}\right)\cdot\frac{3}{4} + \left(\sqrt{1+2^3}\right)\cdot 1 \approx 4.6424 \quad \text{(this is choice A)}$$

and the right-hand sum

$$\left(\sqrt{1+\left(\frac{5}{4}\right)^3}\right)\cdot\frac{1}{4} + \left(\sqrt{1+2^3}\right)\cdot\frac{3}{4} + \left(\sqrt{1+3^3}\right)\cdot 1 \approx 7.9711 \quad \text{(this is choice C)}$$

and take their average. The average of the left-hand sum and right-hand sum gives the trapezoidal sum even when the partition is not uniform.

As we have mentioned earlier, if the integrand is increasing, then the left-hand sum is less than the definite integral and the right-hand sum is greater than the definite integral. The trapezoidal sum always gives a better approximation than at least one of the corresponding one-side sums, left-hand or right-hand. For a trapezoidal sum, you need to consider the concavity of the graph of the integrand to determine whether the estimate is too big or too small.

If the graph of the integrand is concave up (its second derivative is positive), then any trapezoidal sum <u>overestimates</u> the integral; when the graph of the integrand is concave down, any trapezoidal sum <u>underestimates</u> the integral (Figure 5-7).

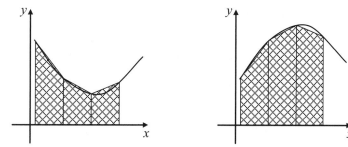

Figure 5-7.   Trapezoidal sums overestimate the definite integral of a concave-up function and underestimate the integral of a concave-down function

 Do not confuse a trapezoidal sum and a midpoint sum. They are the same only when the graph of the function is a straight line!

In fact, midpoint sum errors behave in the opposite way from trapezoidal sum errors. When the graph is concave up, midpoint rectangles give an estimate for the integral that is too small, and when the graph is concave down, midpoint rectangles give an estimate that is too big.

5.4.   Finding Limits of Sums Using Definite Integrals

Occasionally, an AP exam question may ask you to find what looks like the limit of some obscure sum. To solve such a problem, you need to guess that the sum is actually a Riemann sum in disguise for some function and the limit of the sum is actually the definite integral of that function. All you have to do is to reconstruct the function and the limits of integration.

9

What is $\displaystyle \lim_{n \to \infty} \sum_{k=1}^{n} \frac{5}{n} \ln\left(2 + \frac{5k}{n}\right)$?

(A) 0   (B) 4.047   (C) 7.235   (D) 17.235   (E) Nonexistent

We have a limit of the sum of n products, which is one hint that this is really a definite integral in disguise. The missing first term in the sum, when $k = 0$, is $\dfrac{5}{n}\ln 2$. The last term, when $k = n$, is $\dfrac{5}{n}\ln 7$. This is a hint that the limits of integration are 2 and 7. If n is the number of subintervals of equal length in a Riemann sum, then each subinterval would have length $\Delta x = \dfrac{7-2}{n} = \dfrac{5}{n}$, which conveniently happens to be one of the factors in our sum. The other factor is $\ln\left(2 + \dfrac{5k}{n}\right)$. The sum turns out to be the right-hand Riemann sum for $\ln x$ on [2, 7] and its limit is $\displaystyle\int_2^7 \ln x\, dx \approx 7.235$. The answer is C.

10

What is $\displaystyle\lim_{n\to\infty} \frac{1}{n}\left(\frac{1}{1} + \frac{1}{1+\dfrac{1}{n}} + \frac{1}{1+\dfrac{2}{n}} \cdots + \frac{1}{1+\dfrac{n-1}{n}}\right)$?

(A) -0.5 (B) 0 (C) .404 (D) .693 (E) Nonexistent

We can see that the sum is actually the left-hand Riemann sum for the function $f(x) = \dfrac{1}{x}$ on the interval [1, 2] partitioned into n subintervals of length $\Delta x = \dfrac{1}{n}$. Therefore,

$$\lim_{n\to\infty} \frac{1}{n}\left(\frac{1}{1} + \frac{1}{1+\dfrac{1}{n}} + \frac{1}{1+\dfrac{2}{n}} \cdots + \frac{1}{1+\dfrac{n-1}{n}}\right) = \int_1^2 \frac{1}{x}\, dx.$$

$\ln x$ is an antiderivative of $\dfrac{1}{x}$ for $x > 0$, in particular on [1, 2]. From the FTC,

$$\lim_{n\to\infty} \frac{1}{n}\left(\frac{1}{1} + \frac{1}{1+\dfrac{1}{n}} + \frac{1}{1+\dfrac{2}{n}} \cdots + \frac{1}{1+\dfrac{n-1}{n}}\right) = \ln(2) - \ln(1) = \ln(2).$$

The answer is D.

5.5.   Properties of Definite Integrals

The key properties of definite integrals are summarized in Table 5-1. You have to be able to use these properties very fluently and to figure out quickly which property is useful in a particular situation.

Integral over zero length $$\int_a^a f(x)\,dx = 0$$	The integral over a zero-length interval is 0.
Integral of a constant $$\int_a^b c\,dx = c\cdot(b-a)$$	The integral of a constant is equal to that constant times the length of the interval.
Constant Factor property $$\int_a^b k\cdot f(x)\,dx = k\cdot\int_a^b f(x)\,dx$$	Constant factors "slide through" an integration operation, just as they slide through a differentiation operation.
Integral of a sum or difference $$\int_a^b \left[f(x)\pm g(x)\right]dx = \int_a^b f(x)\,dx \pm \int_a^b g(x)\,dx$$	The integral of a sum of functions is the sum of the integrals (for the same limits of integration).
Limits Reversal property $$\int_a^b f(x)\,dx = -\int_b^a f(x)\,dx$$	Swapping the integration limits negates the integral.
Limits Addition property $$\int_a^c f(x)\,dx + \int_c^b f(x)\,dx = \int_a^b f(x)\,dx$$	We can combine two intervals of integration that share an endpoint into one (c <u>does not</u> need to be between a and b).
Comparison property If $f(x)\le g(x)$ for $a\le x\le b$, then $$\int_a^b f(x)\,dx \le \int_a^b g(x)\,dx$$	If a function is less than or equal to another, its integral is less than or equal to the other's integral.

Table 5-1.   Properties of definite integrals

11

If $\int_4^7 g(t)\,dt = 5$, what is $\int_4^7 \left(2 \cdot g(t) + 6\right) dt$?

(A) 15   (B) 16   (C) 22   (D) 23   (E) 28

$\int_4^7 \left(2 \cdot g(t) + 6\right) dt = \int_4^7 2 \cdot g(t)\,dt + \int_4^7 6\,dt = 2 \cdot \int_4^7 g(t)\,dt + 6 \cdot (7 - 4) = 2 \cdot 5 + 18 = 28$. The answer is E.

12

Given that f is an even function and $\int_0^2 f(t)\,dt = 17$, what is $\int_{-2}^2 f(t)\,dt$?

(A) –34   (B) –17   (C) 17   (D) 34   (E) 68

Since f is even, its graph is symmetric over the *y*-axis, making $\int_{-2}^0 f(t)\,dt = \int_0^2 f(t)\,dt$.

Therefore, $\int_{-2}^2 f(t)\,dt = \int_{-2}^0 f(t)\,dt + \int_0^2 f(t)\,dt = 2 \cdot \int_0^2 f(t)\,dt = 2 \cdot 17 = 34$. The answer is D.

13

If $\int_4^{-10} h(w)\,dw = -3$ and $\int_4^6 h(w)\,dw = 5$, what is $\int_{-10}^6 h(w)\,dw$?

(A) 2   (B) –2   (C) –8   (D) 8   (E) 16

In this type of question we have to represent the interval [–10, 6] as a combination of two intervals on which the integrals are known. Swapping the limits of integration on the first integral achieves that: $\int_{-10}^4 h(w)\,dw = -\int_4^{-10} h(w)\,dw = 3$ and we get

$\int_{-10}^6 h(w)\,dw = \int_{-10}^4 h(w)\,dw + \int_4^6 h(w)\,dw = 3 + 5 = 8$. The answer is D.

14

Show that $\int_{-1}^{5} \sin^2\left(\sqrt{t^4+2}\right) dt < 6$.

The integrand, $\sin^2\left(\sqrt{t^4+2}\right)$, is always less than or equal to 1. Therefore,

$\int_{-1}^{5} \sin^2\left(\sqrt{t^4+2}\right) dt \le \int_{-1}^{5} 1\, dt = 6$. Since in fact $\sin^2(x)$ is less than 1 almost everywhere,

$\int_{-1}^{5} \sin^2\left(\sqrt{t^4+2}\right) dt$ can't be equal to 6.

5.6. Calculating Definite Integrals Using Geometry

When areas of regions between the graph and the x-axis can be calculated using formulas from geometry, you can evaluate the integral without doing any fancy approximations or finding an antiderivative. A definite integral is a combination of areas of regions between the graph of the integrand and the x-axis. The areas of regions above the x-axis are added to the total and the areas of the regions below the x-axis are subtracted from the total.

15

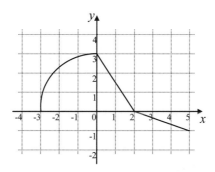

The function $y = f(x)$, graphed above, consists of a quarter circle and two line segments.

What is $\int_{-3}^{5} f(x)\, dx$?

(A) $\dfrac{9\pi}{4} + \dfrac{3}{2}$ (B) $\dfrac{9\pi}{4} + \dfrac{9}{2}$ (C) $\dfrac{9\pi}{2} + \dfrac{3}{2}$ (D) $\dfrac{9\pi}{2} + \dfrac{9}{2}$ (E) 8

Split the integral up into three pieces: $\int_{-3}^{5} f(x)\,dx = \int_{-3}^{0} f(x)\,dx + \int_{0}^{2} f(x)\,dx + \int_{2}^{5} f(x)\,dx$.

The first integral is the area of a quarter circle with radius 3; it is $\dfrac{\pi \cdot 3^2}{4} = \dfrac{9\pi}{4}$. The

second integral is the area of a triangle with base 2 and height 3; it is $\dfrac{1}{2} \cdot 2 \cdot 3 = 3$. The

third integral comes from a triangle with base 3 and height 1. Its area is $\dfrac{1}{2} \cdot 3 \cdot 1 = \dfrac{3}{2}$, but

its contribution to the integral is negative since the graph of f is below the x-axis from 2 to 5. Thus,

$$\int_{-3}^{5} f(x)\,dx = \frac{9\pi}{4} + 3 - \frac{3}{2} = \frac{9\pi}{4} + \frac{3}{2}.$$ The answer is A.

16

What is $\displaystyle\int_{-5}^{5} \sqrt{25 - x^2}\, dx$?

(A) 0   (B) 5   (C) $\dfrac{25\pi}{2}$ (D) 25π (E) 50π

The graph of $y = \sqrt{25 - x^2}$ is a semicircle with radius 5. Indeed, if $y = \sqrt{25 - x^2}$, then $x^2 + y^2 = 25$ (with $y \ge 0$). The value of the integral is therefore the same as the area of a semicircle with radius 5: $\displaystyle\int_{-5}^{5} \sqrt{25 - x^2}\, dx = \frac{1}{2}\pi \cdot 5^2 = \frac{25\pi}{2}$. The answer is C.

5.7.   Calculating Definite Integrals Using the FTC

When you can find an antiderivative for the integrand, you can evaluate a definite integral using the Fundamental Theorem of Calculus. It states that if $f(x)$ is continuous on the closed interval $[a, b]$, and $F(x)$ is an antiderivative of $f(x)$, then

$$\int_a^b f(x)\,dx = F(b) - F(a).$$

17

$$\int_1^4 2x\,dx =$$

(A) 0   (B) 7   (C) 15   (D) 16   (E) 17

Since $\dfrac{d}{dx}x^2 = 2x$, x^2 is an antiderivative for $2x$. So, $\int_1^4 2x\,dx = x^2\Big|_1^4 = 4^2 - 1^2 = 15$. The answer is C.

Note that this integral could also be calculated using geometry. The graph of this function is a straight line and the region under it is a trapezoid with

height $4-1 = 3$ and bases 2 and 8. Its area is $\left(\dfrac{2+8}{2}\right)\cdot 3 = 15$.

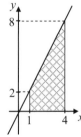

18

$$\int_0^{\frac{\pi}{3}} \sin t\,dt =$$

(A) $\dfrac{1}{2}$ (B) $-\dfrac{1}{2}$ (C) $\dfrac{\sqrt{3}}{2}$ (D) $-\dfrac{\sqrt{3}}{2}$ (E) $\dfrac{\sqrt{3}}{2}-1$

An antiderivative for $\sin x$ is $-\cos x$. So,

$$\int_0^{\frac{\pi}{3}} \sin t\,dt = -\cos x\Big|_0^{\frac{\pi}{3}} = -\cos\frac{\pi}{3} - \left(-\cos 0\right) = -\frac{1}{2} + 1 = \frac{1}{2}.$$ The answer is A.

When evaluating integrals with a lower limit of 0, you may be tempted just to assume that the antiderivative at 0 evaluates to 0. As the above example shows, this is not always the case.

19

Suppose F and G are continuous functions with continuous derivatives of all orders, and $\dfrac{d}{dx}G(x) = F(x)$. Several values for F, G, F', and G' are given in the table below.

x	$F(x)$	$G(x)$	$F'(x)$	$G'(x)$
1	4	−1	0	4
2	−2	2	9	−2
3	5	7	−4	5

Find $\displaystyle\int_1^3 F(t)\,dt$, $\displaystyle\int_1^2 G'(t)\,dt$ and $\displaystyle\int_1^3 F''(t)\,dt$.

Since $\dfrac{d}{dx}G(x) = F(x)$, G is an antiderivative of F. So

$$\int_1^3 F(t)\,dt = G(3) - G(1) = 7 - (-1) = 8.$$

By definition, $G(t)$ is an antiderivative of $G'(t)$. So

$$\int_1^2 G'(t)\,dt = G(2) - G(1) = 2 - (-1) = 3.$$

By definition, $F'(t)$ is an antiderivative of $F''(t)$. So

$$\int_1^3 F''(t)\,dt = F'(3) - F'(1) = -4 - 0 = -4.$$

If the integrand has a non-removable discontinuity somewhere in the interval on which you are integrating, then the FTC <u>cannot</u> be used to evaluate the integral.

To do so will almost certainly give a wrong answer.

20

What is $\int_{-1}^{1} x^{-2} \, dx$?

Where defined, an antiderivative of x^{-2} is $-x^{-1}$. But if you tried using the FTC, you'd get $\int_{-1}^{1} x^{-2} \, dx = -x^{-1} \Big|_{-1}^{1} = (-1) - 1 = -2$. This is clearly wrong since the integrand is always positive. The discontinuity at $x = 0$ prevents us from using the FTC to evaluate this integral. ⌈ This is an ***improper integral***, a BC-only topic discussed in Section 5.10. ⌋ This definite integral does not exist.

5.8.   Calculating Integrals with a Calculator

You should be prepared to use a calculator to evaluate definite integrals on the open calculator parts of the AP exam. Make sure you know how to use your calculator to evaluate an integral.

Most graphing calculators require that you input four arguments to the command that evaluates an integral:

- the integrand function
- the name of the independent variable in the integrand
- the lower limit of integration
- the upper limit of integration

21

Find the area of the region in the first quadrant bounded by the graph of $h(x) = 2\sin(x^2) - x^2$ and the x-axis.

> **Store the integrand as one of the built-in functions in the calculator, such as $\mathtt{Y_1}$, then graph it.**

This will give you an idea what the graph of the integrand looks like and allow you to assess the reasonableness of your answer. It may take you a couple of tries to get an appropriate viewing window. The important thing is that all the relevant behavior be within view. Here, we need to see the entire region in question.

The following TI-83 screen shots illustrate the steps for plotting $y = h(x)$ in the [0, 2] by [−1, 1] window:

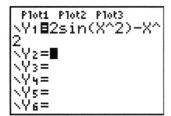  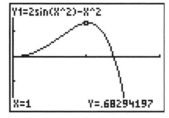

> **When a limit of integration comes from solving an equation, use a variable on your calculator to store it. Then refer to the stored variable when you use that limit.**

This will reduce the chance of making a typing error when you enter the limit, and also will make your answer more accurate. On free response questions, you can use a named variable in your setups, as long as you write down the value of that variable in the exam booklet.

In this example, you need to find the positive *x*-intercept of the function. Use the `zero` option on the CALC menu of a TI-83 to do this (or use the corresponding command on a different model):

 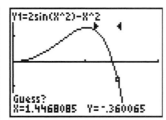 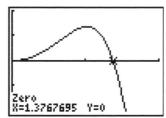

Then return to the HOME screen and store the last answer to a named variable (on TI-83 press ANS then STO➔ and A):

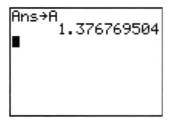

> Some calculators may give less accurate results when you do calculations from the graph screen. The TI-83, for example, computes integrals on the graph screen to an accuracy of only 10^{-3}. Know your calculator! If it is such a machine, use the HOME screen environment rather than the graphing environment to evaluate integrals.

Although in this example it doesn't matter, it is better to be safe than sorry.

Refer to the stored function and the stored integration limits when calculating the integral:

```
Ans→A
            1.376769504
fnInt(Y1,X,0,A)
            .4713616718
■
```

> To receive credit on the free response section of the exam, it is crucial that you use mathematical, not calculator, notation to show your setups.

For example, write the previous calculations as

$$a = 1.3767695$$
$$\text{Area} = \int_0^a \left[2\sin(x^2) - x^2 \right] dx \approx 0.471$$

You will <u>not</u> get full credit for writing `fnInt(2sin(X²)-X²,X,0,A)`.

> If you evaluate the integral of a function that involves an absolute value, be prepared to wait some time for the calculator to give you the answer. Calculators are slower when integrating absolute values. Use the time the calculator is working to think about the question, or do some other work.

5.9. Finding Antiderivatives

As we saw in the previous section, we can use the FTC to evaluate a definite integral. The first step is to find an antiderivative for the integrand. The following sections describe several methods for doing that.

> **But remember: there is no reason to use antiderivatives and the FTC to evaluate an integral on the open calculator part of the AP exam. Just use your calculator!**

22

If $f'(x) = \cos(x^3)$ and $f(5) = 2$, find $f(4)$.

If this type of question appeared on the closed calculator part of the exam, you would have to first find an antiderivative of $f'(x)$, $f(x) = F(x) + C$, then find C from the given initial condition $f(5) = 2$, and finally find $f(4)$. In this question, however, an attempt to simplify $\int \cos(x^3)\,dx$ would be futile, because it can't be expressed by an algebraic formula. Use the FTC and your calculator instead:

$$f(4) = f(5) - \int_4^5 \cos(x^3)\,dx \approx 2 - (-0.0166) \approx 2.017.$$

> **This is a preferred way of tackling a question of this type on the open calculator part of the exam, even if it is possible to find an antiderivative for $f'(x)$ by analytical means.**

Choosing which particular antidifferentiation (integration) method to try can be difficult: there is no precise recipe or algorithm for finding antiderivatives. However, as you practice, you should find that the process gets easier. You'll become familiar with many different forms and patterns, and understand when each of the various methods fails. There's really no substitute for this practice. The Integration Worksheet at the end of this chapter provides two dozen antidifferentiation problems with solutions. Their difficulty is consistent with those you might see on the exam.

The notation for the antiderivative of a function f is $\int f(x)\,dx$. This is similar to the definite integral notation, $\int_a^b f(x)\,dx$, but the meaning here is quite different.

$\int_a^b f(x)\,dx$ **represents a <u>number</u>, while** $\int f(x)\,dx$ **represents a <u>family of</u> <u>functions</u>. If** $F(x)$ **is an antiderivative of** $f(x)$**, then** $F(x)+C$ **is also an antiderivative, for any constant** C**.**

You <u>must</u> include a constant (usually C) in answers to antidifferentiation problems. For example: $\int \dfrac{1}{x}\,dx = \ln|x| + C$.

5.9.1.   General Antidifferentiation Rules

Antidifferentiation is a <u>linear</u> operation, meaning that the antiderivative of a sum (or difference) is the sum (or difference) of the antiderivatives, and a constant factor can be taken out of the integral:

$$\int \big[f(x)+g(x)\big]\,dx = \int f(x)\,dx + \int g(x)\,dx$$

$$\int \big[f(x)-g(x)\big]\,dx = \int f(x)\,dx - \int g(x)\,dx$$

$$\int \big[k \cdot f(x)\big]\,dx = k \cdot \int f(x)\,dx \text{, where } k \text{ is a constant.}$$

There is also a simple rule for the antiderivative of $f(ax+b)$, but you have to be a little careful. First,

if $\int f(x)\,dx = F(x)+C$ **, then** $\int f(x+b)\,dx = F(x+b)+C$ **.**

For example, $\int (x+5)^2\,dx = \dfrac{(x+5)^3}{3}+C$ because $\int x^2\,dx = \dfrac{x^3}{3}+C$.

When $a \neq 1$ an antiderivative needs an additional factor $\dfrac{1}{a}$:

if $\int f(x)\,dx = F(x)+C$ **, then** $\int f(ax+b)\,dx = \dfrac{1}{a}F(ax+b)+C$ **. Be careful not to forget the factor** $\dfrac{1}{a}$ **.**

23

Evaluate $\int \cos(4x - 7)dx$.

The answer is $\frac{1}{4}\sin(4x - 7) + C$. We have used the fact that $\cos u$ is an antiderivative of $\sin u$.

24

Evaluate $\int \frac{1}{3x - 1}dx$.

The answer is $\frac{1}{3}\ln|3x - 1| + C$. We have used the fact that $\int \frac{1}{u}du = \ln|u| + C$. We know it because $\frac{d}{du}\ln u = \frac{1}{u}$.

> **Caution: avoid the mistake of "Universal Logarithmic Antidifferentiation"!**
> **It is tempting for beginning calculus students to think that**
> $\int \frac{1}{something}dx = \ln|something| + C$. **This is one of the most common errors on**
> **AP exams.**

In the above example we had to introduce the factor $\frac{1}{3}$ to find $\int \frac{1}{3x - 1}dx$ correctly. As we will see in Section 5.9.3, in the more general case you need an integral in the form $\int \frac{f'(x)}{f(x)}dx$ to get $\ln|f(x)| + C$.

5.9.2. Antiderivatives from Known Derivatives

For every derivative formula you know, you get its antiderivative counterpart as a bonus. AP exam takers need to memorize the dozen or so antiderivatives listed in Table 5-2, but you should already know them from the tables of derivatives in Chapter 3.

$\dfrac{d}{dx}kx = k$	$\Rightarrow$	$\displaystyle\int k\,dx = kx + C$
$\dfrac{d}{dx}\left(2\sqrt{x}\right) = \dfrac{1}{\sqrt{x}}$	$\Rightarrow$	$\displaystyle\int \dfrac{1}{\sqrt{x}}\,dx = 2\sqrt{x} + C$
$\dfrac{d}{dx}\left(\dfrac{x^{n+1}}{n+1}\right) = x^n$	$\Rightarrow$	$\displaystyle\int x^n\,dx = \dfrac{x^{n+1}}{n+1} + C \ \ (\text{for } n \neq -1)$
$\dfrac{d}{dx}\ln|x| = \dfrac{1}{x}$	$\Rightarrow$	$\displaystyle\int \dfrac{1}{x}\,dx = \ln|x| + C$
$\dfrac{d}{dx}e^x = e^x$	$\Rightarrow$	$\displaystyle\int e^x\,dx = e^x + C$
$\dfrac{d}{dx}\left(\dfrac{a^x}{\ln a}\right) = a^x$	$\Rightarrow$	$\displaystyle\int a^x\,dx = \dfrac{a^x}{\ln a} + C$
$\dfrac{d}{dx}\sin x = \cos x$	$\Rightarrow$	$\displaystyle\int \cos x\,dx = \sin x + C$
$\dfrac{d}{dx}\left(-\cos x\right) = \sin x$	$\Rightarrow$	$\displaystyle\int \sin x\,dx = -\cos x + C$
$\dfrac{d}{dx}\tan x = \sec^2 x$	$\Rightarrow$	$\displaystyle\int \sec^2 x\,dx = \tan x + C$
$\dfrac{d}{dx}\left(-\cot x\right) = \csc^2 x$	$\Rightarrow$	$\displaystyle\int \csc^2 x\,dx = -\cot x + C$
$\dfrac{d}{dx}\sec x = \sec x \cdot \tan x$	$\Rightarrow$	$\displaystyle\int \sec x \cdot \tan x\,dx = \sec x + C$
$\dfrac{d}{dx}\left(-\csc x\right) = \csc x \cdot \cot x$	$\Rightarrow$	$\displaystyle\int \csc x \cdot \cot x\,dx = -\csc x + C$
$\dfrac{d}{dx}\arctan x = \dfrac{1}{1+x^2}$	$\Rightarrow$	$\displaystyle\int \dfrac{1}{1+x^2}\,dx = \arctan x + C$
$\dfrac{d}{dx}\arcsin x = \dfrac{1}{\sqrt{1-x^2}}$	$\Rightarrow$	$\displaystyle\int \dfrac{1}{\sqrt{1-x^2}}\,dx = \arcsin x + C$

Table 5-2.   Common antiderivatives

Many textbooks treat $\int \dfrac{1}{\sqrt{x}} dx$ as a special case of $\int x^n dx$ for $n = -\dfrac{1}{2}$:

$$\int \frac{1}{\sqrt{x}} dx = \int x^{-\frac{1}{2}} dx = \frac{x^{-\frac{1}{2}+1}}{-\frac{1}{2}+1} + C = \frac{x^{\frac{1}{2}}}{\frac{1}{2}} + C = 2\sqrt{x} + C .$$

It works, of course, but it is tedious and error-prone. For good AP test results it is useful to memorize this integral. Also memorize:

$$\int \tan x \, dx = \ln|\sec x| + C$$

$$\int \cot x \, dx = -\ln|\csc x| + C$$

$\int \cot x \, dx$ can be also written as $\ln|\sin x| + C$, but you might find it easier to remember it as $-\ln|\csc x| + C$. In general,

> **in derivatives and integrals, tangents and secants "go together," and cotangents and cosecants also "go together."**

If you are not sure whether your formula for an antiderivative is correct, you can always quickly verify it by taking derivatives of both sides.

For example, $\int \cos\left(2x + \dfrac{\pi}{6}\right) dx = \textit{something} \cdot \sin\left(2x + \dfrac{\pi}{6}\right) + C$. Differentiating both sides (in your head) you get $\cos\left(2x + \dfrac{\pi}{6}\right) = \textit{something} \cdot 2\cos\left(2x + \dfrac{\pi}{6}\right)$. So $\textit{something} = \dfrac{1}{2}$.

We need this factor to compensate for the 2 that pops out when we differentiate $\cos\left(2x + \dfrac{\pi}{6}\right)$.

The formulas in Table 5-2 and other known antiderivatives, combined with the general antidifferentiation rules from Section 5.9.1 and simple algebra, provide answers to many antidifferentiation problems.

The following three questions could appear on the no-calculator section of the exam.

25

$$\int_0^1 (3x+1)^2 \, dx =$$

(A) $\dfrac{7}{3}$ (B) 7   (C) $\dfrac{64}{9}$ (D) 18   (E) 21

$\int (3x+1)^2 \, dx = \dfrac{1}{3} \cdot \dfrac{(3x+1)^3}{3} + C$. By the FTC, $\int_0^1 (3x+1)^2 \, dx = \dfrac{(3x+1)^3}{9} \bigg|_0^1 = \dfrac{64}{9} - \dfrac{1}{9} = 7$.

The answer is B.

26

$$\int_0^{12} \dfrac{1}{\sqrt{1+2z}} \, dz =$$

(A) $-\dfrac{4}{5}$ (B) 2   (C) 4   (D) 8   (E) 16

Recall that $\int \dfrac{1}{\sqrt{u}} \, du = 2\sqrt{u} + C$ and that $\int f(ax+b) \, dx = \dfrac{1}{a} \int f(u) \, du$, where $u = ax+b$.

Therefore, $\int_0^{12} \dfrac{1}{\sqrt{1+2z}} \, dz = \dfrac{1}{2} \cdot 2 \cdot \sqrt{1+2z} \bigg|_0^{12} = 5-1 = 4$. The first factor, $\dfrac{1}{2}$, comes from integrating $f(1+2z)$ rather than $f(z)$. The second factor, 2, comes from the formula for the integral of $\dfrac{1}{\sqrt{u}}$. The answer is C.

27

Evaluate $\int_0^{\pi} \sin\left(3t - \dfrac{\pi}{3}\right) dt$.

$\int_0^{\pi} \sin\left(3t - \dfrac{\pi}{3}\right) dt = \dfrac{1}{3} \cdot \left(-\cos\left(3t - \dfrac{\pi}{3}\right)\right) \bigg|_0^{\pi} = -\dfrac{1}{3}\left[\cos\left(3\pi - \dfrac{\pi}{3}\right) - \cos\left(-\dfrac{\pi}{3}\right)\right] = \dfrac{1}{3}$.

5.9.3.   *u*-Substitution (a.k.a. "Change of Variable")

This antidifferentiation method is inverse to the Chain Rule for derivatives. Since $\frac{d}{dx}F(u(x)) = F'(u(x)) \cdot u'(x)$, the inverse is true for the antiderivatives:

$$\int f(u(x)) \cdot u'(x)\, dx = \int f(u)\, du .$$

For example, if you see $\int \frac{2x}{\sqrt{1+x^2}}\, dx$, you might notice right away that $2x = \frac{d}{dx}\left(1+x^2\right)$ and that we know how to integrate $\frac{1}{\sqrt{u}}$: $\int \frac{1}{\sqrt{u}}\, du = 2\sqrt{u} + C$. Combining these two observations we get $\int \frac{2x}{\sqrt{1+x^2}}\, dx = 2\sqrt{1+x^2} + C$.

What we did here amounts to $\int \frac{2x}{\sqrt{1+x^2}}\, dx = \int \frac{1}{\sqrt{1+x^2}}\, d\left(1+x^2\right)$, but this is very informal, to say the least. Our idea is that $2x \cdot dx = d\left(1+x^2\right)$ and we can pretend that we are integrating by a new variable, $u = 1+x^2$. But none of this should be written in your exam booklet. Do it in your head and then just write down the answer.

To see this approach another way, consider this integral: $\int \left(e^{-2t}+7\right)^3 e^{-2t}\, dt$. Think about doing a *u*-substitution in your head, with $u = e^{-2t}+7$. With that in mind, all we'd need is $du = -2e^{-2t}dt$, which we <u>almost</u> have! We can "manufacture" the *du* we need by introducing a constant factor of -2, and compensating for it by multiplying the integral by $-\frac{1}{2}$. That is,

$\int \left(e^{-2t}+7\right)^3 e^{-2t}\, dt = -\frac{1}{2}\int \left(e^{-2t}+7\right)^3 \left(-2e^{-2t}\right)dt$. Now you can see that we have

$-\frac{1}{2}\int u^3 du$. The result is $\int \left(e^{-2t}+7\right)^3 e^{-2t}\, dt = -\frac{1}{2} \cdot \frac{\left(e^{-2t}+7\right)^4}{4} + C = -\frac{\left(e^{-2t}+7\right)^4}{8} + C$.

This type of shortcut may seem hard at first, especially if your textbook and class work have used a more formal *u*-substitution technique. For our first example, it goes like this:

1. Let $u = 1 + x^2$

2. $\dfrac{du}{dx} = 2x$

3. $\displaystyle\int \frac{2x}{\sqrt{1+x^2}}\,dx = \int \frac{du}{dx}\frac{1}{\sqrt{u}}\,dx = \int \frac{1}{\sqrt{u}}\,du = 2\sqrt{u} + C = 2\sqrt{1+x^2} + C.$

It appears that dx got "cancelled" in $\dfrac{du}{dx}\cdot dx$ and, for practical purposes, it is OK to think

of it that way.

This more methodical approach may be safer, but it takes more time. Judge for yourself whether you have learned to recognize the *u*-substitution pattern correctly and are ready to safely use the shortcut in your head when appropriate. Learning to do that will make your life easier (at least as far as AP Calculus is concerned).

You may also encounter situations where formal *u*-substitution is unavoidable, as in the following example.

28

Find $\displaystyle\int x\sqrt{x-1}\,dx$.

Let $u = x-1 \Rightarrow du = dx \Rightarrow \displaystyle\int x\sqrt{x-1}\,dx = \int (u+1)\sqrt{u}\,du =$

$\displaystyle\int u^{\frac{3}{2}} + u^{\frac{1}{2}}\,du = \frac{2}{5}u^{\frac{5}{2}} + \frac{2}{3}u^{\frac{3}{2}} + C = \frac{2}{5}(x-1)^{\frac{5}{2}} + \frac{2}{3}(x-1)^{\frac{3}{2}} + C.$

The rule that $\displaystyle\int f(ax+b)\,dx = \frac{1}{a}F(ax+b)$, discussed in Section 5.9.1 above, is simply a

special case of *u*-substitution. Here $ax+b = u$, $\dfrac{du}{dx} = a \Rightarrow dx = \dfrac{du}{a}$, and

$\displaystyle\int f(ax+b)\,dx = \frac{1}{a}\int f(u)\,du$.

u-Substitution in Definite Integrals

> When evaluating definite integrals using *u*-substitution, you can substitute the limits of integration, too, and forget the original variable altogether:
>
> $$\int_a^b u'(x) \cdot f(u(x))\, dx = \int_{u(a)}^{u(b)} f(u)\, du$$

29

Evaluate $\int_0^2 \dfrac{e^{3x}}{1+e^{3x}}\, dx$.

Note that the integrand is in the form $k \cdot \dfrac{du}{dx} \cdot \dfrac{1}{u}$, where $u = 1 + e^{3x}$ and k is some constant

factor. Let's work it out: $\dfrac{du}{dx} = 3e^{3x}$, so $e^{3x}\, dx = \dfrac{1}{3}\, du$. Thus *u*-substitution gives us

$$\int \frac{e^{3x}}{1+e^{3x}}\, dx = \frac{1}{3}\int \frac{1}{u}\, du = \frac{1}{3}\ln|u| + C.$$

Now let's figure out the new limits of integration. $u(0) = 1 + e^{3 \cdot 0} = 2$ and

$u(2) = 1 + e^{3 \cdot 2} = 1 + e^6$. So $\displaystyle\int_0^2 \frac{e^{3x}}{1+e^{3x}}\, dx = \frac{1}{3}\int_2^{1+e^6} \frac{1}{u}\, du = \frac{1}{3}\ln(u)\Big|_2^{1+e^6} = \frac{1}{3}\Big(\ln\left(1+e^6\right) - \ln 2\Big).$

30

If $\int_1^3 g(t)\, dt = 8$, what is $\int_2^6 \left[1 + g\left(\dfrac{t}{2}\right)\right] dt$?

(A) 4   (B) 8   (C) 16   (D) 18   (E) 20

First, $\int_2^6 \left[1 + g\left(\dfrac{t}{2}\right)\right] dt = \int_2^6 1\, dt + \int_2^6 g\left(\dfrac{t}{2}\right) dt$. The first integral is equal to 4. In

$\int_2^6 g\left(\dfrac{t}{2}\right) dt$, let $u = \dfrac{t}{2}$. Then $du = \dfrac{1}{2} dt \Rightarrow dt = 2\, du$. $u(2) = 1$ and $u(6) = 3$. Thus

$\int_2^6 g\left(\dfrac{t}{2}\right) dt = 2\int_1^3 g(u)\, du$. But $2\int_1^3 g(u)\, du = 2\int_1^3 g(t)\, dt$ (u is a "dummy" variable). We

get $\int_2^6 \left[1 + g\left(\dfrac{t}{2}\right)\right] dt = 4 + 16 = 20$. The answer is E.

5.9.4. ⌈ Antidifferentiation by Parts (BC Only) ⌋

This method of finding antiderivatives comes from the Product Rule for derivatives:

$$(uv)' = u \cdot v' + v \cdot u'$$

Integrating both sides, we get

$$u \cdot v = \int u \cdot v'\, dx + \int v \cdot u'\, dx \Rightarrow$$

$$\int u \cdot v'\, dx = u \cdot v - \int v \cdot u'\, dx$$

or

$$\int u \cdot dv = u \cdot v - \int v \cdot du$$

where $dv = v'dx$ and $du = u'dx$.

What good is this? Well, sometimes it may be easier to find $\int v \cdot du$ than $\int u \cdot dv$. For example, if $u \cdot dv = x \cdot \cos x\, dx$, then $v \cdot du = \sin x\, dx$ (assuming $u(x) = x$ and $v(x) = \sin x$).

In antidifferentiation by parts, we look for the pattern $\int u \cdot v'\, dx$.

The method applies when the following conditions are met:

1. The integrand is a product of two factors;

2. You know the antiderivative of one of the factors, so you can think of that factor as v' (the other factor is u);

3. It is easier to find the antiderivative of $v \cdot u'$ than of the original integrand, $u \cdot v'$.

There are several points to keep in mind regarding antidifferentiation by parts:

1. Look for a u-substitution <u>first</u>; resort to antidifferentiation by parts only if that fails.

2. Antidifferentiation by parts is often used on integrals of the form $\int x \cdot f(x)\, dx$, where $f(x)$ is something simple, such as e^{kx}, $\sin(ax+b)$ or $\cos(ax+b)$, $\dfrac{1}{\sqrt{ax+b}}$ or $\sqrt{ax+b}$. In these cases $u(x) = x$, dv is everything else.

3. Antidifferentiation by parts also works for $\int x^2 \cdot e^{kx}\, dx$, $\int x^2 \cdot \sin(ax+b)\, dx$, and so on, with $u(x) = x^2$, but then you need to apply it twice: first reducing x^2 to x, then reducing x to 1.

Antidifferentiation by parts is also used to find $\int \ln x\, dx$, $\int \arcsin x\, dx$, and $\int \arctan x\, dx$.[*]
This is slightly counterintuitive, because the integrand does not appear to be a product of two factors. The technique here is to chose $v'(x) = 1$ (i.e., $dv = dx$). For example:

$$\int \arcsin x\, dx = \int (\arcsin x) \cdot 1\, dx = \arcsin x \cdot x - \int x \cdot \left(\frac{d}{dx} \arcsin x \right) dx =$$

$x \cdot \arcsin x - \int \dfrac{x}{\sqrt{1-x^2}}\, dx$. To evaluate $\int \dfrac{x}{\sqrt{1-x^2}}\, dx$, note that

$$\int \frac{x}{\sqrt{1-x^2}}\, dx = -\int \frac{-2x}{2\sqrt{1-x^2}}\, dx = -\int \frac{1}{2\sqrt{1-x^2}}\, d(1-x^2) = -\left(\sqrt{1-x^2} + C \right).$$ We get:

$\int \arcsin x\, dx = x \cdot \arcsin x + \sqrt{1-x^2} + C$.

$\boxed{31}$

Find $\int \ln(2x)\, dx$.

---

[*] AP exam materials may use both $\sin^{-1} x$ and $\arcsin x$, $\cos^{-1} x$ and $\arccos x$, $\tan^{-1} x$ and $\arctan x$.

Use antidifferentiation by parts with $u(x) = \ln(2x)$ and $dv = dx$. This gives

$$\int \ln(2x) \cdot 1 \, dx = \ln(2x) \cdot x - \int x \cdot \frac{2}{2x} \, dx = x \cdot \ln(2x) - x + C.$$

32

Evaluate $\int_1^2 xe^{-3x} \, dx$.

Let $u = x$, $v' = e^{-3x} \Rightarrow v = \int e^{-3x} dx = \frac{e^{-3x}}{-3}$. Then

$$\int xe^{-3x} dx = x \cdot \frac{e^{-3x}}{-3} - \int \frac{e^{-3x}}{-3} \, dx = -\frac{1}{3}\left(x \cdot e^{-3x} - \int e^{-3x} dx\right) =$$

$$-\frac{1}{3}\left(xe^{-3x} - \frac{e^{-3x}}{-3}\right) + C = -\frac{e^{-3x}}{3}\left(x + \frac{1}{3}\right) + C.$$

Remember that this is just the antiderivative; you still must evaluate the entire thing from

1 to 2: $\int_1^2 xe^{-3x} \, dx = -\frac{e^{-3x}}{3}\left(x + \frac{1}{3}\right)\Big|_1^2 = -\frac{e^{-6}}{3}\left(2 + \frac{1}{3}\right) + \frac{e^{-3}}{3}\left(1 + \frac{1}{3}\right) = -\frac{7}{9}e^{-6} + \frac{4}{9}e^{-3}$. Since

we are evaluating a definite integral, we don't have to worry about the value of C — it is there only to keep the notation precise.

33

Evaluate $\int_{-1}^2 x\cos\left(\frac{\pi x^2}{4}\right) dx$.

This should be done **without** antidifferentiation by parts because

$\frac{d}{dx}\left(\frac{\pi x^2}{4}\right) = x \cdot$ *<some constant>*. Just let $u = \frac{\pi x^2}{4}$, $du = \frac{2\pi x}{4} \, dx$, and $\frac{2}{\pi} du = x \, dx$. You

get $\int_{-1}^2 x\cos\left(\frac{\pi x^2}{4}\right) dx = \frac{2}{\pi} \int_{\frac{\pi}{4}}^{\pi} \cos u \, du = \frac{2}{\pi}\sin u\Big|_{\frac{\pi}{4}}^{\pi} = 0 - \frac{2}{\pi} \cdot \frac{\sqrt{2}}{2} = -\frac{\sqrt{2}}{\pi}$.

34

Simplify $\int x \cos \dfrac{x}{5} \, dx$.

Using antidifferentiation by parts, we get

$$\int x \cos \frac{x}{5} \, dx = x \cdot 5 \sin \frac{x}{5} - \int 5 \sin \frac{x}{5} \, dx = 5x \cdot \sin \frac{x}{5} + 25 \cdot \cos \frac{x}{5} + C.$$

35

Simplify $\int e^t \cos t \, dt$.

This one is a little tricky. Integrating by parts, you get: $\int e^t \cos t \, dt = e^t \sin t - \int e^t \sin t \, dt$.
Now you might think you've made no progress, but the key is to persevere! Apply
antidifferentiation by parts to that last integral: $\int e^t \sin t \, dt = e^t \left(-\cos t \right) + \int e^t \cos t \, dt$.
Putting this result together with the original integral gives
$\int e^t \cos t \, dt = e^t \sin t - \left[e^t \left(-\cos t \right) + \int e^t \cos t \, dt \right] = e^t \sin t + e^t \cos t - \int e^t \cos t \, dt$. Thus you
have $\int e^t \cos t \, dt = e^t \sin t + e^t \cos t - \int e^t \cos t \, dt$ or $2 \int e^t \cos t \, dt = e^t \sin t + e^t \cos t$.

Dividing by 2 you get $\int e^t \cos t \, dt = \dfrac{1}{2} \left(e^t \sin t + e^t \cos t \right) + C$.

5.9.5. ⌜ **Antidifferentiation by Partial Fractions (BC Only)** ⌟

The antidifferentiation method of partial fractions is really just an algebraic technique
applied in the context of calculus. This method is used to find antiderivatives of rational

functions $f(x) = \dfrac{p(x)}{q(x)}$, where $p(x)$ and $q(x)$ are two polynomials. We can assume that

the degree of p is smaller than the degree of q (otherwise we could divide p by q and
work with the remainder at the top of the fraction). The method of partial fractions is
based on the algebraic observation that a ratio of two polynomials can be replaced by a

sum of simpler fractions. For example, $\dfrac{1}{(ax+b)(cx+d)} = \dfrac{A}{ax+b} + \dfrac{B}{cx+d}$, where A and

B are some constants. The same works for three or more factors in the denominator.

> **Only the simplest polynomial fractions appear on the BC exam: ones with non-repeating linear factors in the denominator.**

36

Find a function F such that $F'(x) = \dfrac{1}{x^2 - x - 12}$.

This is just another way of asking for an antiderivative of $\dfrac{1}{x^2 - x - 12}$, that is,

$\displaystyle\int \dfrac{1}{x^2 - x - 12}\, dx$. The first step is to factor the denominator: $\dfrac{1}{x^2 - x - 12} = \dfrac{1}{(x-4)(x+3)}$.

Now we want to replace the integrand with a sum of two fractions, such that

$\dfrac{1}{(x-4)(x+3)} = \dfrac{A}{x-4} + \dfrac{B}{x+3}$. There is a technique for finding A and B. Multiplying

through by $(x-4)(x+3)$ gives $1 = A(x+3) + B(x-4)$. This equation must be true for all

values of x, so we can plug any value of x into the equation. By making judicious choices

for x — choices that make one of the terms on the right-hand side equal to 0 — we can

simplify our work. Plugging in $x = 4$ gives $1 = A \cdot 7$, so $A = \dfrac{1}{7}$. Similarly, plugging in

$x = -3$ we get $1 = B \cdot (-7)$, so $B = -\dfrac{1}{7}$. You may have learned other ways to solve for A

and B, but we recommend the above method, which is most straightforward. Now we

have $F(x) = \displaystyle\int \dfrac{1}{x^2 - x - 12}\, dx = \dfrac{1}{7} \int \left[\dfrac{1}{x-4} - \dfrac{1}{x+3} \right] dx = \dfrac{1}{7} \left(\ln|x-4| - \ln|x+3| \right) + C$.

37

Evaluate $\displaystyle\int_4^5 \dfrac{dy}{y - \dfrac{y^2}{100}}$.

It's easier to rewrite the integral first as $100 \int_4^5 \dfrac{dy}{100y - y^2} = 100 \int_4^5 \dfrac{dy}{y(100 - y)}$. Now work

on $100 \int \dfrac{dy}{y(100 - y)}$. We have $\dfrac{1}{y(100 - y)} = \dfrac{A}{y} + \dfrac{B}{100 - y}$ and $1 = A \cdot (100 - y) + B \cdot y$. Let

$y = 100$ to annihilate the first term, giving $1 = A \cdot 0 + B \cdot 100$ and $B = \dfrac{1}{100}$. Let $y = 0$ to

annihilate the second term, giving $A = \dfrac{1}{100}$. So

$$100 \int \frac{dy}{y(100 - y)} = 100 \cdot \frac{1}{100} \int \left[\frac{1}{y} + \frac{1}{100 - y} \right] dy = \ln|y| - \ln|100 - y| + C.$$

$$100 \int_4^5 \frac{dy}{y(100 - y)} = \left(\ln|y| - \ln|100 - y| \right) \Big|_4^5 = (\ln 5 - \ln 95) - (\ln 4 - \ln 96).$$

Integrals like the last one come up in logistic differential equations, discussed in Chapter 7.

5.10. ⌈ Improper Integrals (BC Only) ⌋

$\int_a^b f(x)\, dx$ represents the area of a region (or a combination of regions above and below

the x-axis with appropriate signs) bounded by the x-axis, the graph of the function f, and the vertical lines $x = a$ and $x = b$. This concept can be extended to improper integrals. An ***improper integral*** represents the area of an unbounded region. Consider, for

example, $f(x) = \dfrac{1}{x^2}$ for $x \geq 1$ (Figure 5-8). $\int_1^b \dfrac{1}{x^2}\, dx = -\dfrac{1}{x} \Big|_1^b = 1 - \dfrac{1}{b}$. Let's take the limit

of this value when $b \to \infty$: $\displaystyle\lim_{b \to \infty} \int_1^b \dfrac{1}{x^2}\, dx = 1$.

This limit is interpreted as the area of the region bounded by the x-axis, the graph of $\dfrac{1}{x^2}$,

and the vertical line $x = 1$ (the right vertical boundary disappears into infinity).

$\displaystyle\lim_{b \to \infty} \int_1^b \dfrac{1}{x^2}\, dx = 1$ is written as $\displaystyle\int_1^\infty \dfrac{1}{x^2}\, dx$.

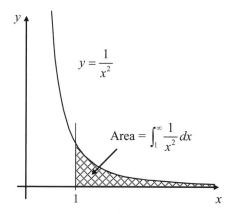

Figure 5-8. The area of the unbounded region for $\displaystyle\int_1^\infty \frac{1}{x^2}\,dx$

There are two types of improper integrals, "horizontal" and "vertical." In a horizontal improper integral, the left limit of integration vanishes into $-\infty$ or the right limit vanishes into ∞, or both limits vanish in respective directions and we integrate over the whole x-axis (Figure 5-9). In a vertical improper integral, we integrate over a closed interval but the function has a vertical asymptote at one or both ends of the interval (Figure 5-10).

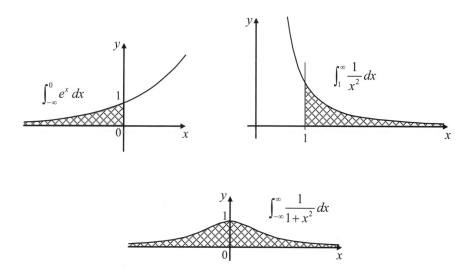

Figure 5-9. "Horizontal" improper integrals

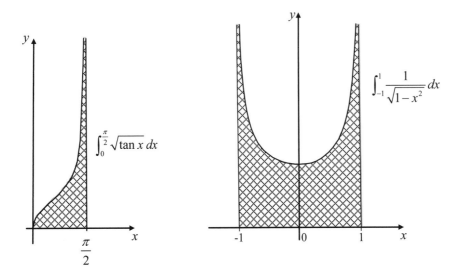

Figure 5-10. "Vertical" improper integrals

We can consider only unbounded regions with one offending end. If we have two offending ends, we represent such an improper integral as the sum of two improper integrals.

> **When the region is unbounded, you <u>cannot</u> directly use the FTC to evaluate the integral, because the FTC requires that the integrand be continuous on a bounded <u>closed</u> interval.**

As we have seen, the technique for evaluating an improper integral "properly" is to evaluate the integral on a bounded closed interval where the function is continuous and the FTC applies — $\int_a^b f(x)\,dx = F(b) - F(a)$ — then take the offending end of the

interval to the limit (e.g., $\int_a^\infty f(x)\,dx = \lim_{b\to\infty}\left[F(b) - F(a)\right]$).

> **If the limit of the integral exists, we say that the integral *converges*. If the limit does not exist, we say that the integral *diverges*.**

Look what happens if we change the integrand from $\dfrac{1}{x^2}$ to $\dfrac{1}{x}$ in the above example:

$\displaystyle\int_1^\infty \frac{1}{x}\,dx = \lim_{b\to\infty}\int_1^b \frac{1}{x}\,dx = \lim_{b\to\infty}\left(\ln|x|\right)\Big|_1^b = \lim_{b\to\infty}\left(\ln b - 0\right) = \lim_{b\to\infty}\ln b$. But $\ln b$ gets arbitrarily

large as b approaches infinity, so $\displaystyle\lim_{b\to\infty}\ln b$ does not exist and the integral diverges.

On the free-response part of the exam, always use the limit notation to evaluate improper integrals. While a shorthand notation, such as

$$\int_1^\infty \frac{1}{x^2}\, dx = \left(-\frac{1}{x} \right)\bigg|_1^\infty = 0 - (-1) = 1 \text{ is expedient, it is mathematically incorrect,}$$

and will result in point deduction.

"Horizontal" Improper Integrals

Evaluate $\displaystyle\int_{-\infty}^0 e^{\frac{x}{3}}\, dx$.

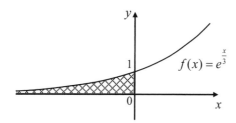

Here we have a region unbounded on the left, so we have to take the limit of the definite integral as the lower limit approaches $-\infty$.

$$\int_{-\infty}^0 e^{\frac{x}{3}}\, dx = \lim_{a\to-\infty} \int_a^0 e^{\frac{x}{3}}\, dx = \lim_{a\to-\infty} 3e^{\frac{x}{3}}\bigg|_a^0 = \lim_{a\to-\infty}\left(3 - 3e^{\frac{a}{3}} \right) = 3 .$$

Find the area of the region under the graph of $y = \dfrac{1}{1+x^2}$.

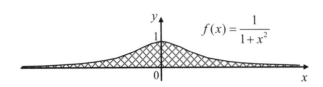

In this example we have a region unbounded on both ends. First let's split the integral into two pieces at $x = 0$, using the Limits Addition property:

$$\int_{-\infty}^{\infty} \frac{1}{1+x^2}\,dx = \int_{-\infty}^{0} \frac{1}{1+x^2}\,dx + \int_{0}^{\infty} \frac{1}{1+x^2}\,dx.$$

Note that the integrand is an even function, so the two integrals in the sum are equal. Let's find one of them:

$$\int_{0}^{\infty} \frac{1}{1+x^2}\,dx = \lim_{b\to\infty} \int_{0}^{b} \frac{1}{1+x^2}\,dx = \lim_{b\to\infty} \left(\arctan x\right)\Big|_{0}^{b} = \lim_{b\to\infty}\left(\arctan b - \arctan 0\right) = \frac{\pi}{2} - 0 = \frac{\pi}{2}.$$

Now we can see that $\displaystyle\int_{-\infty}^{\infty} \frac{1}{1+x^2}\,dx = 2\int_{0}^{\infty} \frac{1}{1+x^2}\,dx = \pi$. Amazing: the area is exactly π.

"Vertical" Improper Integrals

This type of improper integral occurs when the integrand has a vertical asymptote at one of the ends of the integration interval.

40

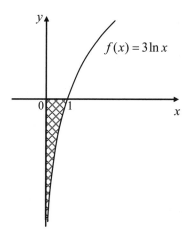

What is the area of the region in the fourth quadrant between the *x*-axis, *y*-axis, and the graph of $y = 3\ln x$?

(A) $\dfrac{1}{3}$ (B) 1   (C) 3   (D) 4   (E) Infinite

As you can see from the picture, you need to evaluate $-\displaystyle\int_0^1 3\ln x\,dx$. Note that

$\displaystyle\lim_{x\to 0^+}\ln x = -\infty$, so this is an improper integral and we need to use the limit of a definite

integral: $-\displaystyle\int_0^1 3\ln x\,dx = -\lim_{a\to 0^+}\int_a^1 3\ln x\,dx =$ (integrating by parts) $-3\lim_{a\to 0^+}\left(x\ln x - x\right)\Big|_a^1 =$

$-3\displaystyle\lim_{a\to 0^+}\left(-1-(a\ln a - a)\right) = 3 + 3\lim_{a\to 0^+}\left(a\ln a\right)$.

From l'Hôpital's Rule, $\displaystyle\lim_{a\to 0^+}\left(a\ln a\right) = \lim_{a\to 0^+}\left(\dfrac{\ln a}{\dfrac{1}{a}}\right) = \lim_{a\to 0^+}\left(\dfrac{\dfrac{1}{a}}{-\dfrac{1}{a^2}}\right) = \lim_{a\to 0^+}\left(-a\right) = 0$.

The answer is C.

Note that the area is the same as in Question 38, because $y = e^{\frac{x}{3}}$ and $y = 3\ln x$ are inverse functions, so the regions are symmetric over the line $y = x$. But the signs of the integrals are different, since the region in this question is below the *x*-axis.

An asymptote can happen somewhere inside the interval of integration. In that case, we have to split the interval into two at the point of the asymptote and consider two improper integrals.

> **When an improper integral is split into two improper integrals, both of them must converge for the sum to converge.**

It is tempting to make symmetry arguments when evaluating improper integrals, but caution must rule the day.

If $\displaystyle\int_{-2}^{2} \frac{1}{x}\,dx$ converges, what is its value?

(A) $-\ln(2)$ (B) 0   (C) $\ln(2)$ (D) 2   (E) It diverges

After looking at the graph, you might think that the integral is zero and choose B. However, to do it properly, you need to split the integral at the discontinuity and look at what happens on each side:

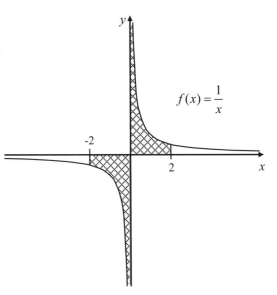

$f(x) = \dfrac{1}{x}$

$$\int_{-2}^{2} \frac{1}{x}\,dx = \lim_{b\to 0^-}\int_{-2}^{b} \frac{1}{x}\,dx + \lim_{a\to 0^+}\int_{a}^{2} \frac{1}{x}\,dx =$$

$$\lim_{b\to 0^-}\left(\ln|x|\Big|_{-2}^{b}\right) + \lim_{a\to 0^+}\left(\ln|x|\Big|_{a}^{2}\right) =$$

$$\lim_{b\to 0^-}\left(\ln|b| - \ln 2\right) + \lim_{a\to 0^+}\left(\ln 2 - \ln|a|\right)$$

Neither of the limits in the last sum exists. Therefore, the original integral does not exist. The correct answer is E.

Integration Worksheet

1. $\displaystyle\int_0^{14} e^{\frac{x}{7}}\,dx$

2. $\displaystyle\int_0^1 \pi + 3x^2\,dx$

3. $\displaystyle\int_1^4 \frac{2x^2 + \sqrt{x}}{x}\,dx$

4. $\displaystyle\int_{\frac{\pi}{3}}^{6} \frac{\sin\left(\frac{3}{t}\right)}{t^2}\,dt$

5. $\displaystyle\int_1^{e^9} \frac{\sqrt{\ln x}}{x}\,dx$

6. $\displaystyle\int_0^{\frac{\pi}{4}} \cos x \sin x\,dx$

7. $\displaystyle\int_0^{\frac{\pi}{12}} \sec 3x \tan 3x\,dx$

8. $\displaystyle\int_{\pi^2}^{4\pi^2} \frac{\cos\sqrt{\frac{w}{4}}}{\sqrt{\frac{w}{4}}}\,dw$

9. $\displaystyle\int_1^2 \frac{dt}{1+t^2}$

10. $\displaystyle\int_{-1}^0 \frac{t\,dt}{\sqrt{4-t^2}}$

11. $\displaystyle\int_{-\frac{1}{2}}^{\frac{1}{2}} \frac{dt}{\sqrt{1-t^2}}$

12. $\displaystyle\int_1^4 |x-3|\,dx$

13. $\displaystyle\int_{\frac{\pi}{12}}^{\frac{\pi}{6}} \csc^2(3s)\,ds$

14. $\displaystyle\int_{-1}^2 \left(3x - x^2\right)^3 (6-4x)\,dx$

15. $\displaystyle\int_0^{\frac{\pi}{4}} \frac{\tan^4 p}{\cos^2 p}\,dp$

16. $\displaystyle\int_{-1}^0 \frac{x}{\left(2+4x^2\right)^2}\,dx$

17. $\displaystyle\int_0^1 \frac{e^{2x}}{e^{2x}+1}\,dx$

18. $\displaystyle\int_0^{\pi/2} \tan\left(\frac{x}{2}\right)\,dx$

19. $\displaystyle\int_0^{e-1} \frac{1}{u+1}\,du$

20. $\displaystyle\int_2^3 e^3\,dt$

21. If the substitution $x = t - 1$ is made in the integral
$\displaystyle\int_1^2 (t-1)^4\,2t\,dt$, what is the resulting integral?

⌈ (BC Only)

22. $\displaystyle\int_1^2 xe^{2x}\,dx$

23. $\displaystyle\int_1^{\sqrt{e}} \ln x\,dx$

24. $\displaystyle\int_1^2 \frac{1}{t^2+2t}\,dt$

25. $\displaystyle\int_3^4 (x-3)^{-\frac{1}{2}}\,dx$

26. $\displaystyle\int_{-\infty}^1 e^{2x}\,dx$

⌋

Worksheet Answers and Solutions

1. $\int_0^{14} e^{\frac{x}{7}} dx = 7 e^{\frac{x}{7}} \Big|_0^{14} = 7\left(e^2 - 1\right).$

2. $\int_0^1 \pi + 3x^2 dx = \left(\pi x + x^3\right)\Big|_0^1 = \pi + 1.$

3. $\int_1^4 \frac{2x^2 + \sqrt{x}}{x} dx = \int_1^4 2x + x^{-\frac{1}{2}} dx = \left(x^2 + 2x^{\frac{1}{2}}\right)\Big|_1^4 = 16 + 4 - (1+2) = 17.$

4.   Let $u = \frac{3}{t}$; $\frac{du}{dt} = -\frac{3}{t^2}$; $-\frac{1}{3} du = \frac{1}{t^2} dt \implies$

   $\int_{\frac{\pi}{3}}^{\frac{6}{\pi}} \frac{\sin\left(\frac{3}{t}\right)}{t^2} dt = -\frac{1}{3} \int_\pi^{\frac{\pi}{2}} \sin u \, du = \frac{1}{3} \cos u \Big|_\pi^{\frac{\pi}{2}} = \frac{1}{3}\left(0 - (-1)\right) = \frac{1}{3}.$

5.   Let $u = \ln x$; $\frac{du}{dx} = \frac{1}{x}$; $du = \frac{1}{x} dx \implies \int_1^{e^9} \frac{\sqrt{\ln x}}{x} dx = \int_0^9 u^{\frac{1}{2}} du = \frac{2}{3} u^{\frac{3}{2}} \Big|_0^9 = \frac{2}{3} \cdot 27 = 18.$

6.   Let $u = \sin x$; $\frac{du}{dx} = \cos x$; $du = \cos x \, dx \implies \int_0^{\frac{\pi}{4}} \cos x \sin x \, dx = \int_0^{\frac{\sqrt{2}}{2}} u \, du = \frac{u^2}{2} \Big|_0^{\frac{\sqrt{2}}{2}} = \frac{1}{4}.$

   (Here you could also choose $u = \cos x$.)

7. $\int_0^{\frac{\pi}{12}} \sec 3x \tan 3x \, dx = \frac{1}{3} \sec 3x \Big|_0^{\frac{\pi}{12}} = \frac{1}{3}\left(\sqrt{2} - 1\right).$

8.   Let $u = \sqrt{\frac{w}{4}}$; $\frac{du}{dw} = \frac{1}{8}\left(\frac{w}{4}\right)^{-\frac{1}{2}}$; $8 du = \frac{1}{\sqrt{\frac{w}{4}}} dw \implies$

   $\int_{\pi^2}^{4\pi^2} \frac{\cos \sqrt{\frac{w}{4}}}{\sqrt{\frac{w}{4}}} dw = 8 \int_{\frac{\pi}{2}}^{\pi} \cos u \, du = 8 \sin u \Big|_{\frac{\pi}{2}}^{\pi} = -8.$

9. $\int_1^2 \frac{dt}{1+t^2} = \arctan t \Big|_1^2 = \arctan 2 - \frac{\pi}{4}.$

10.   Let $u = 4 - t^2$; $\frac{du}{dt} = -2t$; $-\frac{1}{2} du = t \, dt \implies$

   $\int_{-1}^0 \frac{t \, dt}{\sqrt{4 - t^2}} = -\int_3^4 \frac{du}{2\sqrt{u}} = -\sqrt{u} \Big|_3^4 = -\left(2 - \sqrt{3}\right) = \sqrt{3} - 2.$

11. $\int_{-\frac{1}{2}}^{\frac{1}{2}} \dfrac{dt}{\sqrt{1-t^2}} = \arcsin t \Big|_{-\frac{1}{2}}^{\frac{1}{2}} = \dfrac{\pi}{6} - \left(-\dfrac{\pi}{6}\right) = \dfrac{\pi}{3}$

12. $\int_1^4 |x-3|\, dx = \int_1^3 (3-x)\, dx + \int_3^4 (x-3)\, dx = \left(3x - \dfrac{x^2}{2}\right)\Big|_1^3 + \left(\dfrac{x^2}{2} - 3x\right)\Big|_3^4 =$

$\left(9 - \dfrac{9}{2}\right) - \left(3 - \dfrac{1}{2}\right) + (8-12) - \left(\dfrac{9}{2} - 9\right) = \dfrac{5}{2}$

But a geometric solution is shorter: the integral is equal to the sum of the areas of two right triangles.

13. $\int_{\frac{\pi}{12}}^{\frac{\pi}{6}} \csc^2(3s)\, ds = -\dfrac{1}{3}\cot(3s)\Big|_{\frac{\pi}{12}}^{\frac{\pi}{6}} = -\dfrac{1}{3}\cot\left(\dfrac{\pi}{2}\right) + \dfrac{1}{3}\cot\left(\dfrac{\pi}{4}\right) = \dfrac{1}{3}$

14. Let $u = 3x - x^2$; $\dfrac{du}{dx} = 3 - 2x$; $2\,du = (6-4x)\,dx \Rightarrow$

$\int_{-1}^2 (3x - x^2)^3 (6-4x)\, dx = 2\int_{-4}^2 u^3\, du = \dfrac{u^4}{2}\Big|_{-4}^2 = 8 - 128 = -120$.

15. The key is to recognize that $\dfrac{1}{\cos^2 p} = \sec^2 p$. Then let

$u = \tan p$; $du = \sec^2 p\; dp \Rightarrow \int_0^{\frac{\pi}{4}} \dfrac{\tan^4 p}{\cos^2 p}\, dp = \int_0^1 u^4\, du = \dfrac{u^5}{5}\Big|_0^1 = \dfrac{1}{5}$

16. Let $u = 2 + 4x^2$; $\dfrac{du}{dx} = 8x$; $\dfrac{1}{8}\,du = x\,dx \Rightarrow$

$\int_{-1}^0 \dfrac{x}{\left(2+4x^2\right)^2}\, dx = \dfrac{1}{8}\int_6^2 u^{-2}\, du = -\dfrac{1}{8}\cdot\dfrac{1}{u}\Big|_6^2 = -\dfrac{1}{8}\left(\dfrac{1}{2} - \dfrac{1}{6}\right) = -\dfrac{1}{24}$

17. Let $u = e^{2x} + 1$; $\dfrac{du}{dx} = 2e^{2x}$; $\dfrac{1}{2}\,du = e^{2x}\,dx \Rightarrow$

$\int_0^1 \dfrac{e^{2x}}{e^{2x}+1}\, dx = \dfrac{1}{2}\int_2^{e^2+1} \dfrac{du}{u} = \dfrac{1}{2}\ln u \Big|_2^{e^2+1} = \dfrac{1}{2}\left(\ln(e^2+1) - \ln 2\right)$

18. $\int_0^{\pi/2} \tan\left(\dfrac{x}{2}\right) dx = 2\ln\left(\sec\left(\dfrac{x}{2}\right)\right)\Big|_0^{\pi/2} = 2\left(\ln\sqrt{2}\right) = \ln 2$

19. $\int_0^{e-1} \dfrac{1}{u+1}\, du = \ln|u+1|\,\Big\|_0^{e-1} = 1$

20. $\int_2^3 e^3\, dt = e^3 t\Big|_2^3 = 3e^3 - 2e^3 = e^3$. Don't be fooled by the format: e^3 is just a constant.

21. With $x = t - 1$, $dx = dt$ and $t = x + 1$. So,

$\int_1^2 (t-1)^4\, 2t\, dt = 2\int_0^1 x^4 (x+1)\, dx = 2\int_0^1 x^5 + x^4\, dx$

⌐ (BC Only)

22. Integrate by parts. Let $u = x$ and $dv = e^{2x}dx$; $du = dx$; $v = \frac{1}{2}e^{2x}$. So

$\int xe^{2x}dx = \frac{x}{2}e^{2x} - \frac{1}{2}\int e^{2x}dx = \frac{x}{2}e^{2x} - \frac{1}{4}e^{2x}$. Evaluating the definite integral gives

$\int_1^2 xe^{2x}dx = \left(\frac{x}{2}e^{2x} - \frac{1}{4}e^{2x}\right)\Big|_1^2 = \left(e^4 - \frac{1}{4}e^4\right) - \left(\frac{e^2}{2} - \frac{e^2}{4}\right) = \frac{3e^4}{4} - \frac{e^2}{4}$.

23. Integrate by parts. Let $u = \ln x$ and $dv = dx$; $du = \frac{1}{x}dx$; $v = x$. So

$\int \ln x\,dx = x\ln x - \int x\cdot\frac{1}{x}dx = x\ln x - x$. Evaluating the definite integral gives

$\int_1^{\sqrt{e}} \ln x\,dx = x\ln x - x\Big|_1^{\sqrt{e}} = \left(\sqrt{e}\cdot\frac{1}{2} - \sqrt{e}\right) - (0 - 1) = 1 - \frac{\sqrt{e}}{2}$.

24. Use partial fractions: $\frac{1}{t^2 + 2t} = \frac{1}{t(t+2)} = \frac{A}{t} + \frac{B}{t+2}$. Multiplying gives

$1 = A(t+2) + Bt \Rightarrow B = -\frac{1}{2}$ and $A = \frac{1}{2}$. So

$\int_1^2 \frac{1}{t^2 + 2t}dt = \frac{1}{2}\int_1^2 \frac{1}{t}dt - \frac{1}{2}\int_1^2 \frac{1}{t+2}dt = \frac{1}{2}\left[\ln t\Big|_1^2 - \ln(t+2)\Big|_1^2\right] =$

$\frac{1}{2}\left[\ln 2 - (\ln 4 - \ln 3)\right] = \frac{1}{2}\ln\left(\frac{3}{2}\right)$.

25. This is an improper integral, since the integrand has a vertical asymptote at $x = 3$. By definition, $\int_3^4 (x-3)^{-\frac{1}{2}}dx = \lim_{a\to 3}\int_a^4 (x-3)^{-\frac{1}{2}}dx = \lim_{a\to 3} 2(x-3)^{\frac{1}{2}}\Big|_a^4 = 2$.

26. By definition, $\int_{-\infty}^1 e^{2x}dx = \lim_{a\to-\infty}\int_a^1 e^{2x}dx = \lim_{a\to-\infty}\frac{1}{2}e^{2x}\Big|_a^1 = \frac{1}{2}e^2$.

⌐

Chapter 6.   Applications of Integrals

6.1.   Overview

This chapter reviews the following applications of the definite integral:

- Finding the area of a specified region in the coordinate plane;

- Finding the volume of a solid with known cross sections;

- Finding the average value for a function on a specified interval;

- Finding the distance traveled by a particle moving along a line with a given velocity over a specified time interval;

- Other applications that involve finding the net change in a quantity from a known rate of change of that quantity;

- ⌈ Finding the length of an arc described by a given function (BC only). ⌋

Also discussed in this chapter are problems involving accumulation functions. In this type of problem you are given a function $f(x)$ and asked about the properties of the accumulation function $g(x) = \int_a^x f(t)\,dt$. Although they are not applications of the integral as such, problems of this type have become common on recent AP exams, so we discuss them here, too.

⌈ BC-only material also includes finding areas of regions bounded by polar curves and finding the arc length for a curve represented parametrically. These topics are reviewed in Chapter 8. ⌋

6.2.   Finding the Area of a Region

In a typical problem, the region is bounded at the top and bottom by specified curves or lines. One of these lines may be simply the x-axis or another horizontal line, $y = c$. On the left and right, the region may be bounded by vertical lines $x = a$ and $x = b$ (or simply the y-axis). Alternatively, the curves may intersect at a point that becomes one of the boundaries. In the latter case, you may need to find the intersection point first, often using your calculator. Figure 6-1 shows some of the possibilities.

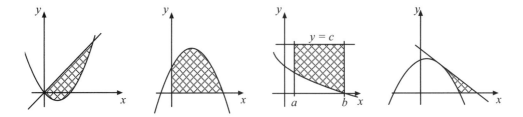

Figure 6-1. Some of the possible situations in "find the area" problems

Every one of these problems is solved using the same basic concept. In every case, think of the area as the sum of many "infinitely thin" rectangles. The width of each rectangle is an increment of the independent variable, often dx, in which case the height of each rectangle is the difference of the two bounding functions: the top function, $f(x)$, minus the bottom function, $g(x)$ (Figure 6-2). In other words,

> **the area bounded by the graphs of *f* and *g* between *x* = *a* and *x* = *b* is equal to**
> $\int_a^b [f(x) - g(x)] dx$**. This works regardless of whether** $f(x)$ **or** $g(x)$ **or both**
> **are above or below the *x*-axis, as long as** $f(x) \geq g(x)$ **for all *x* in [*a*, *b*].**

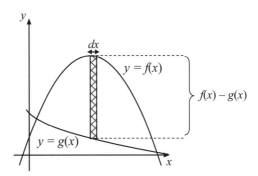

Figure 6-2. Approximating the area using the sum of "infinitely thin" rectangles

If $g(x) = 0$ (i.e., the lower boundary is the *x*-axis), you end up just integrating $f(x)$:
Area = $\int_a^b f(x) dx$. If a function's graph is below the *x*-axis, the "top" curve is simply the line $y = 0$ (that is, the *x*-axis). We subtract from it the "bottom" curve. If $g(x) < 0$, the integrand, $[0 - g(x)] = -g(x)$, is positive (Figure 6-3).

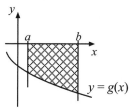

Figure 6-3.   Area of the region below the *x*-axis $= \int_a^b -g(x)\,dx$

1

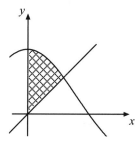

What is the area of the shaded region in the figure above, bounded by the graphs of $y = x$, $y = 2e^{-x^2} - \dfrac{1}{2}$, and the *y*-axis?

Use your calculator to find the *x*-coordinate of the point of intersection of the two graphs $A \approx .7093$. Area $= \int_0^A \left(2e^{-x^2} - \dfrac{1}{2} - x \right) dx \approx 0.606$.

This is a case where you should store the *x*-coordinate of the point of intersection in a calculator variable right after you find it. Then use that variable in your integral expression for the area.

See Section 5.8 for an example.

Sometimes the upper or lower boundary of the region may be made up of pieces of different functions. In that case you need to split the integral into pieces (Figure 6-4).

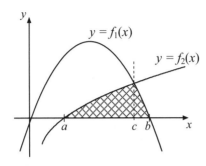

Figure 6-4. The area of the region is $\displaystyle\int_a^c f_2(x)\,dx + \int_c^b f_1(x)\,dx$

2

Consider the same two curves as in Question 1, but this time find the area of the region bounded by their graphs and the *x*-axis.

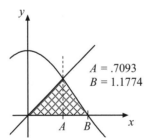

$A = .7093$
$B = 1.1774$

Here, the formula for the top curve "switches" at $x = .7093$. You also need to find the positive *x*-intercept of $y = 2e^{-x^2} - \dfrac{1}{2}$ to use as a limit of integration. The calculator gives it as $x = 1.17741$. With this result stored in *B*, and the *x*-coordinate of the point where the two curves intersect stored in *A*, the area is given by

$$\int_0^A x\,dx + \int_A^B \left(2e^{-x^2} - \frac{1}{2}\right)dx \;\blacksquare\; \approx .251553 + .155745 \approx .407 .$$

3

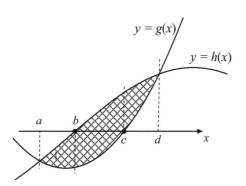

Which of the following represents the area of the shaded region bounded by the graphs of $y = g(x)$ and $y = h(x)$ in the figure above?

(A) $\displaystyle\int_a^d \left[h(x) - g(x) \right] dx$

(B) $\displaystyle\int_a^b \left[-h(x) - g(x) \right] dx + \int_b^d \left[h(x) + g(x) \right] dx$

(C) $\displaystyle\int_a^c -g(x)\, dx + \int_c^d g(x)\, dx + \int_a^b -h(x)\, dx + \int_b^d h(x)\, dx$

(D) $\displaystyle\int_a^b \left[-h(x) - g(x) \right] dx + \int_b^c \left[h(x) + g(x) \right] dx + \int_c^d \left[h(x) - g(x) \right] dx$

(E) $\displaystyle\int_a^b \left[-h(x) - g(x) \right] dx + \int_b^c \left[h(x) - g(x) \right] dx + \int_c^d \left[h(x) + g(x) \right] dx$

The problem is simpler than it might appear. The top curve is the graph of h and the bottom curve is the graph of g. The correct answer is A.

> **In some cases, it might be easier or more prudent to integrate "sideways," taking the independent variable to be y instead of x, and the "heights" of the rectangles to be oriented horizontally rather than vertically.**

4

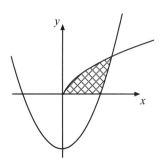

Find the area of the shaded region in the figure above, bounded by the graphs of $y = \sqrt{x}$, $y = 2x^2 - 1$, and the *x*-axis.

The curves intersect at the point (1, 1). You can find the area, of course, as $\int_0^1 \sqrt{x}\, dx - \int_a^1 (2x^2 - 1)\, dx$, where *a* is the *x*-intercept of $y = 2x^2 - 1$. However, it may be easier to look at the problem sideways. In this view, the first curve is $x = y^2$, the second is $x = \left(\dfrac{y+1}{2}\right)^{\frac{1}{2}}$, and the area is $\int_0^1 \left[\left(\dfrac{y+1}{2}\right)^{\frac{1}{2}} - y^2\right] dy = \left[\dfrac{1}{\sqrt{2}} \cdot \dfrac{2}{3}(y+1)^{\frac{3}{2}} - \dfrac{y^3}{3}\right]\Bigg|_0^1 =$

$\left(\dfrac{\sqrt{2}}{3} \cdot 2^{\frac{3}{2}} - \dfrac{1}{3}\right) - \dfrac{\sqrt{2}}{3} = 1 - \dfrac{\sqrt{2}}{3}$.

6.3. Calculating Volumes

Every AB and nearly every BC exam has included a free response question involving the volume of a solid. Beginning in 1998, the only such solids covered on the exams have been those with <u>known cross sections</u>. A technique called "cylindrical shells," still covered in many textbooks, is no longer a required topic in AP Calculus. Of course, you can still use that method to solve a problem on the exam without penalty, but it is unlikely to yield a more economical solution.

There are basically two different ways that solids with known cross sectional area can be described in AP problems:

- Solids of revolution, obtained by rotating a given two-dimensional region around a horizontal or vertical line;

- Solids with cross sections that are explicitly described as some familiar geometric shape.

The volume of a solid like this is found as the integral of the area of a cross section along the width or height of the region: $V = \int_a^b A(x)\,dx$ or $V = \int_c^d A(y)\,dy$.

6.3.1. Volumes of Solids of Revolution

These problems describe a two-dimensional region, usually by giving some functions and lines that define its boundary. The region is then revolved around a horizontal or vertical line, forming a solid. If the axis of revolution serves as one of the boundaries of the region, then each cross section of the resulting solid is a disk. Its area is πR^2, where R is the radius. If the region is not bounded by the axis of revolution on one side, then each cross section is a "washer" (a circular disk with a hole cut out in the middle). Its area is $\pi\left(R^2 - r^2\right)$, where R is the outer radius and r is the inner radius.

Thus, in order to find the area of a cross section, we need to find the radius of the disk, or, if the cross section is a washer, then we need both the outer and the inner radii of the washer. These radii are <u>always</u> the distances from the axis of revolution to the boundary of the region (Figure 6-5). The larger radius, R, is the distance from the axis of revolution to the <u>far</u> side of the region; the smaller radius, r, is the distance from the axis of revolution to the <u>near</u> side of the region.

Volume $= \pi \int_a^b \left[R^2(x) - r^2(x) \right] dx$.

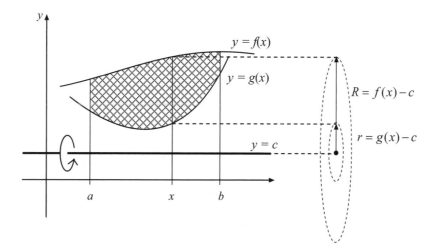

Figure 6-5. The region bounded by $f(x)$ and $g(x)$ is rotated around the horizontal line $y = c$. The inner and outer radii of each "washer" cross section are the distances from the axis of revolution to the boundaries of the region.

What is the volume of the solid generated by revolving the region bounded by the *x*-axis and the graph of $y = 4x - x^2$ about the *x*-axis?

(A) $\dfrac{32\pi}{15}$ (B) $\dfrac{32\pi}{3}$ (C) $\dfrac{256\pi}{15}$ (D) $\dfrac{512\pi}{15}$ (E) $\dfrac{2048\pi}{15}$

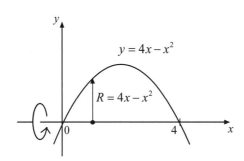

Cross sections perpendicular to the *x*-axis are disks. The radius of each disk is the distance from the curve to the axis of revolution. In this case, that distance is just the *y*-coordinate of a point on the graph of $y = 4x - x^2$. So, the radius is $R(x) = y = 4x - x^2$, and the area of the disk is $A(x) = \pi\left(4x - x^2\right)^2$. The region starts at $x = 0$ and ends at $x = 4$. The volume is given by $V = \pi \int_0^4 \left(4x - x^2\right)^2 dx = \pi \int_0^4 \left(16x^2 - 8x^3 + x^4\right)dx =$

$\pi\left(\dfrac{16}{3}x^3 - \dfrac{8}{4}x^4 + \dfrac{1}{5}x^5\right)\Bigg|_0^4 = \dfrac{512\pi}{15}$. The answer is D.

Consider the region *S* enclosed by the graphs of $y = x^3 - 6x^2 + 9x$ and $y = \dfrac{x}{2}$. Determine which solid has the greater volume, and by how much:

(a)   the solid generated by revolving *S* about the *x*-axis;
(b)   the solid generated by revolving *S* about the line $y = 4$.

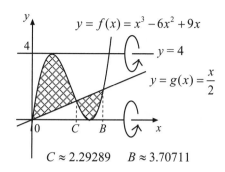

$C \approx 2.29289 \quad B \approx 3.70711$

(a)

It is a good idea to name the two functions to avoid copying their formulas several times in your solution. Let $f(x) = x^3 - 6x^2 + 9x$ and $g(x) = \dfrac{x}{2}$. Also, store these functions in your calculator as Y_1 and Y_2 to avoid entering their formulas several times.

Find the values of x where the two graphs intersect: ▪ $C \approx 2.29289$ and $B \approx 3.70711$, and store them in your calculator. The region S is composed of two pieces. From 0 to C, the cross sections are washers with $R = f(x)$ and $r = g(x)$. From C to B, $R = g(x)$ and $r = f(x)$. The volume is given by

$$V = \pi \int_0^C \left[f(x)\right]^2 - \left[g(x)\right]^2 \, dx + \pi \int_C^B \left[g(x)\right]^2 - \left[f(x)\right]^2 \, dx \; ▪ \approx 61.6324 + 8.10297 \approx 69.735 \,.$$

(b)

When we revolve the region around the line $y = 4$, the cross sections are also washers. From 0 to C, $R = 4 - g(x)$ and $r = 4 - f(x)$. (Recall that R is the distance from the axis of revolution to the far side of the region, while r is the distance from the axis of revolution to the near side of the region.) From C to B, $R = 4 - f(x)$ and $r = 4 - g(x)$. The volume is given by

$$V = \pi \int_0^C \left[4 - g(x)\right]^2 - \left[4 - f(x)\right]^2 \, dx + \pi \int_C^B \left[4 - f(x)\right]^2 - \left[4 - g(x)\right]^2 \, dx \; ▪$$

$\approx 67.6657 + 27.4401 \approx 95.106$. Volume (b) is larger by approximately 25.370 cubic units.

> No matter how tempting, do <u>not</u> try to "simplify" $\pi R^2 - \pi r^2$ as $\pi \left(R - r\right)^2$.
> It's just plain wrong! This is a very common error students make on the AP exam.

6.3.2. Volumes of Solids with Specified Cross Sections

In these problems, the base of the solid is usually described by functions that form its boundary in the *xy*-plane. Each cross section is described as a simple geometrical shape, such as a square, a rectangle, a semicircle, or a triangle, perpendicular to the *x*-axis or the *y*-axis (Figure 6-6).

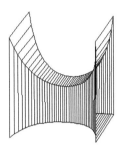

Figure 6-6. A solid with square cross sections

All such volume problems are solved the same way. First, you find a formula for the cross-sectional area of the solid. This is usually a function of *x*, say $A(x)$. Then you integrate the area and get the total volume. Again, $V = \int_a^b A(x)\,dx$. Here *a* and *b* are the *x*-coordinates of the points where the solid starts and stops.

7

The base of a solid is the region *R* in the first quadrant bounded by the graph of $y = 3x^{\frac{1}{2}} - x^{\frac{3}{2}}$ and the *x*-axis. Cross sections of the solid, perpendicular to the *x*-axis, are isosceles right triangles, with one leg in the *xy*-plane. What is the volume of the solid?

(A) $\dfrac{27}{16}$ (B) $\dfrac{27}{8}$ (C) $\dfrac{27}{4}$ (D) $\dfrac{6\sqrt{3}}{5}$ (E) $\dfrac{12\sqrt{3}}{5}$

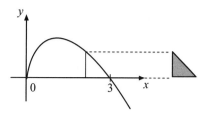

Since each leg of the right triangle has length y, the area of each cross section is

$A = \dfrac{1}{2} y^2 = \dfrac{1}{2} \left(3x^{\frac{1}{2}} - x^{\frac{3}{2}} \right)^2$. The x-intercepts are 0 and 3, so the volume of the solid is

given by $V = \displaystyle\int_0^3 \dfrac{1}{2} \left(3x^{\frac{1}{2}} - x^{\frac{3}{2}} \right)^2 dx$. This integral could be evaluated without a calculator

(after first multiplying out the integrand), so this question could appear on the

no-calculator section of the exam. The volume is $\dfrac{27}{8} = 3.375$. The answer is B.

6.4.   Average Value of a Function

The idea of using an arithmetic average to summarize a collection of data is, of course, very common. An average is easy to calculate: you simply divide the sum of the data values by the number of values. How does this idea translate to finding the average value of a function? A continuous function, after all, has an <u>infinite</u> number of values! How can we average them? Let's start with a finite number of values of f. Suppose the points $x_0, x_1, x_2, \ldots, x_{n-1}, x_n$, where $x_0 = a$ and $x_n = b$, are evenly spaced over the closed interval

$[a, b]$, so that $x_i = x_{i-1} + \Delta x$ ($i = 1, 2, \ldots, n$). $\Delta x = \dfrac{b-a}{n}$. Suppose we say the average

value of f on $[a, b]$ is approximately $\dfrac{f(x_1) + f(x_2) + \ldots + f(x_n)}{n}$. We can rewrite it as

$\dfrac{f(x_1) + f(x_2) + \ldots + f(x_n)}{b-a} \cdot \dfrac{b-a}{n} = \dfrac{1}{b-a} \displaystyle\sum_{i=1}^{n} \Delta x f(x_i)$. The Σ expression is familiar: it is,

of course, a Riemann sum for $\displaystyle\int_a^b f(x)\,dx$. It makes sense to define the average of f on

$[a, b]$ as $\displaystyle\lim_{n \to \infty} \left(\dfrac{f(x_1) + f(x_2) + \ldots + f(x_n)}{n} \right) = \dfrac{1}{b-a} \lim_{\Delta x \to 0} \sum_{i=1}^{n} \Delta x f(x_i) = \dfrac{1}{b-a} \int_a^b f(x)\,dx$.

> **The average value of a function f on a closed interval $[a, b]$ is defined as**
> $\dfrac{1}{b-a} \displaystyle\int_a^b f(x)\,dx$.

Don't forget to divide by $(b-a)$. You can be certain that one of the wrong answers in a multiple choice question will have resulted from just that error!

Recall that, if $f(x) > 0$, $\int_a^b f(x)\,dx$ represents the area of the region under the graph of f between $x = a$ and $x = b$. If you multiply the average by $b-a$, you get the area. Another example: suppose you want to find the average speed at which you traveled during the time period from t_1 to t_2. It makes sense to define it as

$$\text{average speed} = \frac{\text{total distance traveled}}{\text{travel time}} = \frac{\int_{t_1}^{t_2} |v(t)|\,dt}{t_2 - t_1}.$$

8

What is the average vertical distance between the graphs of $y = \cos x$ and $y = \sin x$ on the interval $\left[0, \dfrac{\pi}{4}\right]$?

(A) 0   (B) $\sqrt{2} - 1$ (C) $\dfrac{4}{\pi}\left(\sqrt{2} - 1\right)$

(D) $\dfrac{2\sqrt{2}}{\pi}$ (E) $\dfrac{4}{\pi}$

The vertical distance between the curves is the difference between the y-coordinates of two points with the same x-coordinate. Since $\cos x$ is greater than $\sin x$ on the interval in question, the average distance

$$= \frac{1}{\dfrac{\pi}{4} - 0} \int_0^{\frac{\pi}{4}} [\cos x - \sin x]\,dx = \frac{4}{\pi} \cdot [\sin x + \cos x]\Big|_0^{\frac{\pi}{4}} = \frac{4}{\pi}\left(\sqrt{2} - 1\right).$$

The answer is C.

9

A particle moves along the x-axis so that its position at any time t is given by $x(t) = 2e^{-t^2} - \dfrac{t}{3}$. What is the average distance of the particle from the origin between $t = 0$ and $t = 2$?

(A) 0.549   (B) 0.827   (C) 1.097   (D) 1.315   (E) 1.653

Since you are asked for the average <u>distance</u> from the origin, and distance is always a positive number, you need to find the average of the absolute value of $x(t)$. It is given by $\dfrac{1}{2}\displaystyle\int_0^2 \left| 2e^{-t^2} - \dfrac{t}{3} \right| dt$ ≈ 0.827. The answer is B.

6.5.   Net Change of a Function

If you know the rate of change of some quantity, you can calculate the net change in that quantity over time. If the rate of change is a constant, say r, the net change from t_1 to t_2 is simply $r(t_2 - t_1)$. For a variable rate of change, however, we need an integral. As we know from the Fundamental Theorem, $F(t_2) - F(t_1) = \displaystyle\int_{t_1}^{t_2} F'(t)\,dt$. Therefore, if you know the rate of change $f(t) = F'(t)$, you can find out how much $F(t)$ has changed between two values of the independent variable (here, t_1 and t_2) by integrating $f(t)$ between those two values. In other words,

> **the integral of a rate of change tells you the net change.**

Once you recognize that your answer is an integral, the thing to do is to construct the integral that answers the given question and determine the limits of integration in the context of the problem. Very often, looking at the units can help.

Here are some situations where the integral of the rate of change would be used:

- If you know the rate at which oil is flowing through a pipeline, you can determine how much oil flows through it over a given period of time.

- If you know the rate at which solar collectors are producing electricity, you can determine the total amount of electricity produced over a given period of time.

- If you know the rate at which people are leaving a ballpark, you can determine how many people have left the ballpark over a given period of time.

- If you know the rate at which the water level is rising in a river, you can determine the total change in the water level over a period of time.

- If you know the rate at which the price of milk is rising during a time of economic inflation, you can determine the total price increase over a period of time.

In every case, you would integrate the rate function between two values of time to determine the net change.

During spawning season, the rate of salmon swimming up a river is given by

$S(t) = \dfrac{1500}{1 + \left(\dfrac{t-14}{3}\right)^2}$ salmon/day for $0 \le t \le 28$, where t is the number of days since the

start of spawning season. How many salmon swam up the river during the first two weeks of spawning season?

The total number of salmon that swam up the river is given by $\displaystyle\int_0^{14} \dfrac{1500}{1 + \left(\dfrac{t-14}{3}\right)^2}\, dt$. Using

a calculator to evaluate the integral gives about 6119 salmon.

11

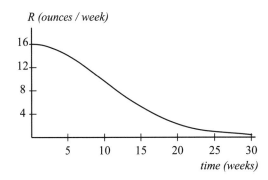

The above graph shows the rate of change in weight of an animal in ounces/week over a 30-week period of time. Which of the following is closest to the total change in weight of the animal in ounces during the 30 weeks?

    (A) $\dfrac{8}{15}$ (B) 16   (C) 30   (D) 240   (E) 480

The area under the graph gives the net change in weight of the animal. It can be approximated by the area of a triangle with base 30 weeks and height 16 ounces per week, giving 240 ounces. The answer is D. This example illustrates how an analysis of the units in the problem can help you reason about its solution.

12

t (hours)	Pump rate, $P(t)$ (gals/hour)
0	250
4	300
8	400
12	500
16	200
20	650
24	450

Fuel is pumped from an underground tank at the rates indicated in the table. The fuel is sold for $1.50/gallon.

(a) Explain the meaning of $\int_0^{24} P(t)\,dt$. Use correct units.

(b) Use a left-hand Riemann sum with 6 subintervals to approximate the total amount of money earned from the sale of fuel over the 24-hour period.

(a) $\int_0^{24} P(t)\,dt$ represents the number of gallons of fuel pumped from the tank during the 24-hour time period.

(b) $\int_0^{24} P(t)\,dt \approx 4 \cdot \left(250 + 300 + 400 + 500 + 200 + 650 \right) = 9200$ gallons.

9200 gallons times $1.50 per gallon gives $13,800.

6.6. Motion of a Particle

One of the classic applications of the integral involves finding either the total distance or the net distance traveled by an object whose velocity is a known function of time. If the velocity v of an object is constant, then over any interval of time, Δt, the net change in position is $\Delta s = v \cdot \Delta t$. If the velocity is not constant, then over an interval of time from t_1 to t_2, the net change in position is $\int_{t_1}^{t_2} v(t)\,dt$.

The velocity of a particle moving along the x-axis is given by $v(t) = \cos t$. If $x(0) = 0$, what is the position of the particle at $t = 2\pi$?

(A) -4 (B) 0 (C) 1 (D) 2 (E) 4

From $t = 0$ to $t = 2\pi$, the particle has exactly the same amount of negative velocity as positive velocity. Thus the net change in position (displacement) is 0. Analytically, we get $x(2\pi) - x(0) = \int_{0}^{2\pi} \cos t\,dt = \sin t \Big|_{0}^{2\pi} = \sin(2\pi) - \sin(0) = 0$. The answer is B.

> **It is very important not to confuse the net change in position of the particle with the total distance traveled.**

If a particle moves along the x- or y-axis, the velocity is positive when the particle is moving in the positive direction (to the right or up) and negative when the particle is moving in the negative direction (to the left or down). In the total distance traveled we count both positive and negative movement. To find the total distance traveled you have to integrate the speed, which is the <u>absolute value</u> of the velocity.

The velocity of a particle moving along a straight line is given by $v(t) = \cos t$. What is the total distance traveled by the particle from $t = 0$ until $t = 2\pi$?

(A) -4 (B) 0 (C) 1 (D) 2 (E) 4

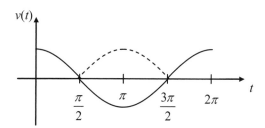

To find the total distance traveled, integrate the speed function (absolute value of velocity). Effectively, this treats motion in the negative direction as contributing positively to the total distance. Thus,

$$\text{Total distance traveled} = \int_0^{2\pi} |\cos t|\, dt = \int_0^{\frac{\pi}{2}} \cos t\, dt + \int_{\frac{\pi}{2}}^{\frac{3\pi}{2}} -\cos(t)\, dt + \int_{\frac{3\pi}{2}}^{2\pi} \cos t\, dt.$$

Notice that when the velocity is negative, we integrate the opposite of the velocity to make it positive. Using symmetry, we can simplify our calculations:

$$\int_0^{2\pi} |\cos t|\, dt = 4\int_0^{\frac{\pi}{2}} \cos t\, dt = 4\sin t \Big|_0^{\frac{\pi}{2}} = 4. \text{ The answer is E.}$$

15

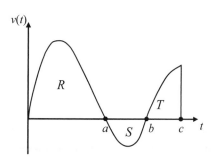

The above graph shows the velocity of a particle moving along the y-axis for $0 \le t \le c$. The particle is at the origin at $t = 0$. The region R has area 8, the region S has area 1.3, and the region T has area 1.7.

(a)   What is the position of the particle at time $t = a$? At $t = b$? At $t = c$?
(b)   At what time t is the particle farthest from the origin?
(c)   What is the total distance traveled by the particle?

(a) The area of R is given by $\int_0^a v(t)\,dt$, which represents the distance traveled from $t = 0$ to $t = a$. Thus the particle is 8 units above the origin at $t = a$. The area of S is given by $\int_a^b -v(t)\,dt$, which represents the distance traveled from $t = a$ to $t = b$.

Note that $\int_a^b v(t)\,dt < 0$, which indicates that the particle is moving <u>down</u> during this time. Thus, the particle is at $8 - 1.3 = 6.7$ units above the origin at $t = b$. The area of T is given by $\int_b^c v(t)\,dt$, which represents the distance traveled from $t = b$ to $t = c$.

The particle is moving up, since its velocity is positive again. Thus the position at $t = c$ is $6.7 + 1.7 = 8.4$.

(b) The particle is farthest from the origin at $t = c$. The important comparison to note here is that there is more area above the x-axis from $t = b$ to $t = c$ than there is area below the x-axis from $t = a$ to $t = b$.

(c) In calculating total distance traveled, add the absolute values of the distances traveled. So, it is $8 + 1.3 + 1.7 = 11$.

6.7.   Accumulation Functions

$g(x) = \int_a^x f(t)\,dt$ (where a is a specified constant) is called the accumulation function for f. An AP question may ask you about the properties of the accumulation function for a given function f, which is often defined graphically (though there are other options, of course).

If the graph of the integrand forms regions whose areas can be calculated using geometry, then you can evaluate g at various values of x. Furthermore, since $g'(x) = f(x)$, you can analyze the derivative of g for extrema and find points of inflection of g.

16

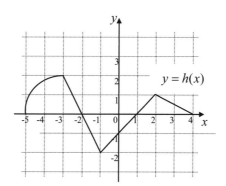

The graph of the function h consists of a quarter-circle and three line segments, as shown above. Let g be the function defined by $g(x) = \int_{-1}^{x} h(t)\, dt$.

(a) Find $g(-5)$.
(b) Find all values of x on the open interval $(-5, 4)$ where g is decreasing. Justify your answer.
(c) Write an equation for the line tangent to the graph of g at $x = -1$.
(d) Find the minimum value of g on the closed interval $[-5, 4]$. Justify your answer.

(a) $g(-5) = \int_{-1}^{-5} h(t)\, dt = -\int_{-5}^{-1} h(t)\, dt = -\left[\int_{-5}^{-3} h(t)\, dt + \int_{-3}^{-1} h(t)\, dt \right]$. $\int_{-5}^{-3} h(t)\, dt$ is the area

of a quarter circle with radius 2. This area is equal to $\frac{1}{4}\pi \cdot 2^2 = \pi$. $\int_{-3}^{-1} h(t)\, dt = 0$,

because it combines the areas of two triangles with equal areas, one above the x-axis, the other below. Therefore, $g(-5) = -\pi$.

(b) From the FTC, we have $g'(x) = h(x)$. Looking at the graph of h, we know $g'(x) < 0$ for $-2 < x < 1$. So g is decreasing on $[-2, 1]$.

(c) $g(-1) = \int_{-1}^{-1} h(t)\, dt = 0$. $g'(-1) = h(-1) = -2$. The equation for the tangent line is $y - 0 = -2(x + 1)$.

(d) The minimum value of g must occur at an endpoint or where $g'(x)$ changes sign from negative to positive. The only candidates are therefore $x = -5$, $x = 1$, and $x = 4$. We can exclude $x = 4$ because the function is increasing when we approach 4 from the left ($g'(x) > 0$). In Part (a) we found $g(-5) = -\pi$. $g(1) = \int_{-1}^{1} h(t)\, dt = -2$. The minimum value of g is $g(-5) = -\pi$.

17

Consider the function g defined by $g(x) = \int_1^x \arctan\left(t^3 - 3t^2 + 2t\right) dt$. What are the x-coordinates of all points of inflection on the graph of g?

    (A) $x = 0$ only   (B) 1.577 only   (C) 0.423 only

    (D) $x = 0.423$ and $x = 1.577$ (E) $x = 0$, $x = 1$, and $x = 2$

As we saw in Chapter 4, a point of inflection on the graph of a function occurs when its derivative changes from increasing to decreasing or vice-versa. From the FTC, $g'(x) = \arctan(x^3 - 3x^2 + 2x)$. Graph $g'(x)$ with the calculator and find its extrema.

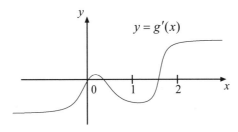

They occur at $x \approx .423$ and $x \approx 1.577$. The answer is D.

If this were a free-response question, you would need to go a step further to justify your answer, because finding extrema is not one of the four allowed calculator operations. To find the extrema of $g'(x)$, you would need to write a formula for $g''(x)$, then use your calculator to find the values of x where $g''(x)$ changes sign.

18

For the function $g(x)$ from the previous question, the number of relative extrema is

    (A) 0   (B) 2   (C) 3   (D) 4   (E) more than 4

We graph $g'(x) = \arctan(x^3 - 3x^2 + x)$ and count the number of times the graph crosses the x-axis. Each time, g has an extremum. There are three such points. The answer is C.

6.8. ⌈ Arc Length (BC Only) ⌋

The basic idea for calculating the length of a curve is to approximate the curve with "infinitely" many straight-line segments that are "infinitely" short (Figure 6-7-a). From the Pythagorean Theorem, the length of each segment is $\sqrt{\Delta x^2 + \Delta y^2}$ (Figure 6-7-b).

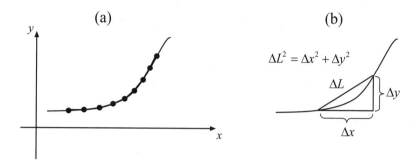

Figure 6-7. Approximating the length of a curve

If we add up all these little lengths, we get $\displaystyle\sum_{i=1}^{n} \sqrt{\Delta x_i^2 + \Delta y_i^2} = \sum_{i=1}^{n} \Delta x_i \sqrt{1 + \left(\frac{\Delta y_i}{\Delta x_i}\right)^2}$. By the

MVT, $\dfrac{\Delta y_i}{\Delta x_i}$ is equal to $\dfrac{dy}{dx}$ for some point inside the i-th subinterval, so the above sum is

a Riemann sum for the integral $\displaystyle\int_a^b \sqrt{1 + \left(\frac{dy}{dx}\right)^2}\, dx$. If the curve is defined with x as the

independent variable, and y as the dependent variable, this is the formula to use:

> **Arc length** $= \displaystyle\int_a^b \sqrt{1 + \left(\frac{dy}{dx}\right)^2}\, dx$.

19

Which of the following expressions represents the length of the curve $y = e^{-x^2}$ for x from 0 to 2?

(A) $\displaystyle\int_0^2 \sqrt{1 + e^{-2x^2}}\, dx$

(B) $\displaystyle\int_0^2 \sqrt{1 + 4x^2 e^{-2x^2}}\, dx$

C) $\displaystyle\int_0^2 \sqrt{1 - e^{-2x^2}}\, dx$

(D) $\displaystyle\int_0^2 \sqrt{1 + 2x e^{-2x^2}}\, dx$

(E) $\displaystyle\pi \int_0^2 e^{-2x^2}\, dx$

If $y = e^{-x^2}$, $\dfrac{dy}{dx} = -2xe^{-x^2}$ and $\left(\dfrac{dy}{dx}\right)^2 = 4x^2 e^{-2x^2}$. The correct answer is B.

20

The three curves, $y = \sin(\pi x)$, $y = 4x - 4x^2$, and $y = 4x^2 - 4x^4$ all go through the points $(0, 0)$ and $(1, 0)$. Which one gives the shortest path between these two points?

The length of the first curve is $L_1 = \displaystyle\int_0^1 \sqrt{1 + \left(\pi\cos(\pi x)\right)^2}\, dx$ ▪ ≈ 2.305.

The length of the second curve is $L_2 = \displaystyle\int_0^1 \sqrt{1 + \left(4 - 8x\right)^2}\, dx$ ▪ ≈ 2.323.

The length of the third curve is $L_3 = \displaystyle\int_0^1 \sqrt{1 + \left(8x - 16x^3\right)^2}\, dx$ ▪ ≈ 2.340.

The first curve is the shortest.

Chapter 7.   Differential Equations

7.1.   What is a Differential Equation?

Prior to studying calculus, you have probably seen equations mainly involving numbers and variables. A solution to such an equation is a number. You probably also came across equalities involving functions. An equality must hold for every x in the common domain of all the functions involved. For example, $\dfrac{1}{x(3-x)} = \dfrac{1}{3}\left(\dfrac{1}{x} + \dfrac{1}{3-x}\right)$. In calculus, we often write equations for <u>functions</u>. A solution to such an equation is a function that satisfies the equation for every x in the domain. For example, the solution for the equation $(x+1)^2 = y + 2x + 1$, when solved for y, is the function $y = x^2$.

> **A *differential equation* is an equation that includes one or more derivatives.**

For example, $\dfrac{dy}{dx} = -xy$. Note that there are many functions that satisfy this differential equation. Among these are $y = e^{-\frac{x^2}{2}}$, $y = -3e^{-\frac{x^2}{2}}$, and $y = \dfrac{1}{\sqrt{2\pi}}e^{-\frac{x^2}{2}}$. In fact, any function of the form $y = Ce^{-\frac{x^2}{2}}$, where C is an arbitrary constant, satisfies this differential equation. A solution that describes the family of all functions that satisfy the differential equation is called the ***general solution*** of that differential equation. A single member of the family of all solutions is called a ***particular solution***. One way to specify which particular solution we are interested in is to specify an ***initial condition***, which is one point on the graph of the solution. For example, the particular solution of the differential equation $\dfrac{dy}{dx} = -xy$ with the initial condition that $y = 3$ when $x = 2$ (i.e., $y(2) = 3$) is

$$y = 3e^2 e^{-\frac{x^2}{2}}.$$

Which of the following functions satisfies the differential equation $\dfrac{dy}{dx} = 3^{x^2} + 2$ with

$y(-1) = 7$?

(A) $y(x) = 2x \cdot \ln 3 \cdot 3^{x^2} + 6\ln 3 + 7$

(B) $y(x) = 3^{x^2} + 2x + 6$

(C) $y(x) = \dfrac{3^{x^2+1}}{x^2+1} + 2x + \dfrac{9}{2}$

(D) $y(x) = 7 + \displaystyle\int_{-1}^{x} \left[3^{t^2} + 2 \right] dt$

(E) $y(x) = 7 + \displaystyle\int_{0}^{x} \left[3^{t^2} + 2 \right] dt$

We don't need to actually solve this differential equation. We can simply look at the five choices and see which one satisfies the conditions in our problem. The solution must satisfy both the equation $\dfrac{dy}{dx} = 3^{x^2} + 2$ and the initial condition $y(-1) = 7$. The

Fundamental Theorem of Calculus tells us that if $y(x) = 7 + \displaystyle\int_{a}^{x} \left[3^{t^2} + 2 \right] dt$, as in Choices

D and E, then $y'(x) = \dfrac{dy}{dx} = 3^{x^2} + 2$. Note that substituting $x = -1$ into the function in

Choice D gives $y(-1) = 7 + \displaystyle\int_{-1}^{-1} \left[3^{t^2} + 2 \right] dt = 7 + 0$. The answer is D.

7.2.   Slope Fields

A slope field is a two-dimensional plot used to visualize the graphs of the solutions to a differential equation.

You don't have to solve the differential equation to create a slope field.

Creating a slope field from a given differential equation is straightforward. You draw (or you are given) a lattice (usually rectangular) of points in the plane. The points in the lattice will often have integer coordinates, but they don't have to. Typically, the equation will be given in the form $\dfrac{dy}{dx} = F(x, y)$, where F is an expression that may include *x* and *y*. Take the *x* and *y* coordinates for each point in the lattice and substitute them into *F* to get a number. This number represents the slope of the solution curve at that point. Then graph a little piece of a straight line centered on the lattice point with the slope you calculated. This straight line represents the tangent line to the solution curve. When viewed "close up," a solution curve is close to the tangent line near the point of tangency.

Suppose we want to draw a slope field for the differential equation $\dfrac{dy}{dx} = \dfrac{x}{2} - \dfrac{y}{4}$ using the lattice points indicated on the coordinate plane below.

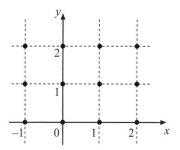

Think of the differential equation as a "slope machine." For each ordered pair, (x, y), indicated in the grid, we substitute *x* and *y* into the slope machine, $\dfrac{dy}{dx} = \dfrac{x}{2} - \dfrac{y}{4}$. The value we get for $\dfrac{dy}{dx}$ tells us the slope of the solution curve at the point we substituted.

We then draw a short line segment with that slope, centered on the grid point.

The table below shows the slopes at the 12 indicated points; the corresponding slope field is shown on the right.

(x, y)	$\dfrac{dy}{dx} = \dfrac{x}{2} - \dfrac{y}{4}$
$(-1, 0)$	$-1/2$
$(-1, 1)$	$-3/4$
$(-1, 2)$	-1
$(0, 0)$	0
$(0, 1)$	$-1/4$
$(0, 2)$	$-1/2$
$(1, 0)$	$1/2$
$(1, 1)$	$1/4$
$(1, 2)$	0
$(2, 0)$	1
$(2, 1)$	$3/4$
$(2, 2)$	$1/2$

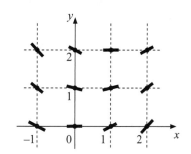

The slope field shown above is really just a small sampling of slopes. For this differential equation, a slope is defined at every point in the coordinate plane. Making a grid with more points, as shown in Figure 7-1, gives a much better picture of how the solution curves behave. This is just a different visualization of the same underlying "field of slopes," which is actually a continuous field with an infinite number of points in it.

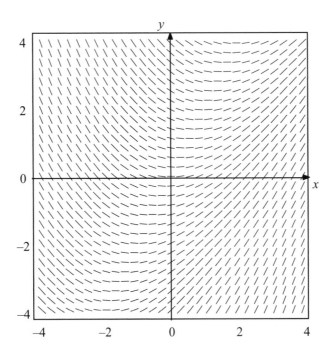

Figure 7-1. A slope field for $\dfrac{dy}{dx} = \dfrac{x}{2} - \dfrac{y}{4}$

Note that along the line $y = 2x$, the slopes are zero because on that line $\dfrac{dy}{dx} = \dfrac{x}{2} - \dfrac{y}{4} = 0$.

This type of analysis can help you answer questions that ask you to match a given differential equation to the corresponding slope field, or vice-versa.

2

Which of the following is a slope field for the differential equation $\dfrac{dy}{dx} = 2x + y$?

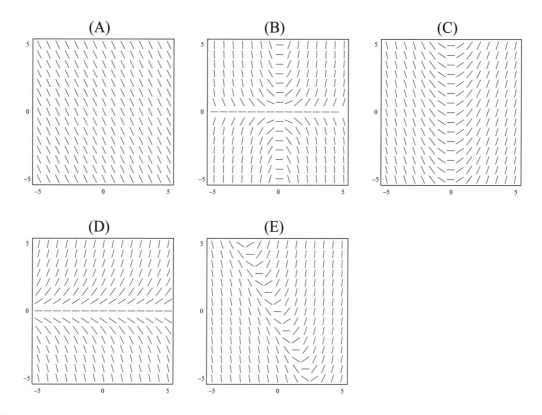

In Choice A, the slopes are equal for all rows and columns in the grid, indicating that the slope is independent of x and y, while our differential equation involves both x and y. In Choice C, the slopes are constant as you move vertically in a column, indicating that slope is not a function of y. In Choice D, the slopes are constant as you move across the same row in the grid, indicating that slope does not depend on x. The only slope fields for which slope is a function of both x and y are in Choices B and E. How can we tell which of these is the right one? One way is to notice that in the given differential equation, when $y = -2x$, the slope is 0. Only in Choice E do we see that fact realized, with horizontal line segments along the line $y = -2x$. The answer is E.

We can also decide between B and E by looking at the signs of the slopes: in the third quadrant, when both x and y are negative, the slopes must be negative, which is not the case in Choice B.

To tell which slope field goes with a given differential equation, try one or more of the following strategies:

- Check if the derivative is a function of only *x*, only *y*, both *x* and *y*, or, in the simplest case, neither *x* nor *y*. If the derivative is a function of *x* alone, then the slopes must be the same within the same column of the grid; if the derivative is a function of *y* alone, then the slopes must be the same within the same row.

- Check what condition would make the slope zero. Look in the grid where that condition is met. The segments must have zero slope at those points.

- Look at what happens to the slope when either *x* is 0 (that is, along the *y*-axis) or when *y* is 0 (that is, along the *x*-axis). Make sure the slopes agree with the prediction.

- Consider what must happen to the slope within certain quadrants of the plane. This strategy is particularly effective when the slope expression is a product or a quotient of functions of *x* and/or *y*.

- Consider what happens to the slope as *x* gets large (that is as you move to the right in the field), or as *y* gets large (as you move up in the field).

- Look at the differential equation to see whether the slopes should be symmetric with respect to the *x*- or *y*-axis, or both.

3

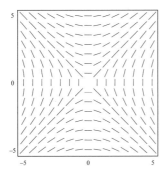

Which differential equation corresponds to the slope field shown above?

(A) $\dfrac{dy}{dx} = x^2 - y^2$ (B) $\dfrac{dy}{dx} = y^2 - x^2$ (C) $\dfrac{dy}{dx} = \dfrac{x}{y}$

(D) $\dfrac{dy}{dx} = \dfrac{y}{x}$ (E) $\dfrac{dy}{dx} = -\dfrac{x}{y}$

All the choices show that the slope is a function of both x and y, as indicated by slopes varying within rows and columns of the grid. From Choices A and B, we'd expect that when $y = \pm x$, the slope would be 0. In the given slope field, this is not the case, eliminating those choices. (Those two choices could also be eliminated by considering symmetry, e.g., the slope at $-x$ must be the same as at x.) From Choice D, we'd expect slopes to be 0 along the x-axis (where $y = 0$) and infinite along the y-axis (where $x = 0$). Again, the slope field is not consistent with these observations. From Choice E, we'd expect the slopes to be positive in the second and fourth quadrants (where x and y have opposite signs), and negative in the first and third quadrants (where x and y have the same sign). The given slope field is just the opposite. The answer is C.

7.3.   Separation of Variables

One method of solving differential equations analytically is called ***separation of variables***. There are more advanced techniques for solving differential equations when variables are not separable, but separating variables is the only method required in AP Calculus.

> **Questions about differential equations where variables are not separable might appear on the AP exams. Such questions may involve a slope field ⌈ or, for the BC exam only, Euler's method ⌋. But if the variables are not separable, the question won't ask for an analytical solution.**

A separable differential equation can be reduced to the form $F(y)\dfrac{dy}{dx} = G(x)$. For example, $yy' = \sin x \cos y \Rightarrow \dfrac{yy'}{\cos y} = \sin x$. $y' + y = x$ is an example of a non-separable differential equation.

To solve a differential equation using separation of variables, perform the following steps:

1. Use algebra to collect like variables on each side of the equation.

2. Antidifferentiate both sides of the resulting equation, taking care to include a constant of integration on one side; call it C.

3. Use the given initial condition, if any, to solve for the constant C.

4. Isolate the dependent variable.

4

Which of the following is a solution of the differential equation $\dfrac{dy}{dx} = 3x^2 e^{-2y}$ with $y(1) = 4$?

(A) $y = \dfrac{1}{2}\ln\left(2x^3\right)$ 　　　　(B) $y = \dfrac{1}{2}\ln\left(2x^3\right) - \ln\sqrt{2} + 4$

(C) $y = \dfrac{1}{2}\ln\left(\dfrac{1}{2}x^3 + e^8 - \dfrac{1}{2}\right)$ 　(D) $y = \dfrac{1}{2}\ln\left(2x^3 + e^8 - 2\right)$

(E) $y = \dfrac{1}{2}\ln\left(x^3 + e^8 - 1\right)$

Follow the steps outlined above:

1. Use algebra to collect like variables on each side of the equation. $e^{2y}\,dy = 3x^2\,dx$. Although $\dfrac{dy}{dx}$ is not a quotient, treating it like one when separating variables always works.

2. Antidifferentiate both sides of the resulting equation, taking care to include a constant of integration. $\displaystyle\int e^{2y}\,dy = \int 3x^2\,dx \;\Rightarrow\; \dfrac{e^{2y}}{2} = x^3 + C$.*

3. Use the given initial condition to solve for the constant C.
    $\dfrac{e^{2\cdot4}}{2} = 1^3 + C \;\Rightarrow\; C = \dfrac{e^8}{2} - 1$.

4. Isolate the dependent variable. $\dfrac{e^{2y}}{2} = x^3 + \dfrac{e^8}{2} - 1 \;\Rightarrow\; e^{2y} = 2x^3 + e^8 - 2 \;\Rightarrow$
    $2y = \ln\left(2x^3 + e^8 - 2\right) \;\Rightarrow\; y = \dfrac{1}{2}\ln\left(2x^3 + e^8 - 2\right)$. The answer is D.

---

* Steps 1 and 2 make an acceptable shortcut for solving a separable differential equation. Actually, what we are doing is this: $e^{2y}\dfrac{dy}{dx} = 3x^2 \;\Rightarrow\; \displaystyle\int e^{2y(x)}y'\,dx = \int 3x^2\,dx$; then, using u-substitution

$u = y(x),\; du = y'dx$, we get $\displaystyle\int e^{2u}\,du = \int 3x^2\,dx \;\Rightarrow\; \dfrac{e^{2u}}{2} = x^3 + C \;\Rightarrow\; \dfrac{e^{2y}}{2} = x^3 + C$.

The Fundamental Theorem can come to the rescue when antidifferentiation using elementary functions is not possible.

5

If $\dfrac{dx}{dt} = \sqrt{1 + \sin\left(t^2\right)}$ and $x(0) = 3$, what is $x(2)$?

(A) 0.349   (B) 0.588   (C) 2.333   (D) 2.413   (E) 5.333

$x(2) = x(0) + \displaystyle\int_0^2 x'(t)\,dt = \;\; x(2) = 3 + \int_0^2 \sqrt{1 + \sin\left(t^2\right)}\,dt = \blacksquare \approx 5.333$. The answer is E.

Using the Fundamental Theorem in this way has become fairly common on AP exams.

6

If $\dfrac{dx}{dt} = 2x\cos\left(t^2\right)$ and $x(-2) = e^2$, what is $x(-3)$?

This problem uses the same ideas as the last example, but it is a little harder. Let us follow the four steps as far as we can:

1. Separate variables. $\dfrac{dx}{2x} = \cos\left(t^2\right)\,dt$. (It doesn't matter which side we put the 2 on.)

2. Antidifferentiate. Here, we construct an antiderivative of the right-hand side using the FTC: $\dfrac{1}{2}\displaystyle\int \dfrac{dx}{x} = \int\cos\left(t^2\right)\,dt \;\Rightarrow\; \dfrac{1}{2}\ln|x| = \int_a^t \cos\left(u^2\right)\,du + C$.

3. Use the initial condition to find C. This gets a little tricky. It appears we have two constants to worry about. But if we are clever, we can manage! If we let $a = -2$, the integral's value is 0 when $t = -2$. Using the initial condition with this choice for a gives $\dfrac{1}{2}\ln\left|e^2\right| = \displaystyle\int_{-2}^{-2}\cos\left(u^2\right)\,du + C$. Solving for C gives $C = 1$. We actually could have used any constant for a. Our choice of $a = -2$ merely simplifies the calculations.

4. Isolate the dependent variable. $\ln|x| = 2\int_{-2}^{t}\cos\left(u^2\right)du + 2 \Rightarrow |x| = e^{2+2\int_{-2}^{t}\cos\left(u^2\right)du}$.

Since raising e to any exponent gives a positive result, and we know that $x(-2) > 0$,

we can safely remove the absolute value signs: $x(t) = e^{2+2\int_{-2}^{t}\cos\left(u^2\right)du}$.

Now we can answer the question! $x(-3) = e^{2+2\int_{-2}^{-3}\cos\left(u^2\right)du}$ ≈ 4.559.

This is certainly an unusual question. But understanding its methods should serve you well.

7.4. The Exponential Model

Differential equations of the form $\dfrac{dy}{dt} = ky$ have solutions that are exponential functions. They arise in so many physical settings that they merit mention in the topic outline for AP Calculus as a model that students should be familiar with.

Applications of the exponential model include:

- population growth in the presence of unlimited nutrients;

- growth of an investment when interest is compounded continuously;

- radioactive decay;

- Newton's Law of Heating and Cooling.

> **The exponential model with $k > 0$ models growth, and the exponential model with $k < 0$ models attrition or decay.**

Once you've solved the differential equation $\dfrac{dy}{dt} = ky$, you've solved all differential equations of that form! The solution follows:

$$\frac{dy}{y} = kdt \Rightarrow \int\frac{dy}{y} = \int k\,dt \Rightarrow \ln|y| = kt + C \Rightarrow |y| = e^{kt+C} \Rightarrow \boxed{y(t) = Ae^{kt}}, \text{ where}$$

$A = \pm e^{C}$. $A = y(0)$ represents the initial value of the modeled variable.

> **Beware! The constant A in the last step could be negative. The absolute values are necessary. You must consider whether y is positive or negative, and calculate A accordingly.**

7

A radioactive substance, such as Thorium-234, decays at a rate proportional to the amount of the substance present at any time t. It takes about 24 days for a sample of Thorium-234 to decay to the point where half of the original amount remains. How long, to the nearest day, will it take for a sample of 10g of Thorium-234 to decay to 1g?

We'll call M the number of grams of Thorium-234 present at time t. The problem states that $\dfrac{dM}{dt} = kM$. Solving this differential equation gives $M(t) = Ae^{kt}$. Since $M(0) = 10$,

$A = 10$. The question tells us that $M(24) = 5$. Thus, $5 = 10e^{k \cdot 24} \implies k \cdot 24 = \ln\left(\dfrac{1}{2}\right)$.

Similarly, since we are looking for t such that $M(t) = 1$, we get

$1 = 10e^{kt} \implies kt = \ln\left(\dfrac{1}{10}\right)$. Dividing one by another we get

$\dfrac{t}{24} = \dfrac{\ln\left(\dfrac{1}{10}\right)}{\ln\left(\dfrac{1}{2}\right)} \implies t = \dfrac{24\ln\left(\dfrac{1}{10}\right)}{\ln\left(\dfrac{1}{2}\right)}$ ≈ 79.726 days. Since the question asks us to round to

the nearest day, the answer is 80 days.

7.5. ⌈ The Logistic Model (BC Only) ⌋

A differential equation of the form $\dfrac{dP}{dt} = kP \cdot \left(1 - \dfrac{P}{L}\right)$, where **k** and **L** are constants, is called a logistic differential equation.

It is used to model how a population grows (or decreases) in the presence of limited resources or space. Note that when P is close to 0, the term $\left(1 - \dfrac{P}{L}\right)$ is close to 1, so the

differential equation becomes approximately $\dfrac{dP}{dt} \approx kP$, which is the exponential model.

This model, discussed in Section 7.4 above, models unrestricted growth or decay.

However, as $t \to \infty$, $P \to L$, so the factor $\left(1 - \dfrac{P}{L}\right) \to 0$, and growth or decay slows

nearly to a halt.

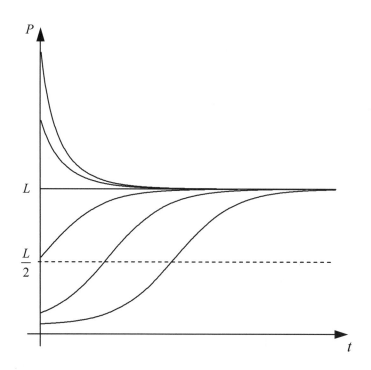

**Figure 7-2.   Several solutions of a logistic differential
equation with different initial conditions**

Figure 7-2 shows several solutions of a logistic equation with different initial values
$P(0)$.   One solution is simply the horizontal line (a constant) $P = L$. This is called the
equilibrium solution. For all other solutions, the line $P = L$ is an asymptote. If $P(0) > L$,
then $\dfrac{dP}{dt}$ starts negative and always remains negative. Such a solution is a decreasing

function, so it actually models attrition, not growth. If $P(0) < L$, then $\dfrac{dP}{dt}$ starts positive
and always remains positive. Such a solution is an increasing function, so it models
growth. But this is bounded growth, because $P(t)$ always remains below L. The
constant L is sometimes called the ***carrying capacity*** of the environment.

Graphing $\dfrac{dP}{dt}$ as a function of P reveals even more about the behavior of the logistic
curves (Figure 7-3).

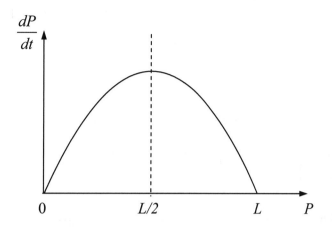

Figure 7-3. $\dfrac{dP}{dt}$ **as a function of** P

This is a parabola pointing down; $\dfrac{dP}{dt} = 0$ when $P = 0$ and $P = L$. As we see from the

graph, $\dfrac{dP}{dt}$ changes from increasing to decreasing precisely at the instant, if any, when

the population is half the carrying capacity, that is, when $P = \dfrac{L}{2}$. So a logistic curve that

crosses the line $P = \dfrac{L}{2}$ has an inflection point at the intersection. $P(t)$ is increasing the

fastest at that time, since $\dfrac{dP}{dt}$ is at its maximum.

[8]

The rate of change in the number of coyotes in a population is given by $\dfrac{dC}{dt} = 9C - 3C^2$,
where C is the number of thousands of coyotes t years after January 1, 2000. For what
value of C is the number of coyotes increasing the fastest?

(A) $C = 0$ (B) $C = 1.5$ (C) $C = 3$ (D) $C = 6$ (E) $C = 9$

Factoring gives $\dfrac{dC}{dt} = 9C \cdot \left(1 - \dfrac{C}{3}\right)$, so the carrying capacity for this equation is $L = 3$.

The population increases the fastest at $C = \dfrac{L}{2}$. The answer is B.

A logistic differential equation can be solved using the method of Partial Fractions:
$$\dfrac{dP}{dt} = kP\left(1 - \dfrac{P}{L}\right) \;\Rightarrow\; \dfrac{dP}{P\left(1 - \dfrac{P}{L}\right)} = \left(\dfrac{1}{P} + \dfrac{1}{L-P}\right)dP = kdt \;\Rightarrow$$

$\ln P - \ln|L-P| = -\ln\dfrac{|L-P|}{P} = kt + C \;\Rightarrow\; \dfrac{|L-P|}{P} = e^{-kt-C} \;\Rightarrow\; \dfrac{L-P}{P} = Ae^{-kt}$, where A can

be any constant, positive, negative, or zero. In any case, the solution is $\boxed{P(t) = \dfrac{L}{1 + Ae^{-kt}}}$,

where $A = \dfrac{L - P(0)}{P(0)}$. But most AP exam questions can be answered using your

familiarity with the qualitative behavior of the logistic curves, without going through the rigors of solving the logistic equation analytically. Nonetheless, knowing this form of the solution certainly can't hurt!

9

If $N(0) = 2000$ and $\dfrac{dN}{dt} = 32000N - 20N^2$, what is $\lim\limits_{t\to\infty} N(t)$?

    (A) 0   (B) 800   (C) 1600   (D) 2000   (E) nonexistent

Factoring $\dfrac{dN}{dt}$ reveals that this is a logistic model with $\dfrac{dN}{dt} = 32000N \cdot \left(1 - \dfrac{N}{1600}\right)$.

$\dfrac{dN}{dt} = 0$ when $N = 0$ and $N = 1600$. $N = 1600$ is the carrying capacity of the environment,

and the value of $\lim\limits_{t\to\infty} N(t)$. The answer is C.

7.6. $\lceil$ **Euler's Method (BC Only)** $\rfloor$

It is not always possible to solve a differential equation analytically. In such cases, numerical approaches are used, and the simplest of these is Euler's method.

Euler's method is an ***iterative*** process. Starting at an initial point, you use the slope information to approximate a point on the solution curve that is close to the initial point. Then you repeat the process at the new point. The process is repeated as many times as needed.

Suppose our differential equation is $\dfrac{dy}{dx} = F(x, y)$ and the point (x_0, y_0) is on the graph of the solution $y = f(x)$. An equation of the tangent line to the graph of $y = f(x)$ at (x_0, y_0) is $y - y_0 = F(x_0, y_0) \cdot (x - x_0)$. If we isolate y, we get $y = y_0 + F(x_0, y_0) \cdot (x - x_0)$.

Remember, this is an equation for the <u>tangent line</u>. Let's pick a point on it, call it (x_1, y_1). There's only one point on the tangent line that you can be sure is on the graph of f, the point (x_0, y_0). So, in general, $y_1 \neq f(x_1)$. However, if x_1 is "close" to x_0 (say, $x_1 = x_0 + h$ for some small number h), then (x_1, y_1) will be close to the point $(x_1, f(x_1))$.

So far we have performed the first step of Euler's method: we have calculated $x_1 = x_0 + h$ and $y_1 = y_0 + F(x_0, y_0) \cdot (x_1 - x_0)$. Since $x_1 - x_0 = h$, this can be simplified to $y_1 = y_0 + F(x_0, y_0) \cdot h$. Note that (x_1, y_1) is not necessarily on the solution curve we are looking for. But we hope it is close to our solution.

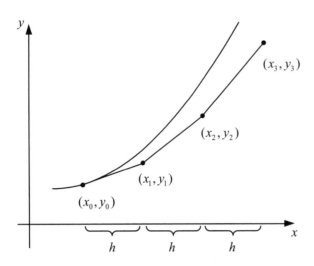

**Figure 7-4.   First 3 steps of Euler's method.   The slope of the
segment from (x_n, y_n) to (x_{n+1}, y_{n+1}) is equal to $F(x_n, y_n)$.**

How do you determine the next point? We repeat the previous step now going from the
point (x_1, y_1) and obtain $x_2 = x_1 + h$ and $y_2 = y_1 + F(x_1, y_1) \cdot h$. We can repeat this
process, getting $x_{n+1} = x_n + h$ and $y_{n+1} = y_n + F(x_n, y_n) \cdot h$ (Figure 7-4).

So, in Euler's method,

$$x_{n+1} = x_n + h \, ; \; y_{n+1} = y_n + F(x_n, y_n) \cdot h \text{, where } \frac{dy}{dx} = F(x, y)$$

If you have ever programmed a computer or a calculator, you can see that such a process
is easy to implement in an iterative procedure. (But no AP question requires such a
program, and even if you have one in your calculator, it won't help you much at the exam
because you will have to show all of your work anyway.)

10

The function $y = f(x)$ is the solution to the differential equation $\dfrac{dy}{dx} = \dfrac{x}{2} - \dfrac{y}{5}$ with the
initial condition $f(2) = 0$. What is the approximation for $f(3)$ if Euler's method is used,
starting at $x = 2$ with a step size of 0.5?

(A) 0.5   (B) 1   (C) 1.075   (D) 1.150   (E) 1.717

Starting with $x_0 = 2, y_0 = 0$, we get $\dfrac{dy}{dx} = \dfrac{2}{2} - \dfrac{0}{5} = 1$. Our first step gets us to

$x_1 = 2.5, y_1 = 0 + 1 \cdot 0.5 = 0.5$. With $x = 2.5$ and $y = 0.5$, we get $\dfrac{dy}{dx} = \dfrac{2.5}{2} - \dfrac{0.5}{5} = 1.15$. We

can compute $x_2 = 3, y_2 = 0.5 + 1.15 \cdot 0.5 = 1.075$. This is a far as we need to go in this question. The answer is C.

Suppose you have a differential equation $\dfrac{dy}{dx} = F(x)$ (the derivative depends only on x,

and not on y). Suppose the initial condition is $\big(a, \, y(a)\big)$ and we want to use Euler's

method to approximate $y(b)$. If $F(x)$ is increasing on the closed interval $[a, b]$, then $\dfrac{dy}{dx}$

is increasing, so the graph of the solution $y(x)$ is concave up. In this case, any approximation of $y(b)$ using Euler's method, no matter the step size, will always underestimate it, because tangent lines are underneath the function graph when the graph is concave up. Similarly, if $F(x)$ is decreasing, the solution graph is concave down on $[a, b]$. Then Euler's method will overestimate $y(b)$.

11

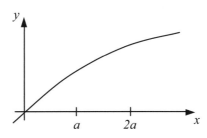

The graph above shows a solution of the differential equation $\dfrac{dy}{dx} = F(x)$ with the initial

condition $(0, 0)$. Suppose we use Euler's method with step size a, starting at $(0, 0)$ and obtain approximations $E(a)$ and $E(2a)$ of $y(a)$ and $y(2a)$ respectively. Which of the following inequalities must be true?

(A) $E(2a) > y(2a) > y(a)$ (B) $y(2a) > E(2a) > y(a)$

(C) $y(2a) > y(a) > E(2a)$ (D) $y(2a) > y(a) > E(a)$

(E) $E(a) > y(a) > E(2a)$

The graph of $y(x)$ is concave down on $[0, 2a]$, so every Euler's method approximation on that interval will overestimate $y(x)$. Since the function $y(x)$ is increasing on the interval, its derivative is positive, so $F(x)$ is positive and Euler's method approximations will also increase. The answer is A. Choices B, C, and D all have an Euler approximation smaller than the corresponding function output. Choice E has $E(a) > E(2a)$.

The next example uses all three representations, graphic, numeric, and analytic, to explore the solution to a differential equation.

12

Consider a differential equation $\dfrac{dy}{dx} = \dfrac{y}{2}\cos(\pi x)$ with the initial condition $y(0) = 2$.

(a)   Sketch the slope field for this equation at the 9 points indicated on the grid below.

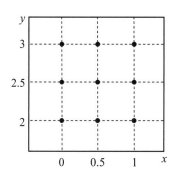

(b)   Starting at $(0, 2)$, use Euler's method with a step of 0.25 to approximate $f(1)$.

(c)   Find $f(x)$ and evaluate $f(1)$.

(a)

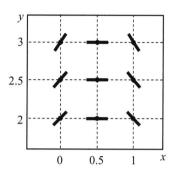

When $x = 0$, the slopes at $y = 2$, 2.5, and 3 are 1, 1.25, and 1.5, respectively. When $x = 0.5$, all slopes are 0. When $x = 1$, the slopes at $y = 2$, 2.5, and 3 are -1, -1.25, and -1.5.

(b)

At $(0,2)$, the slope is $\dfrac{2}{2}\cos(\pi \cdot 0) = 1$. So, $x_1 = \dfrac{1}{4}$ and $y_1 = 2 + 1 \cdot \dfrac{1}{4} = \dfrac{9}{4}$.

At $\left(\dfrac{1}{4}, \dfrac{9}{4}\right)$, the slope is $\dfrac{9}{8}\cos\left(\dfrac{\pi}{4}\right) = \dfrac{9\sqrt{2}}{16}$. So, $x_2 = \dfrac{1}{2}$ and $y_2 = \dfrac{9}{4} + \dfrac{9\sqrt{2}}{16} \cdot \dfrac{1}{4} = \dfrac{9}{4} + \dfrac{9\sqrt{2}}{64}$.

At $\left(\dfrac{1}{2}, \dfrac{9}{4} + \dfrac{9\sqrt{2}}{64}\right)$, the slope is $\left(\dfrac{9}{8} + \dfrac{9\sqrt{2}}{128}\right)\cos\left(\dfrac{\pi}{2}\right) = 0$. So,

$x_3 = \dfrac{3}{4}$ and $y_3 = \dfrac{9}{4} + \dfrac{9\sqrt{2}}{64} + 0 \cdot \dfrac{1}{4} = \dfrac{9}{4} + \dfrac{9\sqrt{2}}{64}$.

At $\left(\dfrac{3}{4}, \dfrac{9}{4} + \dfrac{9\sqrt{2}}{64}\right)$, the slope is $\left(\dfrac{9}{8} + \dfrac{9\sqrt{2}}{128}\right)\cos\left(\dfrac{3\pi}{4}\right) = \left(\dfrac{9}{8} + \dfrac{9\sqrt{2}}{128}\right)\left(-\dfrac{\sqrt{2}}{2}\right) = -\dfrac{9\sqrt{2}}{16} - \dfrac{9}{128}$.

So, $x_4 = 1$ and $y_4 = \dfrac{9}{4} + \dfrac{9\sqrt{2}}{64} + \left(-\dfrac{9\sqrt{2}}{16} - \dfrac{9}{128}\right) \cdot \dfrac{1}{4} = \dfrac{9}{4} + \dfrac{9\sqrt{2}}{64} - \dfrac{9\sqrt{2}}{64} - \dfrac{9}{512} \approx 2.232$.

This calculation may be a little too long for an actual AP question — we just wanted to make sure you got the idea. The AP exam could require you to complete as few as two steps of the solution, especially if a calculator is not permitted.

(c)

Separate variables: $\dfrac{dy}{y} = \dfrac{1}{2}\cos(\pi x)\,dx$.

Antidifferentiate: $\displaystyle\int\dfrac{dy}{y} = \dfrac{1}{2}\int\cos(\pi x)\,dx \Rightarrow \ln|y| = \dfrac{1}{2\pi}\sin(\pi x) + C$.

Use the initial condition to find C: $\ln 2 = \dfrac{1}{2\pi}\sin(\pi\cdot 0) + C \Rightarrow C = \ln 2$.

Solve for y: $y = e^{\frac{1}{2\pi}\sin(\pi x) + \ln 2} = 2e^{\frac{1}{2\pi}\sin(\pi x)}$.

Evaluate at $x = 1$: $y(1) = 2e^{\frac{1}{2\pi}\sin(\pi)} = 2$.

It turns out our Euler's method approximation, $y(1) \approx 2.232$, wasn't too bad. In the graph below, you see a slope field (with many more points in the grid), the solution curve, and an Euler's method approximation with four steps.

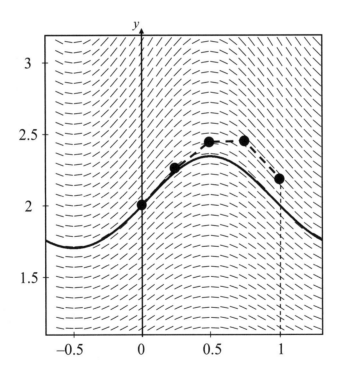

Chapter 8. ⌐ Parametric, Vector, and Polar Functions (BC Only) ⌐

8.1. Parametric Functions

When a particle moves in the *xy*-plane, the parametric equations $x = f(t)$ and $y = g(t)$ can be used to represent the location of the particle and the path that it traces in the plane.

The distinctive characteristic of all parametric curves is that both the *x*- and *y*-coordinates of a point on the curve are defined in terms of another variable, the parameter *t*. Unlike Cartesian curves where the independent variable is commonly named *x* and the dependent variable *y*, on a parametric curve, the parameter *t* is the independent variable, while both *x* and *y* are dependent variables. When a curve is defined parametrically, equations for *x*(*t*) and *y*(*t*) are given. In addition, a domain of values for *t* is specified.

Parametric equations can be used to describe the motion of a particle in the coordinate plane. In this context, we think of the parameter *t* as representing time. Often the motion starts at $t = 0$ (or at some point $t > 0$) and proceeds in the direction of increasing *t*.

Before we get into the calculus of parametrically defined curves, let us look at a few examples that illustrate some important skills you need to have.

1

Find a parametrization for the line segment with endpoints (2, –3) and (4, 1).

There are many ways to solve this problem. One is to first find a Cartesian equation for the line containing the segment. The slope is $\dfrac{1-(-3)}{4-2} = 2$, and an equation for the line is $y - (-3) = 2(x - 2)$ or $y = 2x - 7$. With this equation, creating the parametric equations is easy. Just define $x(t) = t$ and $y(t) = 2t - 7$. Since the smaller value of *x* is 2 and the larger is 4, choose values of *t* between 2 and 4, inclusive. Try graphing this with your calculator in parametric mode.

Another possible solution is to pretend that a particle is moving with a constant speed from the point (2, –3) straight to the point (4, 1). Suppose the particle starts at $t = 0$ and reaches the destination point at $t = 1$. Then the equations should be $x(t) = 2 + 2t$ and $y(t) = -3 + 4t$.

> **A graph of any function** $y = f(x)$ **can be parametrized by setting** $x(t) = t$ **and** $y(t) = f(t)$.

For example, a graph of the function $f(x) = \ln x$ can be parametrized with $x(t) = t$ and $y(t) = \ln t$, for $t > 0$.

Of course, there are other ways to create parametric curves. One of the most basic parametrized curves is the unit circle, which we can represent as
$x = \cos t$, $y = \sin t$, $0 \le t < 2\pi$.

Intersections Versus Collisions

If we follow the movement of two particles simultaneously, their paths can cross any number of times. Such points are called ***intersection*** points. It is further possible that the particles will occupy the same point in the plane at the same value of *t*. Such points are called ***collision*** points.

2

The motion of Particle 1 in the plane for $0 \le t \le 10$ is described by the equations $x_1(t) = t + 1$ and $y_1(t) = t^3 - 5t^2 + 7t + 3$. The motion of Particle 2 over the same time interval is described by the equations $x_2(t) = 7e^t - 6$ and $y_2(t) = 49e^t - 46$. Find all points where the paths of the particles intersect and all points where the particles collide, if any.

For the paths to intersect, we need $x_1(t_1) = x_2(t_2)$ and $y_1(t_1) = y_2(t_2)$. Here we get $t_1 + 1 = 7e^{t_2} - 6$ and $t_1^3 - 5t_1^2 + 7t_1 + 3 = 49e^{t_2} - 46$. Multiplying the first equation by 7 and subtracting it from the second to eliminate e^{t_2}, we get
$t_1^3 - 5t_1^2 + 7t_1 + 3 - 7(t_1 + 1) = -4 \Rightarrow t_1^3 - 5t_1^2 = 0 \Rightarrow t_1^2(t_1 - 5) = 0$. Therefore, the intersection points are at $t_1 = 0$ and $t_1 = 5$. These points are (1, 3) and (6, 38). At these points, $t_2 = 0$ and $t_2 = \ln \dfrac{12}{7}$, respectively. As we can see, only the starting point (1, 3) at $t = t_1 = t_2 = 0$ is a collision point. At $t_1 = 5$ and $t_2 = \ln \dfrac{2}{7}$, respectively, the particles pass through the same intersection point (6, 38), but they pass through it at different times and do not collide.

Tangent Lines

A parametrized curve $x = x(t)$, $y = y(t)$ **is differentiable at** $t = t_0$ **if** x **and** y **are differentiable at** $t = t_0$. **The curve is differentiable on an open interval** $t_1 < t < t_2$ **if** $x'(t)$ **and** $y'(t)$ **exist for all values of** t **in the interval.**

If both $x'(t)$ and $y'(t)$ exist and are not simultaneously zero, then the curve has a tangent line at t.

The slope of that tangent line is $\dfrac{dy}{dx} = \dfrac{dy}{dt} \bigg/ \dfrac{dx}{dt}$.

This follows from the Chain Rule: if $y = y(x(t))$ then $\dfrac{dy}{dt} = \dfrac{dy}{dx} \cdot \dfrac{dx}{dt}$. Note that the tangent line is vertical where $\dfrac{dx}{dt} = 0$ and $\dfrac{dy}{dt} \neq 0$.

In our earlier example of the unit circle $x = \cos t$, $y = \sin t$; $0 \leq t < 2\pi$, $\dfrac{dx}{dt} = -\sin t$ and $\dfrac{dy}{dt} = \cos t$. The slope of the curve is $\dfrac{dy}{dx} = -\dfrac{\cos t}{\sin t}$, which is defined for all points at which $\sin t \neq 0$. If $\sin t = 0$, then the y-coordinate of the point on the circle is 0, and the circle has a vertical tangent.

$\boxed{3}$

A curve is defined by the parametric equations $x = 5t^2 - 7t$, $y = \dfrac{1}{3}t^3 + 6t$. What is the slope of the tangent line to this curve at the point where $t = 2$?

(A) $\dfrac{10}{13}$ (B) $\dfrac{13}{10}$ (C) 10   (D) 13   (E) 23

☞

$\dfrac{dy}{dx} = \dfrac{dy}{dt} \bigg/ \dfrac{dx}{dt} = \dfrac{t^2 + 6}{10t - 7}$. $\dfrac{dy}{dx}\bigg|_{t=2} = \dfrac{2^2 + 6}{10(2) - 7} = \dfrac{10}{13}$. The answer is A.

☜

What is an equation for the line tangent to the curve with parametric equations $x = \dfrac{1}{t}$, $y = \sqrt{t+1}$ at the point where $t = 3$?

(A) $y - 2 = -\dfrac{4}{9}\left(x - \dfrac{1}{3}\right)$ (B) $y - 2 = \dfrac{1}{4}\left(x - \dfrac{1}{3}\right)$ (C) $y - 2 = -\dfrac{9}{4}\left(x - \dfrac{1}{3}\right)$

(D) $y - \dfrac{1}{4} = -\dfrac{4}{9}\left(x + \dfrac{1}{9}\right)$ (E) $y - \dfrac{1}{4} = -\dfrac{9}{4}\left(x + \dfrac{1}{9}\right)$

The point of tangency is $\left(x(3),\, y(3)\right) = \left(\dfrac{1}{3},\, 2\right)$. The slope of the tangent is

$\dfrac{dy}{dx} = \dfrac{dy}{dt} \Big/ \dfrac{dx}{dt} = \dfrac{\frac{1}{2\sqrt{t+1}}}{\frac{-1}{t^2}} \Rightarrow \left.\dfrac{dy}{dx}\right|_{t=3} = \dfrac{\frac{1}{4}}{\frac{-1}{9}} = -\dfrac{9}{4}$. Putting these two results together into a

linear equation, we get $y - 2 = -\dfrac{9}{4}\left(x - \dfrac{1}{3}\right)$, answer C.

A particle moves along a curve in the xy-plane according to the equations $x = t^3 - 3t^2$, $y = \dfrac{1}{3}t^3 - 4t$. What are all the values of t for which the tangent line to the particle's path is vertical?

(A) 0 only   (B) 2 only   (C) 0 and 2 only
(D) –2 and 2 only   (E) –2, 0, and 2

We must find the points where $\dfrac{dx}{dt} = 0$ and $\dfrac{dy}{dt} \neq 0$. $\dfrac{dx}{dt} = 3t^2 - 6t = 0 \Rightarrow t = 0, 2$ and

$\dfrac{dy}{dt} = t^2 - 4 \neq 0 \Rightarrow t \neq -2, 2$. The value $t = 2$ would make the slope indeterminate, but

the curve does not have a vertical tangent at that point because

$\lim\limits_{t \to 2} \dfrac{t^2 - 4}{3t^2 - 6t} = \left.\dfrac{t+2}{3t}\right|_{t=2} = \dfrac{2}{3}$, and thus is finite. Therefore, the only value for which the

tangent line is vertical is $t = 0$. The answer is A.

A point where the slope is indeterminate, that is, where both $\dfrac{dy}{dt}$ and $\dfrac{dx}{dt}$ are zero, may be

at a cusp on the curve. If both $\dfrac{dy}{dt}$ and $\dfrac{dx}{dt}$ are continuous, the curve still has a tangent

line at the cusp point and we can find its slope by taking the limit of the slope when t approaches the value that produces the cusp. In the above example, the slope of the

tangent line at $t = 2$ can be found as $\displaystyle\lim_{t \to 2}\dfrac{dy}{dx} = \lim_{t \to 2}\dfrac{t^2 - 4}{3t^2 - 6t} = \dfrac{2}{3}$. The point on the curve

where $t = 2$ is $\left(x(2),\, y(2)\right) = \left(-4,\, -\dfrac{16}{3}\right)$, so the tangent line at that point has an equation

$y + \dfrac{16}{3} = \dfrac{2}{3}(x + 4)$ (Figure 8-1). The tangent at $t = T$ can be vertical, too, if $\displaystyle\lim_{t \to T}\dfrac{dy}{dx} = \infty$

(e.g., $x = t^5$, $y = t^3$, $\displaystyle\lim_{t \to 0}\dfrac{dy}{dx} = \lim_{t \to 0}\dfrac{3t^2}{5t^4} = \infty$).

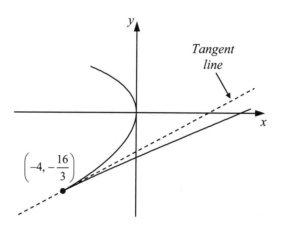

Figure 8-1.   The tangent line to a parametric curve at a cusp

To find $\dfrac{d^2 y}{dx^2}$ for a parametrically defined curve, you need to carefully use the Chain Rule. Try the following example.

$\boxed{6}$

If $x(t) = t^2 - t$, $y(t) = e^{2t}$, determine $\dfrac{d^2 y}{dx^2}$ at the point where $t = 2$.

$$\frac{dx}{dt} = 2t - 1 \text{ and } \frac{dy}{dt} = 2e^{2t} \Rightarrow \frac{dy}{dx} = \frac{2e^{2t}}{2t - 1}.$$

$$\frac{d^2y}{dx^2} = \frac{d}{dx}\left(\frac{dy}{dx}\right) = \frac{d}{dx}\left(\frac{2e^{2t}}{2t-1}\right) = \frac{d}{dt}\left(\frac{2e^{2t}}{2t-1}\right)\bigg/\frac{dx}{dt} = \frac{\left(\dfrac{(2t-1)4e^{2t} - 2e^{2t} \cdot 2}{(2t-1)^2}\right)}{(2t-1)} = \frac{8te^{2t} - 8e^{2t}}{(2t-1)^3}.$$

At $t = 2$, $\dfrac{d^2y}{dx^2} = \dfrac{16e^4 - 8e^4}{(2 \cdot 2 - 1)^3} = \dfrac{8e^4}{27}.$

Length of a Parametric Curve

In Chapter 6 we reviewed the methods for finding the arc length of a curve when y is a function of x. The arc length for a parametric curve can be found with a similar method.

> **If a smooth curve** $x = x(t)$, $y = y(t)$, $a \le t \le b$, **is traced exactly once as t increases from a to b, then the arc length of the curve is**
>
> $$L = \int_a^b \sqrt{\left(\frac{dx}{dt}\right)^2 + \left(\frac{dy}{dt}\right)^2}\, dt.$$

The formula in Chapter 6 can be obtained from the this one by parametrizing the curve as $x(t) = t$, $y = y(x(t))$. Then $\dfrac{dx}{dt} = 1$, $\dfrac{dy}{dt} = \dfrac{dy}{dx}$, and $L = \int_a^b \sqrt{1 + \left(\dfrac{dy}{dx}\right)^2}\, dt$.

> **Be careful: particle's distance from its initial position —**
> $$\sqrt{(x(t) - x(0))^2 + (y(t) - y(0))^2}$$ **— is usually not the same as the length of particle's path when the particle travels from 0 to t.**

7

What is the length of the path described by the parametric equations
$x = \cos(3t)$, $y = \sin t$; $0 \le t \le 2\pi$?

(A) 4.382   (B) 6.049   (C) 6.533   (D) 13.065   (E) 31.416

$\dfrac{dx}{dt} = -3\sin(3t)$ and $\dfrac{dy}{dt} = \cos t$. The arc length formula gives

$L = \displaystyle\int_0^{2\pi} \sqrt{\left(-3\sin(3t)\right)^2 + \left(\cos t\right)^2}\, dt \ \blacksquare \approx 13.065$, answer D.

Integrals for arc length usually cannot be evaluated by finding an antiderivative. Even when they can, it is often a very tedious task. If the problem is on the calculator portion of the exam, be prepared to use the numerical integrator on your calculator. Since the formula is complicated, double-check your input as to the number and placement of parentheses, exponents, and so on. Your calculator may take some time to complete the integral; be patient and start the next problem while waiting for your calculator to finish.

8

The length of the curve described by the parametric equations

$x = \left(\ln t\right)^2$, $y = \dfrac{1}{3}\left(3t+1\right)^3$, $1 \le t \le 2$ is given by

(A) $\displaystyle\int_1^2 \sqrt{1 + \left(\dfrac{\left(3t+1\right)^2}{2\ln t}\right)^2}\, dt$

(B) $\displaystyle\int_1^2 \sqrt{\dfrac{2\ln t}{t} + 3\left(3t+1\right)^2}\, dt$

(C) $\displaystyle\int_1^2 \sqrt{\left(2\ln t\right)^2 + \left(3t+1\right)^4}\, dt$

(D) $\displaystyle\int_1^2 \sqrt{\left(\dfrac{1}{t^2}\right)^2 + \left(3t+1\right)^4}\, dt$

(E) $\displaystyle\int_1^2 \sqrt{\left(\dfrac{2\ln t}{t}\right)^2 + 9\left(3t+1\right)^4}\, dt$

$\dfrac{dx}{dt} = \dfrac{2\ln t}{t}$ and $\dfrac{dy}{dt} = \dfrac{1}{3}\cdot 3\left(3t+1\right)^2 \cdot 3 = 3\left(3t+1\right)^2$. The arc length formula gives

$L = \displaystyle\int_1^2 \sqrt{\left(\dfrac{2\ln t}{t}\right)^2 + \left(3\left(3t+1\right)^2\right)^2}\, dt$, which is equivalent to answer E.

9

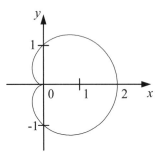

Find the perimeter of the cardioid given by

$$x(t) = (1 + \cos t)\cos t \, , \;\; y(t) = (1 + \cos t)\sin t$$

$$\frac{dx}{dt} = (1 + \cos t)(-\sin t) + (\cos t)(-\sin t) = (-\sin t)(1 + 2\cos t) \;\; \text{and}$$

$$\frac{dy}{dt} = (1 + \cos t)(\cos t) + (\sin t)(-\sin t) = \cos t + \cos^2 t - \sin^2 t \, .$$ One pass around the curve

starts at $t = 0$ and stops at $t = 2\pi$. The arc length is

$$L = \int_0^{2\pi} \sqrt{\left((-\sin t)(1 + 2\cos t)\right)^2 + \left(\cos t + \cos^2 t - \sin^2 t\right)^2} \, dt = 8 \, .$$

8.2.   Vector Functions

Algebraically, an *n*-dimensional vector is an ordered set of *n* numbers. A two-dimensional vector is a pair of numbers (x, y) called components of the vector. We can add two vectors and multiply a vector by a real number. These operations are performed on respective components:

$$\left(x_1, \, y_1\right) + \left(x_2, \, y_2\right) = \left(x_1 + x_2, \, y_1 + y_2\right)$$

$$c\left(x, \, y\right) = \left(cx, \, cy\right)$$

$(0, 0)$ is the zero vector.

Geometrically, a two-dimensional vector is a directed segment between two points on the plane. Two vectors are deemed equal if they have the same length and direction (Figure 8-2).

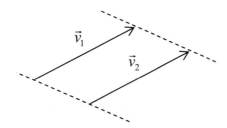

Figure 8-2 Two equal vectors on the plane

A vector (x, y) can be drawn as a directed segment that goes from the origin to the point on the plane with coordinates x and y. Thus, there is a one-to-one correspondence between two-dimensional vectors and points on the plane. But a vector can be drawn between any two points on the plane as needed for better visualization. For example, when a particle is moving on the plane, its position can be described as a vector (x, y). This vector connects the origin to the point at which the particle is located. The velocity of the particle is also a vector. It is often convenient to draw the velocity vector from the point of the current location. The acceleration of the particle is also a vector, but its geometric representation is rarely useful.

> **The length (or *magnitude*) of a vector (x, y) is given by $\sqrt{x^2 + y^2}$ and its direction θ is described by $\tan \theta = \dfrac{y}{x}$.**

The length of a vector $\vec{v}$ is represented by $\left| \vec{v} \right|$.

(Some textbooks use the vector notations $\langle x, y \rangle$ or $x\mathbf{i} + y\mathbf{j}$. You are allowed to use both of these standard notations in your answers on the AP exam, but AP questions use parentheses, and in this book we do the same.)

When a particle moves on the xy-plane, the coordinates of its position can be given as parametric functions $x = x(t)$ and $y = y(t)$ for some time interval $a \le t \le b$. The particle's position vector $\vec{s}(t) = \big(x(t),\ y(t) \big)$ is a ***vector function*** of t. The coordinates of the position at time t are the same as the components of $\vec{s}$ at time t. Therefore,

> **a vector function is essentially a different notation for a parametric function.**

A vector function $\vec{s}(t) = (x(t), y(t))$ is differentiable at t if x and y have derivatives at t. The derivative of $\vec{s}(t)$, $\dfrac{d}{dt}\vec{s}(t)$, is defined as the vector $(x'(t), y'(t))$. A vector function is said to be differentiable if it is differentiable at every point in its domain.

> **If $(x(t), y(t))$ is the position vector of a particle moving along a smooth curve in the *xy*-plane, then, at any time *t*,**
>
> 1. **The particle's velocity vector $\vec{v}(t)$ is $(x'(t), y'(t))$; if drawn from the position point, it is tangent to the curve.**
>
> 2. **The particle's speed along the curve is the length of the velocity vector,**
> $$|\vec{v}(t)| = \sqrt{(x'(t))^2 + (y'(t))^2}.$$
>
> 3. **The particle's acceleration vector $\vec{a}(t)$ is $(x''(t), y''(t))$, is the derivative of the velocity vector and the second derivative of the position vector.**

10

If the position vector of a particle moving along a curve is defined by $\vec{s}(t) = (e^{-t}, -\sin t)$, then the velocity vector of the particle at $t = 0$ is

    (A) $(-1, 0)$ (B) $(1, 0)$ (C) $(-1, 1)$ (D) $(-1, -1)$ (E) $(1, -1)$

$\vec{v}(t) = (-e^{-t}, -\cos t) \Rightarrow \vec{v}(0) = (-e^{-0}, -\cos 0) = (-1, -1)$. The answer is D.

11

A particle moves along a curve so that $x(t) = 6t^2 + \ln t$ and $\dfrac{dy}{dt} = \sin t + 2$. What is the speed of the particle when $t = 2$?

    (A) 5.235   (B) 24.504   (C) 24.672   (D) 24.864   (E) 25.085

The particle's speed along the curve is

$$\left| \vec{v}(t) \right| = \sqrt{\left(\frac{dx}{dt}\right)^2 + \left(\frac{dy}{dt}\right)^2} = \sqrt{\left(12t + \frac{1}{t}\right)^2 + \left(\sin t + 2\right)^2}$$. When $t = 2$, the speed is

$$\sqrt{\left(12 \cdot 2 + \frac{1}{2}\right)^2 + \left(\sin 2 + 2\right)^2} \ \blacksquare \approx 24.672$$, answer C.

12

A particle moves along the graph of $y = x^2 + x$, with its x-coordinate changing at the rate

of $\dfrac{dx}{dt} = \dfrac{2}{t^3}$ for $t > 0$. If $x(1) = 1$, find

(a) the particle's position at $t = 2$;
(b) the speed of the particle at $t = 2$.

(a)

$$x(t) = \int \frac{dx}{dt} \, dt = \int \frac{2}{t^3} \, dt = -\frac{1}{t^2} + C$$. Since $x(1) = 1$, $C = 2$. Now $x = -\dfrac{1}{t^2} + 2$ and

$x(2) = \dfrac{7}{4}$. Substituting the value of $x(2) = \dfrac{7}{4}$ into the given equation for y, we get

$y = \left(\dfrac{7}{4}\right)^2 + \dfrac{7}{4} = \dfrac{77}{16}$. So the particle's position is $\left(\dfrac{7}{4}, \dfrac{77}{16}\right)$.

(b)

The speed of the particle at $t = 2$ is $\left| \vec{v}(2) \right| = \sqrt{\left(\frac{dx}{dt}\right)^2 \Big|_{t=2} + \left(\frac{dy}{dt}\right)^2 \Big|_{t=2}}$. Since y is given as a

function of x, we use the Chain Rule to get $\dfrac{dy}{dt}$, as follows: $\dfrac{dy}{dt} = \dfrac{dy}{dx} \cdot \dfrac{dx}{dt} = (2x + 1)\left(\dfrac{2}{t^3}\right)$.

When $t = 2$ and $x = \dfrac{7}{4}$, $\dfrac{dy}{dt} = \left(\dfrac{9}{2}\right)\left(\dfrac{1}{4}\right) = \dfrac{9}{8}$. So, the speed is

$$\left| \vec{v}(2) \right| = \sqrt{\left(\frac{1}{4}\right)^2 + \left(\frac{9}{8}\right)^2} \ \blacksquare \approx 1.152$$.

An alternative method in both Parts (a) and (b) is to find $x(t)$, then a formula for y in

terms of t: $y = x^2 + x = \left(-\dfrac{1}{t^2}+2\right)^2 + \left(-\dfrac{1}{t^2}+2\right) = \dfrac{1}{t^4} - \dfrac{5}{t^2} + 6$. Then we can use it to find

the position and the speed: $y(2) = \dfrac{1}{16} - \dfrac{5}{4} + 6 = \dfrac{77}{16}$ and $\left.\dfrac{dy}{dt}\right|_{t=2} = \left(-\dfrac{4}{t^5} + \dfrac{10}{t^3}\right)\Big|_{t=2} = \dfrac{9}{8}$.

13

A particle moves along a curve with its position vector given by $\left(3\cos\left(\dfrac{t}{2}\right), 5\sin\left(\dfrac{t}{4}\right)\right)$

for $t \ge 0$. Find its acceleration vector when the particle is at rest for the first time.

When the particle is at rest, the velocity vector, $\left(-\dfrac{3}{2}\sin\left(\dfrac{t}{2}\right), \dfrac{5}{4}\cos\left(\dfrac{t}{4}\right)\right)$, must be equal

to the zero vector $(0,0)$. $-\dfrac{3}{2}\sin\left(\dfrac{t}{2}\right) = 0 \Rightarrow t = 0, 2\pi, 4\pi,\ldots$ and

$\dfrac{5}{4}\cos\left(\dfrac{t}{4}\right) = 0 \Rightarrow t = 2\pi, 6\pi, 10\pi, \ldots$, so the first value for t when the particle is at rest is

2π. The acceleration vector is $\left(-\dfrac{3}{4}\cos\left(\dfrac{t}{2}\right), -\dfrac{5}{16}\sin\left(\dfrac{t}{4}\right)\right)$ and at $t = 2\pi$, the

acceleration vector is $\left(\dfrac{3}{4}, -\dfrac{5}{16}\right)$.

14

A roller coaster track has an inverted loop as a portion of its course. The position of the
car on the track (in feet) at time t seconds, $0 \le t \le 7$, is given by the equations

$$x = 5t - 12\sin(t+2) + 10$$
$$y = 15 + 12\cos(t+2)$$

When is the car at the top of the loop and what is its speed at that time?

y reaches maximum when $\cos(t+2)$ is equal to 1. This occurs when $t+2=2\pi n \Rightarrow t=2\pi n-2$, where n is any integer. $t=2\pi-2$ is the only such value between 0 and 7. The speed at $t=2\pi-2$ is $\sqrt{\left(x'(t)\right)^2+\left(y'(t)\right)^2}=$

$$\sqrt{\left(5-12\cos\left(2\pi\right)\right)^2+\left(-12\sin\left(2\pi\right)\right)^2}=7 \text{ ft/sec.}$$

8.3.   Polar Functions

Polar coordinates allow us to write some very complicated equations in a much simpler form. For example, the Cartesian equation $x^4+2x^3+2x^2y^2+2xy^2-y^2+y^4=0$ is equivalent to the polar equation $r=1-\cos\theta$. Some very simple polar equations have graphs that are very complicated (but beautiful) curves.

The polar coordinates for a point P are $(r,\,\theta)$, where r represents the distance from the origin to the point P and θ is the measure of an angle from the positive x-axis (sometimes called the polar axis) to the ray joining the origin to point P. By convention, angles measured in the counterclockwise direction are positive, and angles measured in the clockwise direction are negative (Figure 8-3).

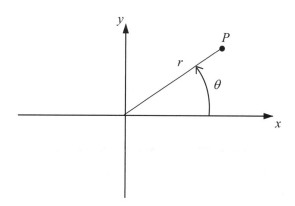

Figure 8-3.   Polar coordinates

In working with polar forms of equations, it is sometimes necessary to convert the coordinates to Cartesian form.

The Cartesian coordinates $(x,\,y)$ and polar coordinates $(r,\,\theta)$ are related by the following equations:

$$x=r\cos\theta,\ \ y=r\sin\theta,\ \ x^2+y^2=r^2,\ \ \tan\theta=\frac{y}{x}$$

Note that in Cartesian coordinates, every point in the plane has exactly one ordered pair that describes it. This is not true in a polar coordinate system, however. In fact, every point has an infinite number of pairs of coordinates, because $(r, \theta + 2\pi n)$, for all integers n, correspond to the same point.

Most graphing calculators have a polar graphing mode, which makes it very easy to draw polar curves. Be sure that you are familiar with the use of this mode in case you need to draw a polar curve. You should also be familiar with the graphs of the most common polar equations, namely circles, lines, cardioids, and three- and four-leaf roses (Figure 8-4), in case there are polar questions on the non-calculator portion of the exam.

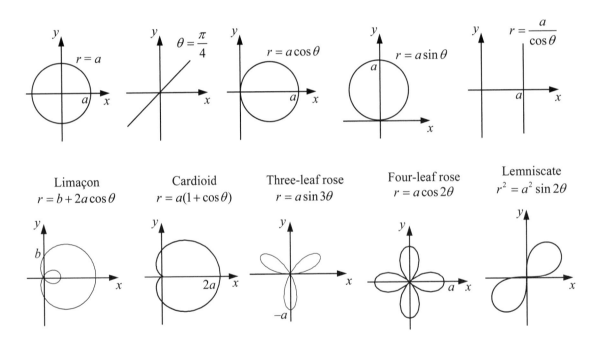

Figure 8-4. Special Polar Graphs

The slope of a tangent line to a polar curve is still found by $\dfrac{dy}{dx}$, <u>not</u> by $\dfrac{dr}{d\theta}$.

We can use the Chain Rule to get $\dfrac{dy}{dx}$ in terms of r and θ:

$$\frac{dy}{dx} = \frac{dy}{d\theta} \bigg/ \frac{dx}{d\theta} = \frac{\dfrac{d}{d\theta}(r\sin\theta)}{\dfrac{d}{d\theta}(r\cos\theta)} = \frac{r'\sin\theta + r\cos\theta}{r'\cos\theta - r\sin\theta}$$

It is not necessary to memorize the last form of this rule: just remember the Chain Rule and the equations relating polar and Cartesian coordinates.

15

For the polar curve $r = \cos(2\theta)$, $0 \le \theta \le 2\pi$, find equations for all the tangent lines at the pole (origin).

The curve is at the pole when $r = 0 \Rightarrow \cos(2\theta) = 0 \Rightarrow \theta = \dfrac{\pi}{4}, \dfrac{3\pi}{4}, \dfrac{5\pi}{4}, \dfrac{7\pi}{4}$.

$$\frac{dy}{dx} = \frac{dy}{d\theta} \bigg/ \frac{dx}{d\theta} = \frac{\dfrac{d}{d\theta}\big(\cos(2\theta)\sin\theta\big)}{\dfrac{d}{d\theta}\big(\cos(2\theta)\cos\theta\big)} = \frac{\cos(2\theta)\cos\theta + \sin\theta\big(-2\sin(2\theta)\big)}{-\cos(2\theta)\sin\theta + \cos\theta\big(-2\sin(2\theta)\big)}.$$

$$\left. \frac{dy}{dx} \right|_{\theta = \frac{\pi}{4}} = \frac{0 \cdot \dfrac{\sqrt{2}}{2} + \dfrac{\sqrt{2}}{2} \cdot (-2) \cdot 1}{-0 \cdot \dfrac{\sqrt{2}}{2} + \dfrac{\sqrt{2}}{2} \cdot (-2) \cdot 1} = 1.$$ When $\theta = \dfrac{\pi}{4}$, the tangent line has equation $y = x$.

Similarly, when $\theta = \dfrac{3\pi}{4}$, the tangent line has equation $y = -x$. When $\theta = \dfrac{5\pi}{4}$, the

tangent line has equation $y = x$ and when $\theta = \dfrac{7\pi}{4}$, the tangent line has equation $y = -x$.

Thus, equations for the tangent lines at the pole are: $y = x$ and $y = -x$.

Area Enclosed by a Polar Curve

Finding the area enclosed by a polar curve, or by a part of a polar curve, offers no new calculus concepts to explore. Rather, it is another setting in which you transform a Riemann sum into a definite integral, and then use the integral to answer questions. Whereas you use rectangles as a first approximation in finding the area under the graph of a function, here you use sectors of a circle (Figure 8-5).

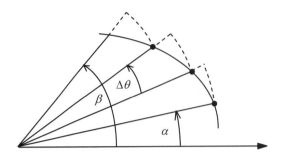

Figure 8-5. Approximating the area of a polar region with sectors

The area of a sector with a small angle $\Delta\theta$ in radians and radius r is $\frac{1}{2}r^2\Delta\theta$.

The area of the polar region from $\theta = \alpha$ to $\theta = \beta$ is given by

$$A = \int_\alpha^\beta \frac{1}{2}\big[r(\theta)\big]^2 \, d\theta.$$

16

The area inside the circle with polar equation $r = 2\sin\theta$ and above the lines with equations $y = x$ and $y = -x$ is given by

(A) $\int_{-\frac{\pi}{4}}^{\frac{\pi}{4}} 2\sin^2\theta \, d\theta$ (B) $\int_{-1}^{1} 2\sin\theta \, d\theta$ (C) $\int_{-1}^{1} \big(2\sin^2\theta - 1\big) d\theta$

(D) $\int_{\frac{\pi}{4}}^{\frac{3\pi}{4}} \sin\theta \, d\theta$ (E) $\int_{\frac{\pi}{4}}^{\frac{3\pi}{4}} 2\sin^2\theta \, d\theta$

The line $y = x$ is equivalent to $\theta = \dfrac{\pi}{4}$, and the line $y = -x$ is

equivalent to $\theta = -\dfrac{\pi}{4}$ or $\theta = \dfrac{3\pi}{4}$. Since the graph of $r = 2\sin\theta$ is

a circle in the first and second quadrants, we want to use $\theta = \dfrac{3\pi}{4}$

for the second line. So we need the portion of the circle between

these two angles. The area is $\int_{\frac{\pi}{4}}^{\frac{3\pi}{4}} \dfrac{1}{2}\big(2\sin\theta\big)^2 d\theta$. The answer is E.

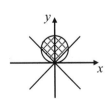

Just as you can find the area between two function graphs, you can find the area between two polar graphs. Since we are looking at sectors of a circle, we must think of the area between two graphs as an outside area minus an inside area. So the area formula becomes $A = \int_{\alpha}^{\beta} \frac{1}{2} \left[\left(r_1(\theta) \right)^2 - \left(r_2(\theta) \right)^2 \right] d\theta$, where $r_1(\theta)$ is the formula for the outside curve and $r_2(\theta)$ is the formula for the inside curve. Sometimes it can be difficult to determine the limits of integration, though.

17

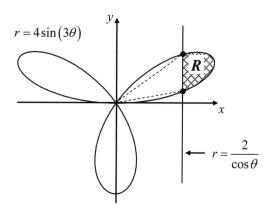

Find the area of the region R enclosed by the graphs of $r = \dfrac{2}{\cos \theta}$ and $r = 4\sin(3\theta)$.

Use the numeric solver on your calculator to find the values of θ where the curves intersect. Solving gives $\alpha \approx 0.1776$ and $\beta \approx 0.7854$. Store these in variables A and B.

The area is given by $\dfrac{1}{2} \int_A^B \left[\left(4\sin(3\theta) \right)^2 - \left(\dfrac{2}{\cos \theta} \right)^2 \right] d\theta \; \blacksquare \approx 2.040$.

Arc Length for Polar Curves

The arc length for a polar curve $r = r(\theta)$ between $\theta = \alpha$ and $\theta = \beta$ is given by

$$L = \int_{\alpha}^{\beta} \sqrt{r^2 + \left(\frac{dr}{d\theta}\right)^2}\, d\theta$$

This topic is <u>not</u> mentioned in the course description. In any event, you can always convert the polar curve to parametric form and use the parametric curve arc length formula.

18

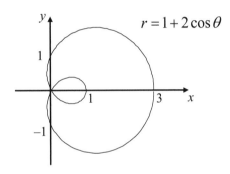

$r = 1 + 2\cos\theta$

Find the total length of the limaçon $r = 1 + 2\cos\theta$ $(0 \le \theta \le 2\pi)$.

$x = r\cos\theta = \cos\theta + 2\cos^2\theta = \cos\theta + \cos 2\theta + 1$.
$y = r\sin\theta = \sin\theta + 2\sin\theta\cos\theta = \sin\theta + \sin 2\theta$.
$\dfrac{dx}{d\theta} = -\sin\theta - 2\sin 2\theta$. $\dfrac{dy}{d\theta} = \cos\theta + 2\cos 2\theta$.

$L = \int_0^{2\pi} \sqrt{\left(\dfrac{dx}{d\theta}\right)^2 + \left(\dfrac{dy}{d\theta}\right)^2}\, d\theta =$

$\int_0^{2\pi} \sqrt{(\sin\theta + 2\sin 2\theta)^2 + (\cos\theta + 2\cos 2\theta)^2}\, dt = \blacksquare \approx 13.365$.

Chapter 9. 「 Series (BC Only) 」

9.1. The Concept of a Series

You probably learned how to write a rational number as a decimal sometime in grade school. For example, you may have seen that $\frac{5}{9} = .55555...$. This was probably one of your first experiences with an ***infinite series***. The decimal on the right-hand side can be rewritten as $.5 + .05 + .005 + .0005 + ...$, a "sum" with an infinite number of terms. This sum ***converges*** to the real number $\frac{5}{9}$.

An ***infinite series*** (or simply a ***series***) is an infinite sum $a_1 + a_2 + ...$ This sum is often written using "sigma" notation:

$$a_1 + a_2 + ... + a_n + ... = \sum_{n=1}^{\infty} a_n$$

To give this expression a precise meaning, we have to be familiar with the concept of a ***sequence*** and the ***limit of a sequence***.

A sequence is a list of numbers, $s_1, s_2, s_3, ...$ (More mathematically, a sequence is defined as a function whose domain is the set of positive integers.)

> **We say that L is the limit of a sequence $s_1, s_2, s_3, ...$ and write $\lim_{n \to \infty} s_n = L$ if, for any sufficiently large n, s_n is close to L.**[*]

This definition is very similar to the definition of the limit of a function as $x \to \infty$ (see Chapter 2).

---

[*] More precisely, $\lim_{n \to \infty} s_n = L$ if for any $\varepsilon > 0$ there exists an $N > 0$ such that $|s_n - L| < \varepsilon$ whenever $n > N$.

The sum of a series is defined as the limit of a sequence, namely the limit of the sequence of the ***partial sums***. The sums

$$s_1 = a_1$$
$$s_2 = a_1 + a_2$$
$$...$$
$$s_n = a_1 + a_2 + ... + a_n$$
$$...$$

are called ***partial sums*** for the series $\sum_{n=1}^{\infty} a_n$.

> **If the limit of the partial sums $\lim_{n\to\infty} s_n = S$ exists, we say that the infinite series**
>
> $\sum_{n=1}^{\infty} a_n$ **converges and call S the sum of the series. Otherwise we say that the**
>
> **series *diverges*.**

9.2.   Geometric Series

$.5 + .05 + .005 + .0005 + ...$ is an example of a ***geometric series***, which is the sum of the terms of a geometric sequence. In a geometric sequence, each term is equal to the previous term multiplied by a constant. The ratio of a term to the previous one is called the ***common ratio*** of the geometric sequence. In the above example, the common ratio is 0.1.

If r is the common ratio and $a = a_1$ is the first term of the sequence, then the formula for the n-th term is $a_n = ar^{n-1}$. For geometric series there is an explicit formula for the

partial sums: $s_n = \sum_{i=1}^{n} ar^{i-1} = a\dfrac{1-r^n}{1-r}$.

> **The geometric series $\sum_{i=1}^{\infty} ar^{i-1}$ converges if and only if $|r| < 1$. If the series**
>
> **converges, its sum is $\dfrac{a}{1-r}$.**

In our example above, $0.55555... = .5 + .05 + .005 + ...$. The common ratio $r = 0.1 < 1$, the series converges, and the sum, not surprisingly, is $\dfrac{0.5}{1-0.1} = \dfrac{5}{9}$.

What is the sum $\dfrac{5}{2}+\dfrac{5}{4}+\dfrac{5}{8}+\dfrac{5}{16}+...$?

 (A) 2 (B) $\dfrac{75}{16}$ (C) $\dfrac{315}{64}$ (D) 5

 (E) This series diverges

This is a geometric series with $a=\dfrac{5}{2}$ and $r=\dfrac{1}{2}$. It converges and the sum is

$\dfrac{5}{2}\cdot\dfrac{1}{1-\dfrac{1}{2}}=5$. The answer is D.

9.3. Tests for Convergence

Not all series converge, of course. This section presents some of the tools that can help determine whether a series converges or not. Table 9-1 summarizes the convergence tests you'll need to know for the AP exam. The following sections discuss these tests in more detail and give a few examples of their use.

You may have studied other tests for convergence. The most common are the Root Test and the Limit Comparison Test. It is OK to use either of these tests on the AP exam, as long as you refer to the test correctly and apply it appropriately. However,

no question will appear on the AP exam that <u>requires</u> the use of a convergence test not listed in Table 9-1.

n-th Term Test	$\displaystyle\sum_{n=1}^{\infty} a_n$ converges $\Rightarrow$ $\displaystyle\lim_{n\to\infty} a_n = 0$. $\displaystyle\lim_{n\to\infty} a_n \neq 0$ $\Rightarrow$ $\displaystyle\sum_{n=1}^{\infty} a_n$ diverges.
Integral Test	$f(x)$ is continuous, positive, and decreasing. $\displaystyle\sum_{n=1}^{\infty} f(n)$ converges $\Leftrightarrow$ $\displaystyle\int_{M}^{\infty} f(x)\,dx$ converges (for some M).
p-Series	$\displaystyle\sum_{n=1}^{\infty} \frac{1}{n^p}$ converges $\Leftrightarrow$ $p > 1$.
Comparison Test	$0 < a_n < b_n$. $\displaystyle\sum_{n=1}^{\infty} b_n$ converges $\Rightarrow$ $\displaystyle\sum_{n=1}^{\infty} a_n$ converges. $\displaystyle\sum_{n=1}^{\infty} a_n$ diverges $\Rightarrow$ $\displaystyle\sum_{n=1}^{\infty} b_n$ diverges.
Ratio Test	$\displaystyle\lim_{n\to\infty}\left\|\frac{a_{n+1}}{a_n}\right\| < 1$ $\Rightarrow$ $\displaystyle\sum_{n=1}^{\infty} a_n$ converges. $\displaystyle\lim_{n\to\infty}\left\|\frac{a_{n+1}}{a_n}\right\| > 1$ $\Rightarrow$ $\displaystyle\sum_{n=1}^{\infty} a_n$ diverges. $\displaystyle\lim_{n\to\infty}\left\|\frac{a_{n+1}}{a_n}\right\| = 1$ $\Rightarrow$ can't tell.
Alternating Series Test	$a_n > 0$, decreasing, $\displaystyle\lim_{n\to\infty} a_n = 0$ $\Rightarrow$ $\displaystyle\sum_{n=1}^{\infty} (-1)^{n-1} a_n$ converges.

Table 9-1. Series Convergence Tests

9.3.1. *n*-th Term Test

If $\lim\limits_{n\to\infty} a_n \neq 0$ then $\sum\limits_{n=1}^{\infty} a_n$ diverges. This test says that for a series to converge, the terms you are adding must get arbitrarily close to 0. Note that this test says that $\lim\limits_{n\to\infty} a_n = 0$ is a necessary condition for convergence of $\sum\limits_{n=1}^{\infty} a_n$. But it is <u>not</u> a sufficient condition. In other words,

just because the *n*-th term goes to zero does not mean the series necessarily converges.

However, if the *n*-th term does not approach zero, the series definitely diverges.

A geometric series with a common ratio greater than 1, such as $1 + 2 + 4 + 8 + 16 + \dots$, obviously has no chance of converging — you keep adding a larger and larger number onto the sum. The series $1 - 1 + 1 - 1 + 1 - 1 + \dots$ also diverges by the *n*-th Term Test. Here the sequence of partial sums is 1, 0, 1, 0, 1, …, which does not converge.

Keep the *n*-th Term Test in mind — it should be the first thing you consider in testing whether a given series diverges.

As we have said, even when the *n*-th term has a limit of 0, the series could still diverge. The classic example is the ***harmonic series***, $1 + \dfrac{1}{2} + \dfrac{1}{3} + \dfrac{1}{4} + \dots + \dfrac{1}{n} + \dots$ Even though $\dfrac{1}{n} \to 0$ as $n \to \infty$, the terms do not approach 0 "fast enough" for this series to converge. You can make $\sum\limits_{n=1}^{N} \dfrac{1}{n}$ as large as you want by choosing a sufficiently large number of terms N. This fact might not be obvious to you; it is explained in the next section.

9.3.2.   Integral Test

Suppose we have a series with the terms given by $a_n = f(n)$, where $f(x)$ is a positive, continuous, and decreasing function for all $x \geq M$. Then the infinite series $\sum\limits_{n=1}^{\infty} a_n$ converges if and only if the improper integral $\int_M^{\infty} f(x)\,dx$ converges.

If $\int_M^{\infty} f(x)\,dx$ diverges, then $\sum\limits_{n=1}^{\infty} a_n$ also diverges.

As an example, let's take the harmonic series discussed in the previous section: $1 + \dfrac{1}{2} + \dfrac{1}{3} + \dfrac{1}{4} + ... + \dfrac{1}{n} + ...$. For this series, $f(x) = \dfrac{1}{x}$. In Chapter 5, you saw that the integral $\int_1^{\infty} \dfrac{1}{x}\,dx$ diverges, since $\ln x$ gets arbitrarily large. Therefore, $\sum\limits_{n=1}^{\infty} \dfrac{1}{n}$ diverges by the Integral Test.

To see why the Integral Test works, note that the n-th partial sum $\sum\limits_{k=1}^{n} \dfrac{1}{k}$ can be viewed as the left-hand Riemann sum for $\int_1^{n+1} \dfrac{1}{x}\,dx$ (Figure 9-1). Since the integrand is decreasing, this left-hand Riemann sum overestimates the integral. Therefore, $\sum\limits_{k=1}^{n} \dfrac{1}{k} > \int_1^{n+1} \dfrac{1}{x}\,dx$. The integral diverges, so the harmonic series diverges, too.

A similar argument works to show that a series converges. Consider, for example, the series $1 + \dfrac{1}{4} + \dfrac{1}{9} + \dfrac{1}{16} + \dfrac{1}{25} ... + \dfrac{1}{n^2} + ... = \sum\limits_{n=1}^{\infty} \dfrac{1}{n^2}$. We can compare it to $\int_1^{\infty} \dfrac{1}{x^2}\,dx$, which we know converges. $\sum\limits_{k=2}^{n} \dfrac{1}{k^2}$ can be viewed as the right-hand Riemann sum for $\int_1^{n} \dfrac{1}{x^2}\,dx$ (Figure 9-2). Since the right-hand sum for a decreasing function underestimates the integral and the integral converges, the partial sums also converge.

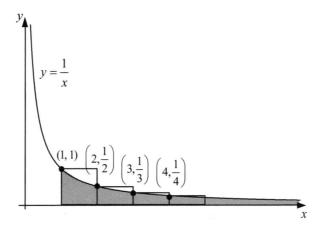

Figure 9-1. $\displaystyle\sum_{k=1}^{n}\frac{1}{k}$ **is a left-hand Riemann sum for** $\displaystyle\int_{1}^{n+1}\frac{1}{x}\,dx$

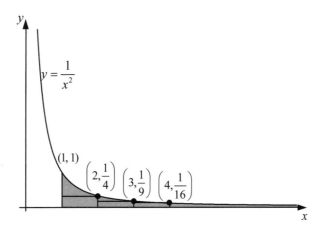

Figure 9-2. $\displaystyle\sum_{k=1}^{n}\frac{1}{k^2}=1+$ **right-hand Riemann sum for** $\displaystyle\int_{1}^{n}\frac{1}{x^2}\,dx$

The need to recognize the connection between a Riemann sum and a series is explicitly mentioned in the College Board's *Topic Outline for Calculus BC*.

9.3.3.   *p*-Series

The Integral Test can be used to determine the convergence of another special kind of series called a **p-series**. This is a series of the form $\sum_{n=1}^{\infty} \dfrac{1}{n^p}$ where p can be any constant. The harmonic series is a special case of a p-series with $p = 1$.

> **A *p*-series converges if $p > 1$ and diverges if $p \le 1$.**

Why? If $p < 0$, then the series diverges by the *n*-th Term Test. If $p > 0$, $f(x) = \dfrac{1}{x^p}$ is positive, continuous, and decreasing for all $x > 0$. So to establish what happens with $\sum_{n=1}^{\infty} \dfrac{1}{n^p}$, we need only to determine what happens with $\int_{M}^{\infty} \dfrac{1}{x^p} dx$. We have considered the case of $p = 1$ in the previous section. For $p \ne 1$, we need to use limits to evaluate this improper integral:

$$\int_{M}^{\infty} \frac{1}{x^p} dx = \lim_{a \to \infty} \int_{M}^{a} x^{-p} dx = \lim_{a \to \infty} \left(\frac{x^{-p+1}}{-p+1} \bigg|_{M}^{a} \right) = \lim_{a \to \infty} \left(\frac{a^{-p+1}}{-p+1} - \frac{M^{-p+1}}{-p+1} \right) = \frac{\lim_{a \to \infty} a^{-p+1}}{-p+1} - \frac{M^{-p+1}}{-p+1} .$$

If $p > 1$, the limit exists because a^{-p+1} goes to 0, and the integral converges; if $p < 1$, the limit does not exist because a^{-p+1} goes to infinity, and the integral diverges. According to the Integral Test, the behavior of the series is the same as the behavior of the integral.

9.3.4.   Comparison Test

The Comparison Test is used to determine whether one series converges or diverges by comparing its terms with the corresponding terms of another series.

> **Suppose $0 \le a_n \le b_n$ for all n greater than some integer M. Then**
>
>   **if $\sum b_n$ converges, then $\sum a_n$ converges;**
>
>   **if $\sum a_n$ diverges, then $\sum b_n$ diverges.**

In other words, if you know the terms of one series are positive and each term is smaller than the corresponding term of another convergent series, then the series converges. If you know the terms of one series are all greater than the terms of another divergent series, then the original series diverges.

9.3.5.   Ratio Test

Another test you need to understand for the AP exam is the powerful Ratio Test. It is actually quite similar to the condition for convergence of a geometric series. In a geometric series, the ratio of consecutive terms, $\dfrac{a_{n+1}}{a_n}$, is constant; a geometric series converges as long as the absolute value of that ratio is less than 1. In many series, while the ratio of consecutive terms is not constant itself, it <u>approaches a constant</u> as n approaches infinity.

> **The Ratio Test says that if** $\lim\limits_{n\to\infty}\left|\dfrac{a_{n+1}}{a_n}\right| < 1$**, then the series converges. If that limit is greater than 1 (or is infinite), then the series diverges. If that limit is exactly 1, the Ratio Test is inconclusive, and the series might converge or diverge.**

In the latter case, you need to apply some other test.

9.3.6.   Alternating Series Test

A series in which the terms alternate in sign between positive and negative is called an **alternating series**. Such series can be represented as

$$a_1 - a_2 + a_3 - a_4 + \ldots + (-1)^{n-1} a_n + \ldots = \sum_{n=1}^{\infty} (-1)^{n-1} a_n \text{ or}$$

$$-a_1 + a_2 - a_3 + a_4 - \ldots + (-1)^{n} a_n + \ldots = \sum_{n=1}^{\infty} (-1)^{n} a_n, \text{ where } a_n > 0. \text{ There's a special test for}$$

convergence of alternating series, appropriately called the Alternating Series Test.

> **If a_n are positive and decreasing and** $\lim\limits_{n\to\infty} a_n = 0$**, then the series** $\sum\limits_{n=1}^{\infty} (-1)^{n-1} a_n$ **converges.**

The series $1 - \dfrac{1}{2} + \dfrac{1}{3} - \dfrac{1}{4} + \ldots + (-1)^{n-1}\dfrac{1}{n} + \ldots$ is called the **alternating harmonic series**. It converges by the Alternating Series Test. All you need to check is that the terms alternate in sign, that they decrease in absolute value, and that the n-th term approaches 0. All three conditions are met, so we know the alternating harmonic series converges. As we will see later in this chapter, this series converges to $\ln 2$.

If an alternating series $\sum_{k=1}^{\infty}(-1)^{k-1}a_k$ (with $0 < a_{k+1} < a_k$ for all k) converges to S, its

partial sums jump around S from side to side with decreasing distances from S (Figure 9-3). As we can see, in this situation $|S - s_n| < |s_{n+1} - s_n| = a_{n+1}$. In other words,

> **if the terms of an alternating series are decreasing and we approximate its sum with the n-th partial sum, the error does not exceed the absolute value of the first omitted term, a_{n+1}.**

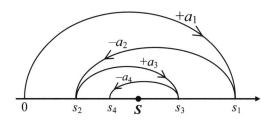

Figure 9-3. Partial sums for an alternating series

9.3.7.   Applying Convergence Tests

While the convergence tests themselves are straightforward, you need some practice to be able to figure out quickly which one to use for a particular series.

Which of the following series converges?

> I. $\displaystyle\sum_{n=1}^{\infty}\frac{1}{n^3}$ II. $\displaystyle\sum_{n=1}^{\infty}\frac{2}{n+1}$ III. $\displaystyle\sum_{n=1}^{\infty}\frac{3^n}{n2^n}$

(A) I only   (B) II only   (C) III only

(D) I and III only   (E) I, II, and III

Series I is a p-series with $p > 1$, therefore the series converges. Series II can be analyzed using the Integral Test, since $f(x) = \dfrac{2}{x+1}$ is continuous, positive, and decreasing.

$\displaystyle\int_1^\infty \frac{2}{x+1}\,dx = \lim_{a\to\infty}\int_1^a \frac{2}{x+1}\,dx = \lim_{a\to\infty}\left(2\ln(x+1)\Big|_1^a\right)$. The integral diverges, so the series

diverges. Series III diverges by the Ratio Test: $\left|\dfrac{a_{n+1}}{a_n}\right| = \dfrac{3}{2}\dfrac{n}{n+1} \to \dfrac{3}{2} > 1$. The answer is A.

The terms of a series may be functions of x. In that case, the series may converge for some values of x and diverge for other values. The set of all values of x for which the series converges is called the ***interval of convergence***. The distance from the midpoint of the interval of convergence to an endpoint is called the ***radius of convergence***. A common example are ***power series***, which have the form $\displaystyle\sum_{n=0}^\infty a_n x^n$ or $\displaystyle\sum_{n=0}^\infty a_n(x-c)^n$. A typical use of the Ratio Test is to find the radius of convergence of a power series.

3

Find the interval and radius of convergence for $\displaystyle\sum_{n=0}^\infty n\left(\frac{2x-5}{3}\right)^n$.

Use the Ratio Test. $\displaystyle\lim_{n\to\infty}\left|\frac{a_{n+1}}{a_n}\right| = \lim_{n\to\infty}\left|\frac{(n+1)\left(\dfrac{2x-5}{3}\right)^{n+1}}{n\left(\dfrac{2x-5}{3}\right)^n}\right| = \lim_{n\to\infty}\left|\frac{n+1}{n}\left(\frac{2x-5}{3}\right)\right| = \left|\frac{2x-5}{3}\right|$. If

$\left|\dfrac{2x-5}{3}\right| < 1$, the series converges, and if $\left|\dfrac{2x-5}{3}\right| > 1$, the series diverges.

$\left|\dfrac{2x-5}{3}\right| < 1 \Leftrightarrow \left|x - \dfrac{5}{2}\right| < \dfrac{3}{2} \Leftrightarrow -\dfrac{3}{2} + \dfrac{5}{2} < x < \dfrac{3}{2} + \dfrac{5}{2} \Leftrightarrow 1 < x < 4$. The radius of

convergence is $\dfrac{3}{2}$ (the distance from the midpoint of this interval to either endpoint).

Since the Ratio Test is inconclusive when the limit is exactly 1, we have to test the endpoints separately. When $x = 1$ or $x = 4$, the series becomes $\displaystyle\sum_{n=0}^\infty n\left(\frac{\pm 3}{3}\right)^n = \sum_{n=0}^\infty (\pm n)$, which diverges by the n-th Term Test. So the interval of convergence is $1 < x < 4$.

4

Determine whether the series $\sum_{n=1}^{\infty} \dfrac{1}{n^2+e^n}$ converges or diverges. Justify your answer.

Compare the given series to $\sum_{n=1}^{\infty} \dfrac{1}{e^n}$, which converges since it is a geometric series with a common ratio $\dfrac{1}{e}<1$. Since every term of $\sum_{n=1}^{\infty} \dfrac{1}{n^2+e^n}$ is less than the corresponding term of $\sum_{n=1}^{\infty} \dfrac{1}{e^n}$, the series $\sum_{n=1}^{\infty} \dfrac{1}{n^2+e^n}$ converges by the Comparison Test.

5

Which of the following series converges for all values of x between -1 and 1?

I. $\sum_{n=1}^{\infty} \dfrac{1}{(3x)^n}$ II. $\sum_{n=1}^{\infty} \dfrac{2^n}{n^3}x^n$ III. $\sum_{n=1}^{\infty} \dfrac{x^n}{(2n)!}$

(A) I only   (B) III only   (C) I and II only

(D) II and III only   (E) I, II, and III

Series I is a geometric series, with a common ratio of $\dfrac{1}{3x}$. It converges as long as $\left|\dfrac{1}{3x}\right|<1$. This is true for $x<-\dfrac{1}{3}$ or $x>\dfrac{1}{3}$, and the series does not converge for all x between -1 and 1.

Applying the Ratio Test to Series II gives $\lim\limits_{n\to\infty}\left|\dfrac{\dfrac{2^{n+1}}{(n+1)^3}x^{n+1}}{\dfrac{2^n}{n^3}x^n}\right|<1$. Simplifying, we get

$\lim\limits_{n\to\infty}|2x|<1$, which is true for $-\dfrac{1}{2}<x<\dfrac{1}{2}$. The series does not converge for all x between -1 and 1.

Applying the Ratio Test to Series III gives $\lim\limits_{n\to\infty} \left| \dfrac{\dfrac{x^{n+1}}{(2n+2)!}}{\dfrac{x^n}{(2n)!}} \right| = \lim\limits_{n\to\infty} \dfrac{x}{(2n+1)(2n+2)} = 0 < 1$.

The series converges for all values of x. The answer is B.

9.4.   Taylor and Maclaurin Polynomials

If f is differentiable at $x = c$, then the linear function $L(x) = f(c) + f'(c)(x - c)$ has the same value as $f(x)$ at $x = c$, that is $f(c) = L(c)$. Also, $L'(c) = f'(c)$. As we know, this is the equation for the tangent line to the graph of $y = f(x)$ at $x = c$. $L(x)$ gives a linear approximation for $f(x)$ in the vicinity of c. If f is differentiable twice at $x = c$, we can take this a step further and create a quadratic approximation for $f(x)$ in the vicinity of c:

$Q(x) = f(c) + f'(c)(x - c) + \dfrac{f''(c)}{2}(x - c)^2$. This quadratic polynomial has the same value, the same first derivative value, and the same second derivative value as f at $x = c$. $Q(x)$ gives a better approximation for $f(x)$ in the vicinity of c than $L(x)$.

If f has n derivatives at $x = c$, we can construct an n-th degree polynomial $T_n(x)$, such that $T_n(c) = f(c)$ and $T_n'(c) = f'(c)$, ..., $T_n^{(k)}(c) = f^{(k)}(c)$, for $k = 1,...,n$. The equation for such a polynomial is

$$T_n(x) = f(c) + f'(c) \cdot (x - c) + \dfrac{f''(c)}{2!}(x - c)^2 + ... + \dfrac{f^{(n)}(c)}{n!}(x - c)^n$$

This polynomial is called an *n-th degree Taylor polynomial* for f at $x = c$. If $c = 0$, such a polynomial is often called a *Maclaurin polynomial*.

For example, the third-degree Maclaurin polynomial for f is given by
$T_3(x) = f(0) + f'(0) \cdot x + \dfrac{f''(0)}{2!}x^2 + \dfrac{f'''(0)}{3!}x^3$.

To understand why a Taylor polynomial looks the way it does, try to calculate the derivatives of $T_n(x)$. The factorials in the coefficients cancel the factors that arise from successively differentiating $(x - c)^n$, and you get $T_n^{(k)}(c) = f^{(k)}(c)$.

What is the third-degree Taylor polynomial for $f(x) = \tan x$ at $x = 0$?

(A) $x - \dfrac{x^3}{3!}$ (B) $x + \dfrac{x^3}{3!}$ (C) $x - \dfrac{x^3}{3}$ (D) $x + \dfrac{x^3}{3}$ (E) $x + \dfrac{2x^3}{3}$

We need the values of the first three derivatives of $\tan x$ at $x = 0$:

$f'(x) = \sec^2 x \implies f'(0) = \sec^2 0 = 1$;

$f''(x) = 2\sec x \cdot \sec x \cdot \tan x = 2\sec^2 x \cdot \tan x \implies f''(0) = 2\sec^2 0 \cdot \tan 0 = 0$;

$f'''(x) = 2\sec^2 x \cdot \sec^2 x + \tan x \cdot 4 \cdot \sec x \cdot \sec x \cdot \tan x \implies f'''(0) = 2$.

The third-degree Taylor polynomial for $\tan x$ at $x = 0$ is $T_3 = 0 + x + 0 + \dfrac{2}{3!}x^3 = x + \dfrac{x^3}{3}$.

The answer is D.

The n-th degree Taylor polynomial approximates $f(x)$ in the vicinity of c better than any other n-th degree polynomial. Note that polynomials require only the three basic arithmetic operations — addition, subtraction, and multiplication — to evaluate. These operations can be performed quickly with a pencil and paper or with a machine. This gives us one way to evaluate transcendental functions. An amazing result with enormous practical importance!

In order to maintain a "level playing field" for students with a CAS-equipped calculator and those without one, some recent AP exams have provided the values of the derivatives and simply asked students to put them together to get a Taylor polynomial. A value for the function and its derivatives can be provided in a table or by a list of formulas for the function and its derivatives.

Suppose g is a function with derivatives of all orders for all real numbers. Assume $g(0) = -4$, $g'(0) = 2$, $g''(0) = 6$, and $g'''(0) = -8$. Write the third-degree Taylor polynomial for g at $x = 0$, and use it to approximate $g(0.5)$.

$$T_3(x) = -4 + 2x + \frac{6}{2!}x^2 - \frac{8}{3!}x^3 = -4 + 2x + 3x^2 - \frac{4}{3}x^3 \, .$$

$g(0.5) \approx -4 + 2 \cdot 0.5 + 3 \cdot 0.5^2 - \dfrac{4 \cdot 0.5^3}{3}$. If this is a free-response question, you do not have

to simplify any more; leave your answer just like that. Also note the use of $\approx$ instead of $=$ on the last line.

> $T_n(x)$ is typically <u>not</u> equal to $f(x)$ when $x \neq c$. It is only an
> approximation.

The only value of g we can be sure about is $g(0) = -4$. Do not use the equal sign improperly.

8

Suppose f is a function whose n-th derivative satisfies $f^{(n)}(x) = (2^x + 1)(n+1)!$ for all x and n. If $f(3) = -2$, what is the fourth-degree Taylor polynomial for f at $x = 3$?

    (A) $-2 + 18 \cdot (x-3) + 27 \cdot (x-3)^2 + 36 \cdot (x-3)^3 + 45 \cdot (x-3)^4$

    (B) $-2 + 18x + 27x^2 + 36x^3 + 45x^4$

    (C) $-2 + 18 \cdot (x-3) + 54 \cdot (x-3)^2 + 216 \cdot (x-3)^3 + 1080 \cdot (x-3)^4$

    (D) $-2 + 18x + 54x^2 + 216x^3 + 1080x^4$

    (E) $-2 + 18 \cdot (x-3) + 27 \cdot (x-3)^2 + 72 \cdot (x-3)^3 + 270 \cdot (x-3)^4$

$$T_4(x) = -2 + \frac{9 \cdot 2!}{1!}(x-3) + \frac{9 \cdot 3!}{2!}(x-3)^2 + \frac{9 \cdot 4!}{3!}(x-3)^3 + \frac{9 \cdot 5!}{4!}(x-3)^4 =$$

$-2 + 18 \cdot (x-3) + 27 \cdot (x-3)^2 + 36 \cdot (x-3)^3 + 45 \cdot (x-3)^4$. The answer is A.

Our last example in this section involves the "traditional" calculation of a Taylor polynomial by hand. Such a problem could appear on the non-calculator section of the AP exam.

9

Which of the following is the fourth-degree Taylor polynomial for $\cos\left(3x - \dfrac{\pi}{6}\right)$ at $x = 0$?

(A) $\dfrac{\sqrt{3}}{2} + \dfrac{3x}{2} - \dfrac{9\sqrt{3}x^2}{4} - \dfrac{9x^3}{4} + \dfrac{27\sqrt{3}x^4}{16}$

(B) $1 - \dfrac{\left(3x - \dfrac{\pi}{6}\right)^2}{2} + \dfrac{\left(3x - \dfrac{\pi}{6}\right)^4}{24}$

(C) $-\dfrac{\sqrt{3}}{2} - \dfrac{3x}{2} + \dfrac{9\sqrt{3}x^2}{4} + \dfrac{9x^3}{4} - \dfrac{27\sqrt{3}x^4}{16}$

(D) $\dfrac{\sqrt{3}}{2} + \dfrac{x}{2} - \dfrac{\sqrt{3}x^2}{4} - \dfrac{x^3}{12} + \dfrac{\sqrt{3}x^4}{48}$

(E) $\dfrac{\sqrt{3}}{2} - \dfrac{x}{2} + \dfrac{\sqrt{3}x^2}{4} + \dfrac{x^3}{12} - \dfrac{\sqrt{3}x^4}{48}$

We need the outputs from the function and its first four derivatives at $x = 0$.

$f(x) = \cos\left(3x - \dfrac{\pi}{6}\right) \Rightarrow f(0) = \dfrac{\sqrt{3}}{2}$

$f'(x) = -3\sin\left(3x - \dfrac{\pi}{6}\right) \Rightarrow f'(0) = \dfrac{3}{2}$

$f''(x) = -9\cos\left(3x - \dfrac{\pi}{6}\right) \Rightarrow f''(0) = -\dfrac{9\sqrt{3}}{2}$

$f'''(x) = 27\sin\left(3x - \dfrac{\pi}{6}\right) \Rightarrow f'''(0) = -\dfrac{27}{2}$

$f^{(4)}(x) = 81\cos\left(3x - \dfrac{\pi}{6}\right) \Rightarrow f^{(4)}(0) = \dfrac{81\sqrt{3}}{2}$

Put these together to form the polynomial:

$T_4(x) = \dfrac{\sqrt{3}}{2} + \dfrac{3}{2}x - \dfrac{9\sqrt{3}}{2 \cdot 2!}x^2 - \dfrac{27}{2 \cdot 3!}x^3 + \dfrac{81\sqrt{3}}{2 \cdot 4!}x^4 = \dfrac{\sqrt{3}}{2} + \dfrac{3x}{2} - \dfrac{9\sqrt{3}x^2}{4} - \dfrac{9x^3}{4} + \dfrac{27\sqrt{3}x^4}{16}$. The answer is A.

9.5.   Taylor and Maclaurin Series

For a function with <u>all</u> derivatives we can construct the Taylor polynomial of any degree. Then

> **we can view the *n*-th degree Taylor polynomial $T_n(x)$ as a partial sum for the series**
>
> $$f(c)+f'(c)\cdot\left(x-c\right)+\frac{f''(c)}{2!}\left(x-c\right)^2+...+\frac{f^{(n)}(c)}{n!}\left(x-c\right)^n+...$$
>
> **This series is called the *Taylor series* for *f* at *x = c*.**

For many functions *f*, their Taylor series converge to $f(x)$ everywhere or for *x* in a certain interval containing *c*. For example, as we will see shortly, the Taylor series for $\sin x$ at $x = 0$ is $x-\frac{x^3}{3!}+\frac{x^5}{5!}-....$ This series converges to $\sin x$ for any real number *x*.

So, we can approximate $\sin(1)$ by evaluating $1-\frac{1^3}{3!}+\frac{1^5}{5!}-...,$ adding as many terms as needed to make the approximation sufficiently accurate.

> **The special case of Taylor series for *f*, when *c = 0*, is called the *Maclaurin series*. It is given by**
>
> $$f(0)+f'(0)\cdot x+\frac{f''(0)}{2!}x^2+\frac{f'''(0)}{3!}x^3+...+\frac{f^{(n)}(0)}{n!}x^n+...=\sum_{n=0}^{\infty}\frac{f^{(n)}(0)}{n!}x^n$$

Table 9-2 lists Maclaurin series that are specifically mentioned in the AP Course Description. You should memorize them.

Series	Converges for:
$\dfrac{1}{1-x} = 1 + x + x^2 + x^3 \ldots + x^n + \ldots = \displaystyle\sum_{n=0}^{\infty} x^n$	$-1 < x < 1$
$e^x = 1 + x + \dfrac{x^2}{2!} + \dfrac{x^3}{3!} + \ldots + \dfrac{x^n}{n!} + \ldots = \displaystyle\sum_{n=0}^{\infty} \dfrac{x^n}{n!}$	all real numbers x
$\sin x = x - \dfrac{x^3}{3!} + \dfrac{x^5}{5!} - \ldots + (-1)^n \dfrac{x^{2n+1}}{(2n+1)!} + \ldots = \displaystyle\sum_{n=0}^{\infty} (-1)^n \dfrac{x^{2n+1}}{(2n+1)!}$	all real numbers x
$\cos x = 1 - \dfrac{x^2}{2!} + \dfrac{x^4}{4!} - \ldots + (-1)^n \dfrac{x^{2n}}{(2n)!} + \ldots = \displaystyle\sum_{n=0}^{\infty} (-1)^n \dfrac{x^{2n}}{(2n)!}$	all real numbers x

Table 9-2. Some of the important Maclaurin series[*]

For the AP exam, you may be required to produce a Maclaurin series by manipulating a known one.

For example, suppose you need to find the Maclaurin series for $x\sin(x^2)$. You could try creating it by calculating all the derivatives for $x\sin(x^2)$ at $x = 0$. However, this becomes overwhelmingly tedious (unless you have a computer algebra system at hand). A much easier way is to substitute x^2 for x in the series for $\sin x$, and then to multiply the resulting series by x:

$$\sin\left(x^2\right) = x^2 - \frac{x^6}{3!} + \frac{x^{10}}{5!} - \ldots + (-1)^n \frac{x^{4n+2}}{(2n+1)!} + \ldots = \sum_{n=0}^{\infty} (-1)^n \frac{x^{4n+2}}{(2n+1)!}. \text{ Multiplying by } x$$

$$\text{gives } x\sin\left(x^2\right) = x^3 - \frac{x^7}{3!} + \frac{x^{11}}{5!} - \ldots + (-1)^n \frac{x^{4n+3}}{(2n+1)!} + \ldots = \sum_{n=0}^{\infty} (-1)^n \frac{x^{4n+3}}{(2n+1)!}.$$

---

[*] If you substitute $i\theta$ for x in the series for e^x (where $i = \sqrt{-1}$), you get

$$e^{i\theta} = 1 + i\theta - \frac{\theta^2}{2!} - \frac{i\theta^3}{3!} + \frac{\theta^4}{4!} + \ldots = \left[1 - \frac{\theta^2}{2!} + \frac{\theta^4}{4!} - \ldots\right] + i\left[\theta - \frac{\theta^3}{3!} + \ldots\right] = \cos\theta + i\sin\theta, \text{ known as Euler's}$$

formula. In particular, for $\theta = \pi$, you get the famous equality (which many mathematicians regard as one of the most beautiful facts in mathematics): $e^{i\pi} = -1$ or $e^{i\pi} + 1 = 0$, an equation with the five most important numbers in mathematics.

10

Find the Maclaurin series for $\dfrac{1}{1+2x}$.

To obtain a series for $\dfrac{1}{1+2x}$, all we need to do is replace x by $-2x$ in the series for $\dfrac{1}{1-x}$. We get: $\dfrac{1}{1+2x} = 1 - 2x + 4x^2 - 8x^3 \ldots + (-2x)^n + \ldots = \displaystyle\sum_{n=0}^{\infty} (-2x)^n$. It converges for $-1 < 2x < 1 \Leftrightarrow -\dfrac{1}{2} < x < \dfrac{1}{2}$.

11

Find the Maclaurin series for $\dfrac{1}{1+x^2}$.

Simply substitute $-x^2$ for x in the series for $\dfrac{1}{1-x}$. You get

$$\dfrac{1}{1+x^2} = 1 - x^2 + x^4 - x^6 + \ldots + (-1)^n x^{2n} + \ldots = \displaystyle\sum_{n=0}^{\infty} (-1)^n x^{2n}.$$

> **By the way, you could well be asked on the exam to give a formula for the *n*-th term of a series. It's a good idea to think about and practice how to construct such a formula (for example, using powers of -1 to make the signs alternate).**

You can also obtain a new series by differentiating or antidifferentiating each term of a known series. There is a theorem (the proof of which goes beyond a typical AP Calculus course) that guarantees that the resulting series has the same center and radius of convergence as the original series (but convergence of the result may differ at the endpoints.)

12

Find the Maclaurin series for $\ln(1+x)$.

To get the Maclaurin series for $\ln(1+x)$ we can take the series for $\dfrac{1}{1+x}$ and

antidifferentiate both sides: $\displaystyle\int\frac{1}{1+x}\,dx = \int\left[1-x+x^2-x^3\ldots+(-x)^n+\ldots\right]dx \Rightarrow$

$\ln(1+x) = C + x - \dfrac{x^2}{2} + \dfrac{x^3}{3} - \dfrac{x^4}{4} + \ldots + \dfrac{(-1)^n x^{n+1}}{n+1} + \ldots$ for $|x| < 1$. Letting $x = 0$, we get

$\ln(1) = C$, so $C = 0$ and $\ln(1+x) = x - \dfrac{x^2}{2} + \dfrac{x^3}{3} - \dfrac{x^4}{4} + \ldots + \dfrac{(-1)^n x^{n+1}}{n+1} + \ldots = \displaystyle\sum_{n=0}^{\infty} \frac{(-1)^n x^{n+1}}{n+1}.$

This series also converges at $x = 1$, by the Alternating Series Test, and we get

$1 - \dfrac{1}{2} + \dfrac{1}{3} - \dfrac{1}{4} + \ldots = \ln 2$.[*] Thus, the alternating harmonic series converges to $\ln 2$.

13

Find the Taylor series for $g(x) = \arctan(x)$ at $x = 0$.

Notice that the series you constructed in one of the previous examples,

$\dfrac{1}{1+x^2} = 1 - x^2 + x^4 - x^6 + \ldots + (-1)^n x^{2n} + \ldots$, is the series for the derivative of $\arctan(x)$.

You can antidifferentiate it to produce this one:

$\arctan(x) = C + x - \dfrac{x^3}{3} + \dfrac{x^5}{5} - \dfrac{x^7}{7} + \ldots + (-1)^n \dfrac{x^{2n+1}}{2n+1} + \ldots$. Since $\arctan(0) = 0$, $C = 0$. The

answer is $\arctan(x) = x - \dfrac{x^3}{3} + \dfrac{x^5}{5} - \dfrac{x^7}{7} + \ldots + (-1)^n \dfrac{x^{2n+1}}{2n+1} + \ldots = \displaystyle\sum_{n=0}^{\infty} (-1)^n \frac{x^{2n+1}}{2n+1}.$

Note that the series for $\dfrac{1}{1+x^2}$ and $\arctan(x)$ have the same center and radius of

convergence, as expected. However, the series for $\dfrac{1}{1+x^2}$ diverges at both endpoints,

$x = -1$ and $x = 1$, while the series for $\arctan(x)$ converges at both endpoints.

---

[*] Strictly speaking, we cannot automatically assume that the series for $\ln(1+x)$ converges to $\ln 2$ at $x = 1$. But, in this case, the series for $\ln(1+x)$ converges for $-1 < x \le 1$ and it is possible to prove that its sum is continuous (from the left) at $x = 1$. In general, the sum may be discontinuous — consider, for example, $\displaystyle\sum_{n=1}^{\infty}\left(x^{n+1} - x^n\right).$

You should be careful when applying Taylor and Maclaurin series to calculating function values. Without more information, there is no guarantee that a Taylor series converges. Moreover, there is no guarantee, even if it does converge, that it converges to the function that generated the series! When can you be sure? Taylor's Theorem answers this question.

9.6.   Taylor's Theorem and Lagrange Remainder

If f has at least $n + 1$ derivatives in an interval around $x = c$, then for any x in that interval,

$$f(x) = f(c) + f'(c)(x-c) + \frac{f''(c)}{2!}(x-c)^2 + \ldots + \frac{f^{(n)}(c)}{n!}(x-c)^n + R_n(x)$$

where $R_n(x) = \dfrac{f^{(n+1)}(\tilde{x})}{(n+1)!}(x-c)^{n+1}$ for some $\tilde{x}$ between c and x.

$R_n(x)$ is called the *Lagrange form of the remainder*.

The practical significance of this theorem is twofold:

1. If you can determine a maximum value of $\left| f^{(n+1)}(\tilde{x}) \right|$ for values $\tilde{x}$ in the closed interval $[c, x]$ (or $[x, c]$), then you can find a bound on the error when you approximate $f(x)$ with the n-th degree Taylor polynomial at $x = c$. Just substitute that maximum for $f^{(n+1)}(c)$ in the formula for $R_n(x)$.

2. If you can show that $R_n(x)$ approaches 0 as n approaches infinity, then you can be sure that the Taylor series converges to $f(x)$.

This is indeed an amazing result.

Let's see this theorem at work. Suppose we use the third-degree Taylor polynomial for $f(x) = \arcsin(x)$ at $x = 0$ to approximate $f\left(\dfrac{1}{2}\right) = \arcsin\left(\dfrac{1}{2}\right)$ (which is $\dfrac{\pi}{6}$). To find the polynomial, we must calculate the function and its first 3 derivatives at $x = 0$. We get

$$f(x) = \arcsin(x) \qquad\qquad\qquad f(0) = 0$$

$$f'(x) = \frac{1}{\sqrt{1-x^2}} = \left(1-x^2\right)^{-\frac{1}{2}} \qquad\qquad f'(0) = 1$$

$$f''(x) = x\left(1-x^2\right)^{-\frac{3}{2}} \qquad\qquad\qquad f''(0) = 0$$

$$f'''(x) = 3x^2\left(1-x^2\right)^{-\frac{5}{2}} + \left(1-x^2\right)^{-\frac{3}{2}} \qquad f'''(0) = 1$$

Thus, $T_3(x) = x + \dfrac{x^3}{3!} = x + \dfrac{x^3}{6}$. Using this polynomial to approximate $\arcsin\left(\dfrac{1}{2}\right)$ gives

$\arcsin\left(\dfrac{1}{2}\right) \approx \dfrac{1}{2} + \dfrac{1}{48} \approx 0.520833$. This is a pretty good estimate for $\dfrac{\pi}{6} \approx 0.523599$.

Figure 9-4 shows the graphs of $y = \arcsin(x)$ and $y = T_3(x)$.

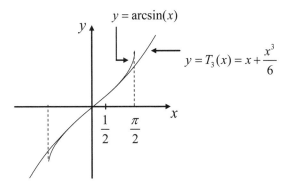

Figure 9-4. Approximation of $f(x) = \arcsin x$ with the third-degree Taylor polynomial

Taylor's Theorem guarantees the existence of a number $\tilde{x}$ between 0 (where our polynomial was created) and $\dfrac{1}{2}$ (where the approximation is made) such that the

Lagrange form of the remainder $R_3\left(\dfrac{1}{2}\right) = \dfrac{f^{(4)}(\tilde{x})}{4!}\left(\dfrac{1}{2}-0\right)^4 = \dfrac{f^{(4)}(\tilde{x})}{24 \cdot 16}$ is equal to the

difference $\arcsin\left(\dfrac{1}{2}\right) - \left(\dfrac{1}{2}+\dfrac{1}{48}\right)$. Usually we are not interested in the exact value of $\tilde{x}$;

we are more interested in finding an upper bound for $R_3\left(\dfrac{1}{2}\right)$. In many cases, such a

bound can be established analytically, as shown below. You could also use a calculator to find a bound for the remainder. Although this might not make sense from a practical

point of view (you could evaluate $\arcsin\left(\dfrac{1}{2}\right)$ itself if you had a calculator!), doing so

may help you see the significance of Taylor's Theorem.

Here $f^{(4)}(x) = \left(1-x^2\right)^{-\frac{5}{2}} 6x + 3x^2\left(1-x^2\right)^{-\frac{7}{2}} 5x + \left(1-x^2\right)^{-\frac{5}{2}} 3x =$

$9x\left(1-x^2\right)^{-\frac{5}{2}} + 15x^3\left(1-x^2\right)^{-\frac{7}{2}}$. For $0 \le x \le \dfrac{1}{2}$, $|x| \le \dfrac{1}{2}$ and $\left|\left(1-x^2\right)^{-1}\right| \le \left|\left(1-\dfrac{1}{4}\right)^{-1}\right| = \dfrac{4}{3}$, so

$\left|f^{(4)}(x)\right| = \left|9x\left(1-x^2\right)^{-\frac{5}{2}} + 15x^3\left(1-x^2\right)^{-\frac{7}{2}}\right| \le 9 \cdot \dfrac{1}{2} \cdot \left(\dfrac{4}{3}\right)^{\frac{5}{2}} + 15 \cdot \dfrac{1}{8} \cdot \left(\dfrac{4}{3}\right)^{\frac{7}{2}} =$

$\dfrac{16\sqrt{3}}{3} + \dfrac{80\sqrt{3}}{27} < 9\sqrt{3} < 18$. Therefore, $\left|R_3\left(\dfrac{1}{2}\right)\right| = \left|\dfrac{f^{(4)}(\tilde{x})}{24 \cdot 16}\right| < \dfrac{18}{24 \cdot 16} = \dfrac{3}{128} < 0.025$. This

estimate is really high: actually $R_3\left(\dfrac{1}{2}\right) = \dfrac{\pi}{6} - \left(\dfrac{1}{2}+\dfrac{1}{48}\right) \approx 0.003$. If you solve

$\dfrac{9\tilde{x}\left(1-\tilde{x}^2\right)^{-\frac{5}{2}} + 15\tilde{x}^3\left(1-\tilde{x}^2\right)^{-\frac{7}{2}}}{24 \cdot 16} = \arcsin\left(\dfrac{1}{2}\right) - \left(\dfrac{1}{2}+\dfrac{1}{48}\right)$ with a calculator, you will find

$\tilde{x} \approx 0.112$.

14

Use the Lagrange error bound to show that an approximation to $\sqrt{e}$ obtained by using the fourth-degree Taylor polynomial for e^x at $x = 0$ is within $.0005$ of the exact value of $\sqrt{e}$.

The fifth derivative of e^x is e^x, and its maximum value on the interval $\left(0, \dfrac{1}{2}\right)$ occurs at $x = \dfrac{1}{2}$ (since e^x is an increasing function). Thus the Lagrange error bound is

$$\frac{e^{\frac{1}{2}}}{5!}\left(\frac{1}{2}\right)^5 \approx .000429\,.$$

Note that, as written, this example isn't terribly practical: to get the error bound, we used the approximate value of $\sqrt{e}$. Actually, we do not need the exact value. Since $e < 3$,

$$\sqrt{e} < 1.8 \text{ and } \frac{e^{\frac{1}{2}}}{5!}\left(\frac{1}{2}\right)^5 < \frac{1.8}{120\cdot 32} < \frac{1.8}{120\cdot 30} = .0005\,.$$

For the curious, $\sqrt{e} = 1.64872127...$ while $1 + \dfrac{1}{2} + \dfrac{\left(\dfrac{1}{2}\right)^2}{2!} + \dfrac{\left(\dfrac{1}{2}\right)^3}{3!} + \dfrac{\left(\dfrac{1}{2}\right)^4}{4!} = 1.6484375$. The actual error is $.00028377...$ Taylor's Theorem doesn't tell us the <u>exact</u> error. Rather, it gives us a <u>bound</u> for the error. The actual error cannot exceed the error bound.

Error Bound for Alternating Taylor Series

As we know from Section 9.3.6, for a converging alternating series with decreasing terms the error for the n-th partial sum approximation does not exceed the absolute value of the first omitted term in the series. Thus, for a converging alternating Taylor series with decreasing terms, the n-th Taylor polynomial approximation is bounded by the first omitted term in the series.

15

If the first 3 non-zero terms of the series for $\arctan(x)$ at $x = 0$ are used to approximate $\arctan(1)$, what is the alternating series error bound for this approximation?

From the series for $\arctan(x)$ derived earlier, $\arctan(1) = 1 - \dfrac{1}{3} + \dfrac{1}{5} - \dfrac{1}{7} + \dots$

The alternating series error bound from using the first three terms $1 - \dfrac{1}{3} + \dfrac{1}{5}$ is just the first omitted term, $\dfrac{1}{7}$.

Since the value of $\arctan(1)$ is $\dfrac{\pi}{4}$, this series can be used to approximate π.

What a powerful and deep topic infinite series are! You have seen that the sum of an infinite number of terms can converge to a real number (often a transcendental number). Here are some examples:

$$1 = \frac{1}{2} + \frac{1}{4} + \frac{1}{8} + \dots$$

$$e = 1 + 1 + \frac{1}{2!} + \frac{1}{3!} + \dots$$

$$\frac{\pi}{4} = \arctan(1) = 1 - \frac{1}{3} + \frac{1}{5} - \frac{1}{7} + \dots$$

In other cases, we can determine that an infinite series converges, and though we don't know what it converges <u>to</u>, we can figure out how close we are to the sum when we stop at some particular term, by using the Lagrange Error Bound or the Alternating Series error bound.

You can just about guarantee that there will be a series question on the free-response part of the BC exam. Here are a couple more examples.

16

A function f has derivatives of all orders for all real numbers. It is known that $f(1) = 5$, $f'(1) = -3$, $f''(1) = 2$, and $f'''(1) = -1$. Which of the following statements are necessarily true?

   I. $f(2)$ can be exactly determined from the given information.
   II. The third-degree Taylor polynomial for f at $x = 1$ can be determined from the given information.
   III. The Taylor series for f at $x = 1$ converges to $f(x)$ for all real numbers x.

   (A) I only   (B) II only
   (C) III only   (D) I and II only
   (E) II and III only

I is false, because the only information given concerns the value of f and its derivatives at $x = 1$. No information can be determined about the value of $f(2)$ (except that it exists).

II is true. We have all we need to write the third-degree Taylor polynomial for f at $x = 1$.
III is not necessarily true, since we don't know anything about the remainder term. The answer is B.

17

The Maclaurin polynomial for $\cos x$ with four non-zero terms is used to approximate $\cos(1)$. Show that the approximation is within .001 of $\cos(1)$.

$\cos x = 1 - \dfrac{x^2}{2!} + \dfrac{x^4}{4!} - \dfrac{x^6}{6!} + \dots$. This is an alternating series with decreasing terms, (i.e., a series that converges by the Alternating Series Test). So the error from using the first n terms is less than the first omitted term (according to the alternating series error bound).

$\cos(1) \approx 1 - \dfrac{1^2}{2!} + \dfrac{1^4}{4!} - \dfrac{1^6}{6!}$ with error less than $\dfrac{1}{8!} = \dfrac{1}{40320} < 0.001$.

Chapter 10. Annotated Solutions to Past Free-Response Questions

The material for this chapter is on our web site:

http://www.skylit.com/calculus/fr.html

That page includes links to free-response questions from recent years and an annotated solution for each question, including Form B exams.

Appendix: Calculator Skills

You should practice several calculator skills prior to the AP Exam. A few examples follow, with calculator-assisted solutions for the TI-83, TI-84, TI-86, and TI-89 models. There are other acceptable calculator methods to solve these problems. If your calculator model does not match one of the models presented, consult your user's manual to solve the examples.

A.1.   Graphing a Function

This is the simplest calculator skill required on the exam. Usually, the hardest part is making sure you enter the function correctly on your calculator, and that you choose a suitable viewing window. Be sure to check that the parentheses that enclose function arguments (as in `sin(X)`) are properly matched.

> **Be sure that your calculator is set to the `Radian` mode when you take the exam. (To set the mode, go to the `MODE` menu.)**

You will see several graphing examples in the following sections.

A.2.   Solving an Equation

Graphing calculators offer several methods for solving an equation. The easiest approach is to enter the functions for each side of the equation, graph the two functions, and use the `Intersect` command from the `GRAPH` screen. However, this method may not give the required accuracy. If you are looking for a root or zero of a function, graph the function and use the `Root` or `Zero` command to find the x-intercept of the graph. The TI-83 and TI-84 come with an `EQUATION SOLVER`, and the TI-86 has a `SOLVER` environment where you can find zeroes using arbitrary names of variables (while in the graphing environment, the independent variable is usually named `X` or `T`).

> **Before using the `EQUATION SOLVER` environment, you should graph the equation in order to get an idea how many roots there are, and where those roots are located. The one you find with the `EQUATION SOLVER` might not be the one you need. Using a "seed" value for the variable that is close to the root you want usually works.**

Example 1

The derivative of a function f is given by $f'(x) = \sin\left(\sqrt{x^2 + 1}\right) - \cos\left(\sqrt{x}\right)$. Find all the values of x in the open interval $(0, 6)$ where f has a local minimum.

TI-83/TI-84 Solution

We need to determine where the derivative changes sign from negative to positive. It is a good idea to graph the function first, especially when you are given a domain of values for the independent variable. Then use the `Zero` command from the `CALC` menu.

Here are the necessary steps. First press `WINDOW` to set your viewing window:

```
WINDOW
 Xmin=0
 Xmax=6
 Xscl=1
 Ymin=-1
 Ymax=1
 Yscl=1
 Xres=1■
```

Note that the choices of `Xmin` and `Xmax` correspond to the interval given in the problem. Press `Y=` and set Y_1 to $f'(x)$, taking care that your parentheses are properly matched:

```
Plot1  Plot2  Plot3
\Y1■sin(√(X²+1))
-cos(√(X))
\Y2=
\Y3=
\Y4=
\Y5=
\Y6=
```

Then press `GRAPH` to plot the graph:

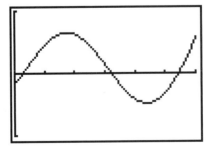

You can see two points where the derivative changes sign from negative to positive. Go to the CALC menu (2nd+TRACE), and choose 2:zero:

You will be prompted to enter a left bound for the zero. You can either use the arrow keys to move the cursor to the left of the leftmost zero, or simply enter a number for *x* using the keypad. You have to be sure the number is to the left of the zero you are looking for (but still in the viewing window). Type in 0 for *x*:

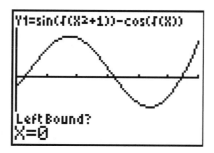

Then either move the cursor over to the right of the zero, or enter a value of *x* that you know is to the right of the zero (but not past the next zero). Suppose you use the cursor (and press ENTER when you are satisfied with its position):

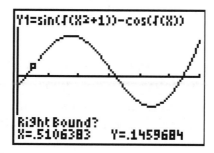

Your calculator then prompts you: Guess? If you've chosen the left and right bounds so that only one root is between them, then just press ENTER. The zero will then be calculated, and stored into the variable X.

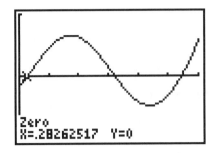

Repeat the process to find the other zero where the derivative changes sign from negative to positive:

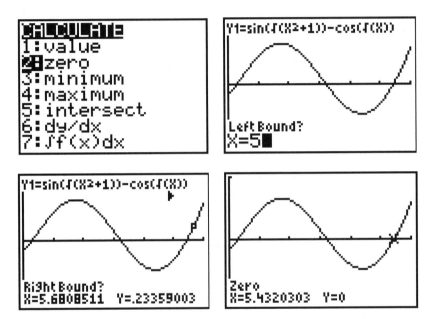

The answers (accurate to three digits to the right of the decimal) are $x \approx .283$ and $x \approx 5.432$.

> **For the exam, you must use the MATH/zero command from the CALC menu on the Graph screen, (or the EQUATION SOLVER environment from the HOME screen discussed below) to find a root or an intersection point. Just tracing along a graph to find roots or intersection points may not give you the required accuracy.**

TI-86 or TI-89 Solution

The same basic procedure works for the TI-86 or TI-89. However, to find a zero of a function on the TI-86, use the Root command on the MATH menu on the GRAPH screen. On the TI-86, the MATH menu is on the second page of commands on the GRAPH screen. On the TI-89, press the F5 key and use the Zero command:

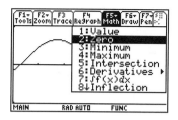

The following example illustrates the use of the EQUATION SOLVER environment on the TI-83 or TI-84.

Example 2

The number of bees in a colony is given by $B(t) = 523e^{.18t}$, where t is the number of days since the colony was established. The derivative of $B(t)$ is given by $B'(t) = 94.14e^{.18t}$. On what day is the number of bees in the colony increasing at the rate of 1000 bees per day?

TI-83 or TI-84 Solution

Press MATH, then 0 to bring up the EQUATION SOLVER. Press the arrow-up key to get into the equation editing mode (identified by the eqn:0= line), if you are not already there.

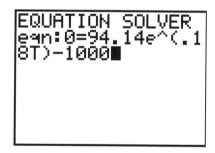

Note that the equation we've entered is equivalent to $0 = B'(t) - 1000$. Solving the equation answers the question posed in the example. Press ENTER to return to the main solver screen. Use the arrow keys to position the cursor on the line that starts with T=, then press ALPHA, then ENTER to solve for T.

> **Be patient: this operation may take a few seconds.**

```
94.14e^(.18T)...=0
 T=13.129984624...
bound={-1E99,1...
```

$T \approx 13.30$; the answer is on the 14th day.

Example 3

The derivative of a function g is given by $g'(x) = \dfrac{x}{2} - \cos\left(x^2\right) + 0.3$. What is the x coordinate of a local maximum point on the graph of g?

TI-83 or TI-84 Solution

Enter $g'(x)$ into Y_1, and graph it in the decimal window.

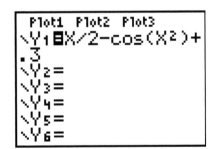   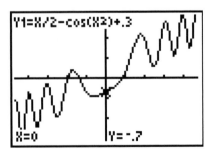

We need to find a zero where $g'(x)$ changes sign from positive to negative. From the graph, this zero appears to be near $x = -1$. So, on the `Solver` menu, enter the equation as shown, then enter a "seed" value of -1 for X.

```
EQUATION SOLVER        Y₁=0
eqn:0=Y₁                 X=-1
                        bound={-1E99,1…
```

Press ALPHA, then ENTER to solve for X.

```
Y₁=0
∎X=-1.409857087…
 bound={-1E99,1…
∎left-rt=0
```

The local maximum occurs at the point where $x = -1.410$.

A.3.   Evaluating a Derivative at a Point

Example 4

Find the slope of the line tangent to the graph of $y = 2x + \sin(1 + x^2) + \cos(1 - x^2)$ at the point where the graph crosses the x-axis.

Solution

First enter the function into Y_1.   Then use the procedure from the previous section to find the zero of a function.

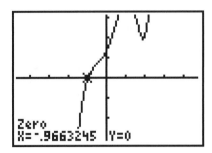

The zero is at $x = -.9663245$. Immediately after finding that zero, return to the HOME screen and use the STO➔ command to store the calculator's current value of the x-coordinate from the graph, which is X on the TI-83 (or x on the TI-86, or xc on the TI-89) into a variable, like A on the TI-83 or TI-86 (or a on the TI-89):

Then calculate the numerical derivative of Y_1 at A. The numerical derivative command is number 8 on the MATH menu:

On the TI-83, the variable Y_1 must be pulled off of the Y-VARS submenu on the VARS menu, Choice 1. On the TI-86 or TI-89, you can just type y1 from the keypad. On the TI-89, you can use the derivative command located at 2nd 8 on the keypad, in conjunction with the | command to evaluate the derivative:

A.4.   Evaluating an Integral Numerically

> **On the open calculator free-response part of the AP exam (Section II, Part A), always use your calculator when you need to evaluate a definite integral. You don't get "extra credit" for evaluating an integral by first finding an antiderivative.**

Example 5

Find the area of the region in the first quadrant bounded by the graphs of $f(x) = \sin(x^2) + \cos(x^2) + x^2$, $g(x) = e^x - 1$, and the y-axis.

Solution

First, take a look at the graphs:

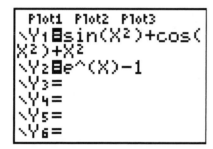

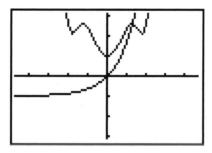

To find the area, we must first find where the curves intersect.

On the TI-83, press CALC (2nd+TRACE), then choose 5 to use the intersect command. Press ENTER three times to select the first curve, second curve, and "guess" for the intersection point:

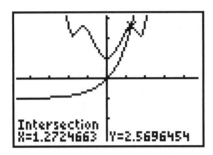

(You can use the up and down arrow keys to select from among several graphed functions the two you want, and the left and right arrow keys to change the "guess" to a value that will result in the intersection you want to find.)

On the TI-86, ISECT is on the second page of the MATH menu. The screen below shows where to find `Intersection` on the TI-89.

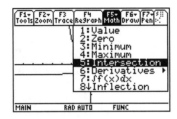

On the TI-89, you need to choose lower and upper bounds for the *x*-coordinate of the point of intersection.

After you've found the intersection, go to the HOME screen. Then immediately store the current value of X into the variable A (on the TI-86 use x, and on the TI-89 use xc) by

pressing X STO➔ A.

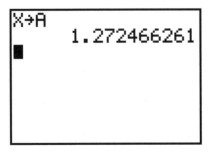

Now you're ready to evaluate an integral to get the area. Area = $\int_0^A Y_1(X) - Y_2(X)\, dx$.

Enter the integral as follows:

On the TI-83, `fnInt` is option 9 in the `MATH` menu, and you must use the `Y-VARS` submenu on the `VARS` menu, Choices 1 and 2, to enter Y_1 and Y_2. On the TI-86, `fnInt` is on the `CALC` menu.

On the TI-89, it is a good idea to clear out all your single letter variables before starting a problem like this. You do that by pressing `2nd F1 ENTER Clear a-z`:

You can use the integral key found on the keypad.

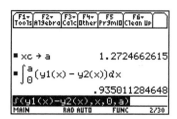

On the TI-89, you should press ◆ `ENTER` after typing the integral to force the calculator to give you a numeric (approximate) result.

Practice Exam AB-1

Calculus AB
Section I, Part A
Time — 55 minutes
Number of questions — 28

No calculator is allowed for these questions.

---

1. If $f(x) = e^{2x} \tan^{-1}(x)$, then $f'(1) =$

   (A) $\dfrac{e^2}{2}$ (B) $\dfrac{e^2 \pi}{4}$ (C) e^2 (D) $\dfrac{e^2 \pi}{2}$ (E) $\dfrac{e^2(\pi+1)}{2}$

---

2. $\displaystyle\int_1^8 x^{-\frac{2}{3}}\, dx =$

   (A) $-\dfrac{31}{48}$ (B) $-\dfrac{1}{4}$ (C) $\dfrac{1}{3}$ (D) 1 (E) 3

---

3. If $f(x) = e^{-x} + \sin x - \cos x$, then $f''(0) =$

   (A) -2 (B) -1 (C) 0 (D) 1 (E) 2

---

4. The slope of the line tangent to the curve $3x^2 - 2xy + y^2 = 11$ at the point $(1, -2)$ is

   (A) $-\dfrac{1}{6}$ (B) 0 (C) 1 (D) $\dfrac{5}{3}$ (E) 10

---

5. $\dfrac{d}{dx}\left[\ln(\sec x)\right] =$

(A) $\cos x$ (B) $\tan x$ (C) $\cos x \cot x$

(D) $\dfrac{\sec x \tan x}{x}$ (E) $\dfrac{\sec x}{x} + \ln\left(\sec x \tan x\right)$

---

6. $\displaystyle\int_{\pi/6}^{\pi/4} 2\sin(2x)\cos(2x)\,dx =$

(A) $-\dfrac{3}{8}$ (B) $\dfrac{1}{8}$ (C) $\dfrac{5\pi^2}{288}$ (D) $\dfrac{1}{4}$ (E) $\dfrac{3}{8}$

---

7. The area of the region bounded by the graphs of $y = x^2$ and $y = \sqrt{x}$ is

(A) $\dfrac{1}{4}$ (B) $\dfrac{1}{3}$ (C) $\dfrac{1}{2}$ (D) $\dfrac{2}{3}$ (E) 1

---

8. If $f(x) = 2x$, $g(x) = x^2$, and $h(x) = 2^x$, which of the following limits is equal to 0?

(A) $\displaystyle\lim_{x\to\infty} \dfrac{g(x)}{f(x)}$ (B) $\displaystyle\lim_{x\to\infty} \dfrac{g(x)}{h(x)}$ (C) $\displaystyle\lim_{x\to\infty} \dfrac{h(x)}{f(x)}$

(D) $\displaystyle\lim_{x\to\infty} \dfrac{h(x)}{g(x)}$ (E) $\displaystyle\lim_{x\to\infty} \dfrac{h(x)}{f(x)g(x)}$

---

9. $\displaystyle\lim_{x\to 5} \dfrac{2x^2 - 50}{x^2 - 15x + 50} =$

(A) -4 (B) -1 (C) 0 (D) 1 (E) 2

10. If $f''(0) < 0$, $f'(0) = 0$, and $f(0) > 0$, which of the following could be the graph of $y = f(x)$?

(A)   (B)   (C)   (D)   (E)

11. If $\int_0^3 e^{\sin x}\, dx = k$, then $\int_1^2 x\, e^{\sin(4-x^2)}\, dx =$

(A) $-\dfrac{k}{2}$ (B) $-\dfrac{k}{3}$ (C) $\dfrac{k}{6}$ (D) $\dfrac{k}{3}$ (E) $\dfrac{k}{2}$

12. An object moves in a straight line with acceleration $a(t) = t^2$ ft/sec^2. If the initial velocity of the object is -72 ft/sec, at what time t does the object change direction?

(A) $t = \sqrt[3]{6}$ (B) $t = 6$ (C) $t = 6\sqrt{2}$

(D) $t = 36$ (E) Never

13. What is the x-coordinate of the point of inflection of the graph of $y = x^3 + 3x^2 - 45x + 81$?

(A) -9 (B) -5 (C) -1 (D) 1 (E) 3

14. Which of the following functions satisfies $0 < f''(x) < f'(x) < f(x)$ for all x?

(A) $f(x) = e^{-x}$ (B) $f(x) = e^{x/2}$ (C) $f(x) = e^x$

(D) $f(x) = e^{2x}$ (E) $f(x) = e^{x^2}$

15. Given that $f(-3) = 4$ and $f'(-3) = 2$, which of the following is the tangent line approximation of $f(-3.1)$?

(A) 3.8   (B) 3.9   (C) 4.0   (D) 4.1   (E) 4.2

16. If $f(x) = \sqrt{x-1}$, then the average value of f over the interval $1 \le x \le 5$ is

(A) $\dfrac{1}{4}$ (B) $\dfrac{1}{2}$ (C) $\dfrac{4}{3}$ (D) $\dfrac{8}{3}$ (E) $\dfrac{16}{3}$

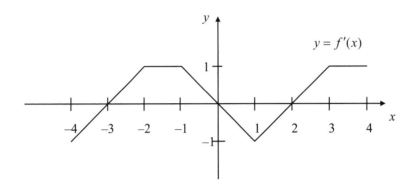

Questions 17-19 refer to the graph of $y = f'(x)$, the <u>derivative</u> of f, shown above. The graph consists of five line segments, two of which are horizontal.

17. At $x = 1$, f has a

(A) point of discontinuity
(B) point of inflection
(C) point of nondifferentiability
(D) local maximum
(E) local minimum

18. Over the interval $-4 < x < 4$, how many local maxima does f have?

(A) One   (B) Two   (C) Three   (D) Four   (E) Infinitely many

19. If $f(2) = 1$, what is the value of $f(-2)$?

(A) $-\dfrac{3}{2}$ (B) $-\dfrac{1}{2}$ (C) $\dfrac{1}{2}$ (D) $\dfrac{3}{2}$ (E) $\dfrac{5}{2}$

20. The midpoint-sum approximation for $\int_{-4}^{2} x^2\, dx$ using three subintervals of equal length is

(A) 11   (B) 14   (C) 22   (D) 24   (E) 28

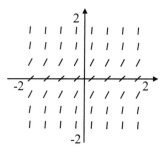

21. Which of the following differential equations could be represented by this slope field?

(A) $\dfrac{dy}{dx} = x^3 + 1$ (B) $\dfrac{dy}{dx} = \tan(x)$ (C) $\dfrac{dy}{dx} = \dfrac{1}{1 + x^2}$

(D) $\dfrac{dy}{dx} = 1 + y^2$ (E) $\dfrac{dy}{dx} = \tan^{-1}(y)$

22. A particle moves in a straight line with velocity $v(t) = 4 - t^2$ feet per second. What is the total distance the particle travels between time $t = 0$ and $t = 3$ seconds?

(A) $\dfrac{7}{3}$ ft   (B) 3 ft   (C) 6 ft   (D) $\dfrac{23}{3}$ ft   (E) $\dfrac{25}{3}$ ft

23. If the region bounded between $y = x^2$ and the horizontal line $y = 1$ is rotated about the x-axis, the volume of the resulting solid of revolution is

(A) $\dfrac{2\pi}{3}$ (B) $\dfrac{4\pi}{5}$ (C) $\dfrac{16\pi}{15}$ (D) $\dfrac{4\pi}{3}$ (E) $\dfrac{8\pi}{5}$

24. If a function $y = f(x)$ satisfies the differential equation $\dfrac{dy}{dx} = -4y$ and $f(0) = 6$, then $f(x) =$

(A) $-2x^2 + 6$ (B) $-\dfrac{x}{4} + 6$ (C) $6e^{-4x}$

(D) $e^{-4x} + 5$ (E) $-\dfrac{1}{4}\ln\left(x + e^{-24}\right)$

---

25. If $F(x) = \displaystyle\int_0^{x^2} \cos(t^3)\,dt$, then $F'(x) =$

(A) $\cos(x^3)$ (B) $\cos(x^6)$ (C) $2x\cos(x^3)$

(D) $2x\cos(x^6)$ (E) $x^2\cos(x^3)$

---

26. $\displaystyle\int_1^e \dfrac{\ln(x)}{x}\,dx =$

(A) $\dfrac{1}{2e^2}$ (B) $\dfrac{1}{e}$ (C) $\dfrac{1}{2}$ (D) 1 (E) e

---

27. If the sum of two positive numbers x and y is 3, then the absolute maximum value of $x^3 + 12xy$ is

(A) 0 (B) 25 (C) 27 (D) 32 (E) 135

---

28. The side of a cube is expanding at a constant rate of 2 centimeters per second. What is the instantaneous rate of change of the surface area of the cube, in cm^2 per second, when its volume is 27 cubic centimeters?

(A) 6 (B) 24 (C) 36 (D) 54 (E) 72

Calculus AB
Section I, Part B
Time — 50 minutes
Number of questions — 17

A graphing calculator is required for some questions.

29. If $f'(x) = \tan^{-1}(x^3 - x)$, at how many points is the tangent line to the graph of $y = f(x)$ parallel to the line $y = 2x$?

(A) None   (B) One   (C) Two   (D) Three   (E) Infinitely many

30. Which of the following functions are continuous but not differentiable at $x = 0$?

I. $f(x) = x^{1/3}$
II. $g(x) = |x|$
III. $h(x) = x \cdot |x|$

(A) I only   (B) II only   (C) I and II
(D) II and III   (E) I, II, and III

31. An object moves along the y-axis with coordinate position $y(t)$ and velocity $v(t) = \sqrt{t} - \cos(e^t)$ for $t \geq 0$. At time $t = 1$, the object is

(A) moving downward with negative acceleration
(B) moving upward with negative acceleration
(C) moving downward with positive acceleration
(D) moving upward with positive acceleration
(E) at rest

32. If $f'(x) = \sin(2^x)$ for all x, then the smallest value of x at which f has a relative minimum is

(A) 0   (B) 0.651   (C) 1.652   (D) 2.236   (E) 2.651

33. The average value of $f(x) = x^3$ over the interval $a \leq x \leq b$ is

(A) $3b + 3a$

(B) $b^2 + ab + a^2$

(C) $\dfrac{b^3 + a^3}{2}$

(D) $\dfrac{b^3 - a^3}{2}$

(E) $\dfrac{(b^4 - a^4)}{4(b-a)}$

34. Given that $F'(x) = f(x)$, the value of $\displaystyle\int_a^b x f(x^2)\,dx$ is

(A) $\dfrac{F(b^2) - F(a^2)}{2}$

(B) $bF(b^2) - aF(a^2)$

(C) $2F(\sqrt{b}) - 2F(\sqrt{a})$

(D) $\dfrac{b^2 F(b^2) - a^2 F(a^2)}{2}$

(E) $2F(b^2) - 2F(a^2)$

35. Which one of the following integral expressions represents the area of the region bounded by the graphs of $y = 2 - x$, $y = \ln(x)$, and the x-axis?

(A) $\displaystyle\int_0^{1.557} (2 - x - \ln(x))\,dx$

(B) $\displaystyle\int_1^{1.557} \ln(x)\,dx + \int_{1.557}^2 (2 - x)\,dx$

(C) $\displaystyle\int_{1.557}^2 (\ln(x) - 2 + x)\,dx$

(D) $\displaystyle\int_0^2 (2 - x)\,dx - \int_1^{1.557} \ln(x)\,dx$

(E) $\displaystyle\int_0^1 (2 - x)\,dx + \int_1^{1.557} (2 - x - \ln(x))\,dx$

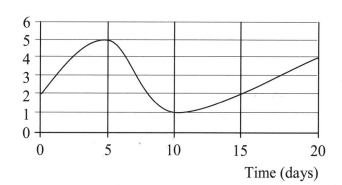

Questions 36-38 refer to the graph above that shows the rate of bamboo growth $g(t)$ in centimeters (cm) per day over a 20-day period.

36. What is the average rate of change of $g(t)$ over the interval $0 \le t \le 5$ days?

(A) $\dfrac{3}{5}$ cm per day per day

(B) 1 cm per day per day

(C) $\dfrac{7}{5}$ cm per day per day

(D) 3 cm per day per day

(E) $\dfrac{7}{2}$ cm per day per day

37. If the bamboo is 60 cm tall at time $t = 10$ days, approximately how tall is it at $t = 20$ days?

(A) 25 cm   (B) 64 cm   (C) 70 cm   (D) 82 cm   (E) 100 cm

38. At which of the following times t is $g'(t)$ at its absolute minimum over the time interval $0 \le t \le 20$?

(A) $t = 0$ (B) $t = 5$ (C) $t = 7$ (D) $t = 10$ (E) $t = 20$

39. A function f is continuous for $0 \leq x \leq 5$ and differentiable for $0 < x < 5$. Given that $f(0) = -2$ and $f(5) = 3$, which of the following statements must be true?

   I. $f'(c) = 1$ for some c such that $0 < c < 5$
   II. $f(c) = 0$ for some c such that $0 < c < 5$
   III. $f(c) = -1$ for some c such that $0 < c < 5$

(A) I only   (B) II only   (C) I and II

(D) II and III   (E) I, II, and III

40. The volume of a sphere of radius r is $\frac{4}{3}\pi r^3$. The volume of a gas-filled spherical balloon increases 6 cubic inches for each degree (Celsius) increase in temperature. If the temperature increases at a constant rate of 2 degrees per minute, then at what rate is the radius of the balloon changing at the instant when the volume is 36π cubic inches?

(A) $\dfrac{1}{\pi}$ inches per minute

(B) $\dfrac{1}{2\pi}$ inches per minute

(C) $\dfrac{1}{3\pi}$ inches per minute

(D) $\dfrac{1}{6\pi}$ inches per minute

(E) $\dfrac{2}{3\pi}$ inches per minute

41. If the region bounded between $y = \dfrac{1}{x}$ and the x-axis between the vertical lines $x = 1$ and $x = e$ is rotated about the line $y = -2$, the volume of the resulting solid of revolution is represented by

(A) $\pi \int_1^e \left(\dfrac{1}{x} + 2\right)^2 dx$ (B) $\pi \int_1^e \left(\dfrac{1}{x^2} + 2\right) dx$ (C) $\pi \int_1^e \left(\dfrac{1}{x} + 2\right) dx$

(D) $\pi \int_1^e \left[\left(\dfrac{1}{x} + 2\right)^2 - 4\right] dx$ (E) $\pi \int_1^e \left[\left(\dfrac{1}{x} + 2\right)^2 + 4\right] dx$

42. The function g is given by the formula $g(x) = \int_0^{x/2} e^{-t^2} dt$. An equation for the tangent line to the graph of g at the point $x = 1$ is

(A) $y - 0.461 = 0.184(x-1)$

(B) $y - 0.461 = 0.368(x-1)$

(C) $y - 0.461 = 0.389(x-1)$

(D) $y - 0.461 = 0.779(x-1)$

(E) $y - 0.461 = 1.558(x-1)$

---

43. The area of the region in the first quadrant bounded between the graphs of $y = \cos(x^3)$, $y = \sin(x^3)$, and the y-axis is

(A) 0.270   (B) 0.692   (C) 0.709   (D) 0.923   (E) 0.983

---

44. A solid has as its base the region bounded by $y = \sqrt{x}$, the x-axis, and the vertical line $x = 4$. Each cross-section of the solid perpendicular to the y-axis is a square. Which one of the following expressions represents the volume of the solid?

(A) $\int_0^4 \sqrt{x}\, dx$ (B) $\int_0^4 x\, dx$ (C) $\int_0^2 (4 - y^2)\, dy$

(D) $\int_0^2 (4 - y)^2\, dy$ (E) $\int_0^2 (4 - y^2)^2\, dy$

---

45. Fluid leaks out of a bottle at the rate of $\ln(1+t) + \cos(t + \sqrt{t})$ milliliters per minute, where t is measured in minutes. If the bottle contains 375 milliliters of fluid at time $t = 0$ minutes, how many milliliters of fluid are in the bottle at time $t = 60$ minutes?

(A) 159.337 (B) 185.427 (C) 187.500 (D) 189.573 (E) 215.663

Practice Exam AB-1

Calculus AB
Section II, Part A
Time — 45 minutes
Number of problems — 3

A graphing calculator is required for some problems or parts of problems.

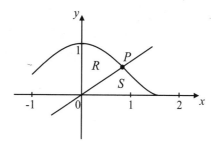

1. Consider the graphs of $y = \dfrac{2}{3}x$ and $y = \sqrt{\cos^3(x)}$ shown in the picture above. The two curves intersect at point P. Region R is bounded by the two curves and the y-axis. Region S is bounded by the two curves and the x-axis.

   (a) Find the slope of the tangent line to $y = \sqrt{\cos^3(x)}$ at point P.

   (b) Find the area of R.

   (c) Find the area of S.

   (d) Find the volume of the solid obtained when region R is rotated about the x-axis.

2. A tortoise runs along a straight track, starting at position $x = 0$ at time $t = 0$. The tortoise has a velocity of $v(t) = \ln(1 + t^2)$ inches per minute, where t is measured in minutes.

    (a) Find the average acceleration of the tortoise over the time interval $0 \le t \le 10$ minutes. Find the instantaneous acceleration of the tortoise at time $t = 10$ minutes. Indicate the correct units.

    (b) How far does the tortoise travel between time $t = 0$ and time $t = 10$ minutes?

    (c) At what time t, $0 < t < 10$, is the tortoise's instantaneous velocity equal to its average velocity over the time interval $0 \le t \le 10$?

    (d) A hare begins running from the same starting position at time $t = 9.5$ minutes with an initial velocity b inches per minute and a constant acceleration of 2 inches per minute per minute; the hare catches up with the tortoise at time $t = 10$ minutes. Find b.

---

3. Methane is produced in a cave at the rate of $r(t) = e^{\sin\left(\frac{\pi}{4}t\right)}$ liters per hour at time t hours. The initial amount of methane in the cave at time $t = 0$ is 20 liters. At time $t = 8$ hours, a pump begins to remove the methane at a constant rate of 1.5 liters per hour.

    (a) At what time t during the time interval $0 \le t \le 8$ hours is the amount of methane increasing most rapidly?

    (b) What is the total amount of methane in the cave at time $t = 8$ hours?

    (c) What is the average rate of methane accumulation in the cave over the time interval $0 \le t \le 24$ hours?

    (d) What is the absolute maximum amount of methane in the cave over the time interval $0 \le t \le 24$ hours?

Calculus AB
Section II, Part B
Time — 45 minutes
Number of problems — 3

No calculator is allowed for these problems.

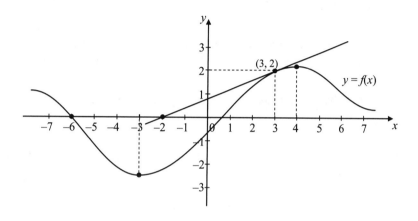

4.   The above picture shows the graph of a differentiable function $y = f(x)$ and the tangent line to that graph at the point (3, 2). $f(x)$ has horizontal tangents at $x = -3$ and $x = 4$. The function $g(x)$ is defined as $g(x) = \int_0^x f(t)\,dt$. Suppose we know that $g(3) = \dfrac{11}{5}$.

(a)   Find $g'(3)$ and $g''(3)$.

(b)   For what values of x in the open interval $-7 < x < 7$ does g have a relative maximum? Justify your answers.

(c)   For what values of x in the open interval $-7 < x < 7$ does g have a point of inflection? Justify your answers.

(d)   Let $C(x)$ be the average value of $f(t)$ over the interval $[0, x]$. Find $C(3)$ and $C'(3)$.

x	-3	-2	-1	0	1	2	3
$f(x)$	5	3	2	-1	1	-4	-7

5.   The table above shows selected values for a twice-differentiable function f.

(a)   Find the midpoint Riemann sum approximation for $\int_{-3}^{3} f(x)\,dx$ using 3 subintervals of equal length.

(b)   Show that the average rate of change of f over the interval $-3 \le x \le 0$ is the same as the average rate of change of f over the interval $0 \le x \le 3$.

(c)   Explain why there must be at least one value c such that $0 < c < 2$ and $f'(c) = 0$.

(d)   Must there be at least one value d such that $-3 < d < 3$ and $f''(d) = 0$? Explain why or why not.

---

6.   A function $y = f(x)$ satisfies the differential equation $\dfrac{dy}{dx} = \dfrac{x^2 - 1}{y}$ with initial condition $f(2) = 1$.

(a)   Sketch the slope field for the given differential equation at the 12 points indicated below.

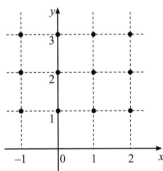

(b)   Find $f''(2)$.

(c)   Use separation of variables to find an expression for $f(x)$.

Practice Exam AB-2

Calculus AB
Section I, Part A
Time — 55 minutes
Number of questions — 28

No calculator is allowed for these questions.

---

1. If $f(x) = x^2 \ln x$, then $f'(x) =$

    (A) 2 (B) $x + 2 \ln x$ (C) $2x \ln x$

    (D) $1 + 2x \ln x$ (E) $x + 2x \ln x$

---

2. $\displaystyle\int_3^{11} \frac{1}{\sqrt{2x + 3}}\, dx =$

    (A) $\ln \dfrac{5}{3}$ (B) 2 (C) 4 (D) $\dfrac{98}{3}$ (E) $\dfrac{196}{3}$

---

3. If $\lim\limits_{x \to c} f(x) = f(c)$ for all values of c, $0 \le c \le 5$, and $f(0) \ne f(5)$, which of the following could be false?

    (A) $f(4)$ exists   (B) $f'(1)$ exists   (C) $\lim\limits_{x \to 2^+} f(x)$ exists

    (D) $\lim\limits_{x \to 3^-} f(x) = \lim\limits_{x \to 3^+} f(x)$ (E) $\lim\limits_{x \to 0^+} f(x) \ne \lim\limits_{x \to 5^-} f(x)$

---

4. $\lim\limits_{x \to 0^+} \dfrac{12}{4 + e^{\csc x}}$ is

    (A) $\dfrac{-12}{5}$ (B) 0 (C) $\dfrac{12}{5}$ (D) 3 (E) nonexistent

---

5.   Which of the following differential equations corresponds to the slope field shown in the figure above?

(A) $\dfrac{dy}{dx} = \dfrac{xy}{2}$ (B) $\dfrac{dy}{dx} = \dfrac{y}{x}$ C) $\dfrac{dy}{dx} = -\dfrac{y}{x}$

(D) $\dfrac{dy}{dx} = \dfrac{x}{y}$ (E) $\dfrac{dy}{dx} = -\dfrac{x}{y}$

6. $\dfrac{d}{dx}\left(2^{\cos x}\right) =$

(A) $-(\sin x)2^{\cos x}$ (B) $(\ln 2)2^{\cos x}$ (C) $-(\ln 2)(\sin x)2^{\cos x}$

(D) $(\sin x)2^{\cos x}$ (E) $(\ln 2)(\cos x)2^{\cos x - 1}$

7. $\displaystyle\int \dfrac{1}{\sqrt{e^x}}\,dx =$

(A) $\sqrt{e^x} + C$ (B) $2\sqrt{e^x} + C$ (C) $-2\sqrt{e^x} + C$

(D) $\dfrac{1}{\sqrt{e^x}} + C$ (E) $-\dfrac{2}{\sqrt{e^x}} + C$

8.   If $f'(x) = x^3(x+2)^2$, then the graph of f has inflection points when $x =$

(A) -2 only   (B) 0 only   (C) -2 and 0

(D) -2 and $-\dfrac{6}{5}$ (E) $-2, -\dfrac{6}{5}$, and 0

9. If $f(x) = \int_{2}^{\sin x} \sqrt{1+t^2}\, dt$, then $f'(x) =$

(A) $(1+x^2)^{\frac{3}{2}}$

(B) $\cos x(1+\sin x)^{\frac{1}{2}}$

(C) $(1+\sin^2 x)^{\frac{1}{2}}$

(D) $\cos x(1+\sin^2 x)^{\frac{1}{2}}$

(E) $\cos x(1+\sin^2 x)^{\frac{3}{2}}$

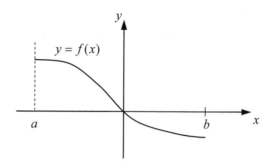

10. Let f be a continuous function as shown in the figure above. The area of the region bounded by $f(x)$, the x-axis, and $x = a$ is 5. If $\int_{a}^{b} f(x)\, dx = 3$, then $\int_{0}^{b} f(x)\, dx$ is

(A) –3

(B) –2

(C) 2

(D) 8

(E) cannot be determined

11. $\lim\limits_{x \to -\infty} \dfrac{2x+3}{\sqrt{x^2+x+1}}$ is

(A) –2

(B) –1

(C) 0

(D) 2

(E) nonexistent

12. The solution to the differential equation $\dfrac{dy}{dx} = \dfrac{x}{\cos y}$ with the initial condition $y(1) = 0$ is

(A) $y = \sin^{-1}\left(\dfrac{x^2-1}{2}\right)$

(B) $y = \sin^{-1}\left(\dfrac{x^2}{2}\right)$

(C) $y = \cos^{-1}\left(x^2-2\right)$

(D) $y = \ln\left[\cos(x-1)\right]$

(E) $y = \ln(\sin x)$

13. For $x > 0$, $\dfrac{d}{dx}\sin^{-1}(\ln x) =$

(A) $\dfrac{1}{x(1+\ln^2 x)}$ (B) $\dfrac{1}{x\sqrt{1-2\ln x}}$ (C) $\dfrac{1}{\sqrt{1-\ln^2 x}}$

(D) $\dfrac{1}{x\sqrt{1-\ln^2 x}}$ (E) $\dfrac{\cos^{-1}(\ln x)}{x}$

14. The function g is given by $g(x) = \dfrac{3x^2}{e^{3x}}$. On which of the following intervals is g increasing?

(A) $(-\infty, 0)$ (B) $\left(-\infty, \dfrac{2}{3}\right)$ (C) $\left(0, \dfrac{2}{3}\right)$ (D) $(0, \infty)$ (E) $\left(\dfrac{2}{3}, \infty\right)$

15. If $x^2 y + y^2 + 4 = 0$, then when $x = 2$, the value of $\dfrac{dy}{dx}$ is

(A) -2 (B) -1 (C) 0 (D) 2 (E) nonexistent

16. The slope to the tangent line to the graph of $y = \tan 2x$ at $x = \dfrac{\pi}{8}$ is

(A) $\dfrac{1}{\sqrt{2}}$ (B) $\sqrt{2}$ (C) 2 (D) $2\sqrt{2}$ (E) 4

17. The area of the region enclosed by the graph of $y = 5 - x^2$ and the line $y = 1$ is

(A) $\dfrac{4}{3}$ (B) $\dfrac{8}{3}$ (C) 4 (D) $\dfrac{16}{3}$ (E) $\dfrac{32}{3}$

18. The average value of e^{3x} on the interval $[0, 4]$ is

(A) $\dfrac{e^{12}-1}{12}$ (B) $\dfrac{e^{12}}{12}$ (C) $\dfrac{e^{12}-1}{4}$ (D) $\dfrac{e^{12}}{4}$ (E) $\dfrac{e^{12}-1}{3}$

19. If the region enclosed by the graphs of $y = \sqrt{x-1}$, $x = 4$, and the x-axis is revolved about the x-axis, the volume of the solid generated is

(A) $2\pi\sqrt{3}$ (B) $\dfrac{7\pi}{2}$ (C) 4π (D) $\dfrac{9\pi}{2}$ (E) 12π

20. If $x^2 - y^2 = 5$, what is the value of $\dfrac{d^2y}{dx^2}$ at the point (3, 2)?

(A) $-\dfrac{13}{8}$ (B) $-\dfrac{11}{8}$ (C) $-\dfrac{7}{8}$ (D) $-\dfrac{5}{8}$ (E) $-\dfrac{1}{4}$

21. The velocity of a particle at time t is given by the function $v(t) = t^3 - \sin t + 2$. What is the acceleration of the particle at time $t = 2\pi$?

(A) $6\pi - 1$ (B) $6\pi + 1$ (C) $12\pi^2 - 1$ (D) $12\pi^2$ (E) $12\pi^2 + 1$

22. $\displaystyle\int_0^{\sqrt{e-1}} \dfrac{x}{x^2+1}\, dx =$

(A) 0   (B) $\dfrac{\ln(e-1)}{2}$ (C) $\dfrac{1}{2}$

(D) 1   (E) $\sqrt{e-1}$

23. If $f'(x) = \tan x$, $-\dfrac{\pi}{2} < x < \dfrac{\pi}{2}$, and $f\left(\dfrac{\pi}{3}\right) = \ln 6$, the value of $f\left(\dfrac{\pi}{4}\right)$ is

(A) $\ln 6 - 2\sqrt{3} + \sqrt{2}$ (B) $\ln 6 - 2$ (C) $\ln\left(3\sqrt{2}\right)$

(D) $\ln\left(4\sqrt{2}\right)$ (E) $\ln\left(12\sqrt{2}\right)$

24. For what value of x is the line tangent to $y = x^2$ parallel to the line tangent to $y = \sqrt{x}$?

(A) 0   (B) $\dfrac{\sqrt[3]{4}}{4}$ (C) $\dfrac{1}{2}$ (D) $\dfrac{\sqrt[3]{2}}{2}$ (E) 1

25. The function $f(x) = \dfrac{\ln x}{x}$, $x > 0$, is decreasing on the interval

(A) $(0, 1)$ (B) $(0, e)$ (C) $(1, e)$ (D) $(1, \infty)$ (E) (e, ∞)

26. $\displaystyle\lim_{x \to 2} \dfrac{e^{2x} - e^4}{x - 2} =$

(A) e (B) $2e$ (C) $2e^2$ (D) e^4 (E) $2e^4$

27. $\displaystyle\int_0^{\frac{\pi^2}{9}} \dfrac{dx}{\sqrt{x} \cos^2\left(\sqrt{x}\right)} =$

(A) $\dfrac{\sqrt{3}}{3}$ (B) $\dfrac{\sqrt{3}}{2}$ (C) $\dfrac{2\sqrt{3}}{3}$ (D) $\sqrt{3}$ (E) $2\sqrt{3}$

28.

x	−1	1	5	8
$f(x)$	5	2	5	−1

Let f be a continuous, differentiable function defined for all real values of x. The table above shows values of f at certain values of x. The function f must have at least

(A) one point of inflection and at least one relative minimum
(B) one point of inflection and at least two relative maxima
(C) one zero and at least two points of inflection
(D) two zeros and at least one relative minimum
(E) two points of inflection and at least one relative maximum

Calculus AB
Section I, Part B
Time — 50 minutes
Number of questions — 17

A graphing calculator is required for some questions.

---

29. What is the area of the region in the first quadrant enclosed by the graphs of $y = 2 - x^2$, $y = 3\sin x$, and the y-axis?

(A) 0.591   (B) 0.604   (C) 0.982   (D) 1.281   (E) 1.924

---

30. Let f and g be differentiable functions with $g(x) \neq 0$ for all x. If $h(x) = \dfrac{f(x)}{g(x)}$ and $h'(x) = \dfrac{f(x)\left[g(x) - g'(x)\right]}{\left[g(x)\right]^2}$, then $f(x)$ could be

(A) e^x (B) $\ln x$ (C) $\sin x$ (D) $\cos x$ (E) $\dfrac{1}{x}$

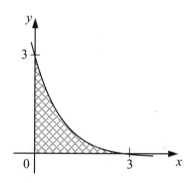

31. The base of a solid is the region in the first quadrant bounded by the x-axis, the y-axis, and the graph of $y = (3 - x)e^{-x}$, as shown in the figure above. If cross sections of the solid perpendicular to the x-axis are squares, what is the volume of the solid?

(A) 2.050   (B) 3.081   (C) 3.249   (D) 10.208   (E) 12.998

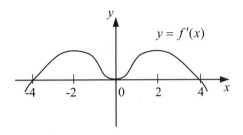

32. The graph of the derivative of *f* is shown in the figure above. Which of the following could be the graph of *f*?

(A)

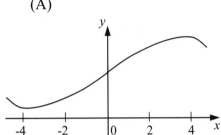

(B)

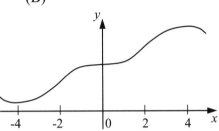

(C)

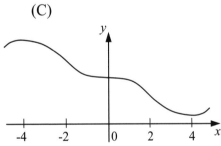

(D)

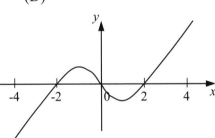

(E)

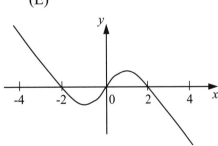

33. If $f(x)$ defines an even polynomial function, which of the following must be equal to the value of $\int_a^b f(x)dx$?

   I. $\int_a^b f(-x)dx$

   II. $\int_a^b |f(x)|dx$

   III. $\int_{-b}^{-a} f(x)dx$

   (A) I only   (B) II only   (C) III only   (D) I and II only   (E) I and III only

---

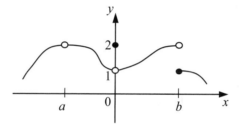

34. The graph of the function f is shown in the figure above. Which of the following statements about f is true?

   (A) $f(a)$ exists

   (B) $\lim_{x \to a} f(x) = 2$

   (C) $\lim_{x \to b} f(x) = 1$

   (D) $\lim_{x \to b^-} f(x) = \lim_{x \to b^+} f(x)$

   (E) f is continuous at $x = 0$

---

35. The rate at which a bacteria population grows is proportional to the number of bacteria present. At time $t = 2$ hours, 500 bacteria were present, and at time $t = 6$ hours, 1500 bacteria were present. Approximately how long does it take for the population to double?

   (A) 2.5 hours   (B) 5.4 hours   (C) 7.6 hours

   (D) 8.4 hours   (E) 9.6 hours

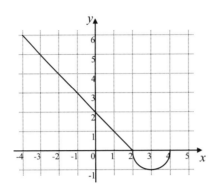

36. The graph of the piecewise-defined function f, for $-4 \leq t \leq 4$, consists of a line segment and a semicircle as shown in the figure above. The function g is defined by $g(x) = \int_{-4}^{x} f(t)\,dt$. What is the value of $g(4)$?

(A) $2 + \pi$ (B) $18 - \pi$ (C) $18 - \dfrac{\pi}{2}$ (D) $18 + \dfrac{\pi}{2}$ (E) $18 + \pi$

37. The first derivative of the function f is given by $f'(x) = \dfrac{8\cos x}{x^2} - \dfrac{1}{8}$. On the open interval $(1, 10)$ the graph of f has

(A) one relative maximum and no relative minima
(B) one relative minimum and no relative maxima
(C) two relative maxima and one relative minimum
(D) two relative minima and one relative maximum
(E) no relative extrema

38.

Time (sec)	0	3	8	10
Rate (gal/sec)	16	10	6	5

Water drains continuously from a tank. The rate (in gallons per second) at which the water drains out is measured at the times (in seconds) given in the table above. What is the trapezoidal approximation, based on all of the data in the table, for the total amount of water that has drained from the tank in the first ten minutes?

(A) 37 gallons   (B) 70 gallons   (C) 79.5 gallons

(D) 90 gallons   (E) 110 gallons

39. At time $t \geq 0$, the position of a particle moving along the x-axis is given by $x(t) = \dfrac{t^3}{3} + 2t + 2$. For what value of t in the interval $[0, 3]$ will the instantaneous velocity of the particle equal the average velocity of the particle from time $t = 0$ to time $t = 3$?

(A) 1   (B) $\sqrt{3}$ (C) $\sqrt{7}$ (D) 3   (E) 5

40. If $f'(x) = \sqrt{1+x^3}$ and $f(1) = 0.5$, then $f(4) =$

(A) 7.562   (B) 8.062   (C) 12.871   (D) 13.371   (E) 17.871

41. Let f be a twice differentiable function of x such that, when $x = c$, f is decreasing, concave up, and has an x-intercept. Which of the following is true?

(A) $f(c) < f'(c) < f''(c)$

(B) $f(c) < f''(c) < f'(c)$

(C) $f'(c) < f(c) < f''(c)$

(D) $f'(c) < f''(c) < f(c)$

(E) $f''(c) < f(c) < f'(c)$

42. Let f be the function given by $f(x) = x^2 e^{-x}$. For what value of x is the slope of the line tangent to the graph of f at $(x, f(x))$ equal to 0.2?

(A) –0.091   (B) 0.112   (C) 0.605   (D) 1.418   (E) 4.708

43. The Riemann sum $\dfrac{\pi}{n}\displaystyle\sum_{i=1}^{n}\sin\left(\dfrac{\pi i}{n}\right)$, where n is any integer, is an approximation for

(A) $\displaystyle\int_0^{\pi} \sin x \, dx$ (B) $\displaystyle\int_0^{\pi} \cos x \, dx$ (C) $\displaystyle\int_0^{1} \sin \pi x \, dx$

(D) $\displaystyle\int_0^{1} \cos \pi x \, dx$ (E) $\displaystyle\int_0^{1} \sin x \, dx$

44. If the region enclosed by the y-axis and the relation $x = 4 - y^2$ is revolved about the y-axis, the volume of the solid generated is

(A) 25.133   (B) 33.510   (C) 53.617   (D) 107.233   (E) 214.466

45. The radius of a spherical balloon is increasing at a constant rate of 5 millimeters per second. What is the rate of increase of the surface area of the balloon at the instant when the volume is 36π cubic millimeters? $(V = \dfrac{4}{3}\pi r^3,\ S = 4\pi r^2.)$

(A) 24π mm^2 / sec   (B) 36π mm^2 / sec   (C) 40π mm^2 / sec

(D) 120π mm^2 / sec   (E) 180π mm^2 / sec

Practice Exam AB-2

A graphing calculator is required for some problems or parts of problems.

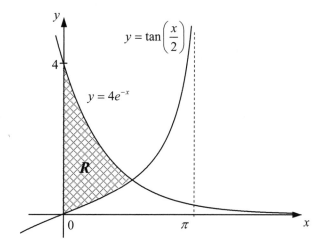

1. Let R be the shaded region in the first quadrant enclosed by the graphs of $y = 4e^{-x}$, $y = \tan\left(\dfrac{x}{2}\right)$, and the y-axis, as shown in the figure above.

    (a) Find the area of the region.

    (b) Find the volume of the solid generated when the region R is revolved about the x-axis.

    (c) The region R is the base of a solid. For this solid, each cross section perpendicular to the x-axis is a semicircle. Find the volume of this solid.

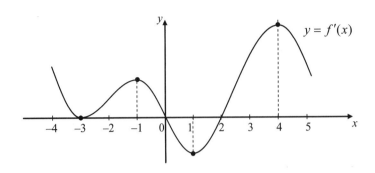

2.   The figure above shows the graph of f', the derivative of the function f, for $-4 \le x \le 5$. The graph of f' has horizontal tangent lines at $x = -3, -1, 1,$ and 4.

(a)   Find all values of x, for $-4 < x < 5$, for which f is decreasing. Justify your answer.

(b)   Find all values of x, for $-4 < x < 5$, at which f attains a relative maximum. Justify your answer.

(c)   Find all values of x, for $-4 < x < 5$, for which the graph of f is concave up.

(d)   Given $f(-4) = -2$, $f(0) = 5$, and $f(5) = 8$, sketch a possible graph of f on the axes below.

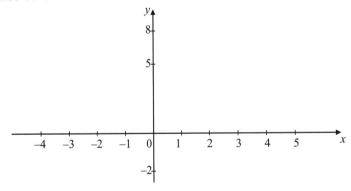

t	0	4	8	12	16	20	24
$R(t)$	320	221	130	82	39	22	11

3. The rate of decay, in grams per minute, of a radioactive substance is a differentiable, decreasing function R of time, t, in minutes. The table above shows the decay rate as recorded every 4 minutes over a 24-minute period.

   (a) Use data from the table to approximate the average decay rate, in grams per minute, of the radioactive substance over the time interval $0 \le t \le 24$ minutes by using a midpoint Riemann sum approximation with 3 subintervals of equal length.

   (b) Approximate $R'(12)$. Show the computations that lead to your answer. Indicate units of measure.

   (c) A physicist proposes the function G, given by $G(t) = 320(0.882)^t$, as a model for the rate of decay of the radioactive substance at time t, where t is measured in minutes and $G(t)$ is measured in grams per minute. Find the average value, in grams per minute, of $G(t)$ over the time interval $0 \le t \le 24$ minutes.

   (d) Use the function G, defined in Part (c), to find $G'(12)$. Using appropriate units, explain the meaning of your answer in terms of the decay rate of the substance.

Calculus AB
Section II, Part B
Time — 45 minutes
Number of problems — 3

No calculator is allowed for these problems.

---

4.   A particle moves along the y-axis so that its position at any time $t \geq 0$ is given by
$y(t) = t^2 - 4\ln(t+1) - 1$.

(a)   Find the velocity $v(t)$ at any time $t \geq 0$.

(b)   For any time $t \geq 0$, find the lowest y value attained by the particle. Justify your answer.

(c)   Find all values of t for which the speed of the particle is increasing. Justify your answer.

(d)   Find the total distance traveled by the particle from time $t = 0$ to time $t = 2$.

---

5.   A function $y = f(x)$ satisfies the differential equation $\dfrac{dy}{dx} = \dfrac{1+x^2}{y}$.

(a)   On the axes provided, sketch the slope field for the given differential equation at the 11 points indicated.

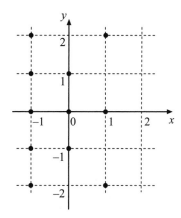

(b)   Use the slope field for the given differential equation to explain why a solution could not have the graph shown below.

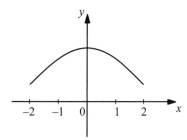

(c)   Find the particular solution $y = f(x)$ to the given differential equation which contains the point $(3, -4)$.

6.   An oil tanker spills 100,000 cubic meters (m^3) of oil, which forms a slick that spreads on the water surface in a shape best modeled by a circular disc of radius r meters and thickness h meters. The radius of the circular disc is increasing at a rate of 3 m/min. At $t = T$, the area of the "circular" slick reaches 100π m^2.

(a)   How fast is the area of the slick increasing at $t = T$?

(b)   How fast is the thickness of the slick decreasing at $t = T$?

(c)   Find the rate of change of the area of the slick with respect to the thickness at $t = T$.

Practice Exam AB-3

Calculus AB
Section I, Part A
Time — 55 minutes
Number of questions — 28

No calculator is allowed for these questions.

---

1. If $f(x) = e^{2\ln x}$, then $f'(3) =$

    (A) 6   (B) 9   (C) e^6 (D) e^9 (E) $\dfrac{e^9}{9}$

---

2. What is the total area enclosed by the curves $f(x) = x - x^2$ and $g(x) = x^3 - x$ between the values of $x = 0$ and $x = 1$?

    (A) $\dfrac{5}{12}$ (B) $\dfrac{7}{12}$ (C) $\dfrac{5}{7}$ (D) 3   (E) 5

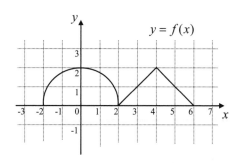

Questions 3 and 4 refer to the graph above.

3. The graph of $f(x)$, shown above, consists of a semicircle and two line segments. $f'(1) =$

    (A) -1 (B) $-\dfrac{1}{\sqrt{3}}$ (C) $\dfrac{1}{\sqrt{3}}$ (D) 1   (E) $\sqrt{3}$

---

4.   For which values of x does $f'(x)=0$?

     (A) 0 only   (B) 2 only   (C) 0 and 4 only

     (D) –2, 2, and 6 only   (E) –2, 0, 2, 4, and 6

---

5.   The position of a particle moving along a straight line at any time t is given by $s(t)=t^3+9t^2-27$. What is the velocity of the particle at the time when the acceleration is zero?

     (A) –27   (B) –3   (C) 18   (D) 27   (E) 81

---

6.   If $[x]$ represents the greatest integer that is less than or equal to x, then $\displaystyle\lim_{x\to 0^-}\frac{1}{[x]}=$

     (A) –2   (B) –1   (C) 0

     (D) 1   (E) the limit does not exist

---

7.   If $\dfrac{dy}{dx}=\sin(3x-3)+4$ and $y(1)=7$, then $y=$

     (A) $-\dfrac{1}{3}\cos(3x-3)+4x+3$

     (B) $-\dfrac{1}{3}\cos(3x-3)+4x+\dfrac{10}{3}$

     (C) $-3\cos(3x-3)+10$

     (D) $-3\cos(3x-3)+4x+3$

     (E) $-3\cos(3x-3)+4x+6$

8. What is an equation for the line tangent to $y = \tan^{-1} x$ at $x = \sqrt{3}$?

(A) $y - \dfrac{\pi}{3} = -\dfrac{1}{2}\left(x - \sqrt{3}\right)$

(B) $y - \dfrac{\pi}{6} = -\dfrac{1}{4}\left(x - \sqrt{3}\right)$

(C) $y - \dfrac{\pi}{3} = -\dfrac{1}{4}\left(x - \sqrt{3}\right)$

(D) $y - \dfrac{\pi}{6} = \dfrac{3}{4}\left(x - \sqrt{3}\right)$

(E) $y - \dfrac{\pi}{3} = \dfrac{1}{4}\left(x - \sqrt{3}\right)$

---

9. If f is a differentiable function and $f(0) = -1$ and $f(4) = 3$, then which of the following must be true?

I.   There exists a c in [0, 4] where $f(c) = 0$.

II. There exists a c in [0, 4] where $f'(c) = 0$.

III. There exists a c in [0, 4] where $f'(c) = 1$.

(A) I only   (B) II only   (C) I and II only
(D) I and III only   (E) I, II, and III

---

10. Suppose $f(x) = x^4 + ax^2$. What is the value of a if f has a local minimum at $x = 2$?

(A) –24   (B) –8   (C) –4   (D) $-\dfrac{1}{2}$ (E) $-\dfrac{1}{6}$

---

11. $\displaystyle \int_{-\pi}^{0} \frac{\sin x}{2 + \cos x}\, dx =$

(A) $-\ln 3$ (B) -1 (C) 1 (D) $\ln 3$ (E) $\dfrac{4}{3}$

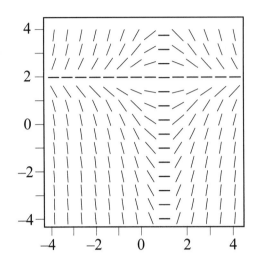

12. The graph above is a slope field for which of the following differential equations?

(A) $\dfrac{dy}{dx} = y - 2x$ (B) $\dfrac{dy}{dx} = 1 + x + y$ (C) $\dfrac{dy}{dx} = (1-x)(y-2)$

(D) $\dfrac{dy}{dx} = xy^2$ (E) $\dfrac{dy}{dx} = (x-1)y^2$

13. If $y = e^{8x^2+1}$, then $\dfrac{dy}{dx} =$

(A) e^{8x^2} (B) e^{8x^2+1} (C) $16xe^{8x^2}$

(D) $16xe^{8x^2+1}$ (E) $(8x^2+1)e^{8x^2}$

14. For what point on the graph of $y = xe^{-2x}$ is the tangent line horizontal?

(A) $\left(-1, -e^2\right)$ (B) $\left(-\dfrac{1}{2}, -\dfrac{e}{2}\right)$ (C) $\left(\dfrac{1}{2}, 0\right)$

(D) $\left(\dfrac{1}{2}, \dfrac{1}{2e}\right)$ (E) $\left(1, \dfrac{1}{e^2}\right)$

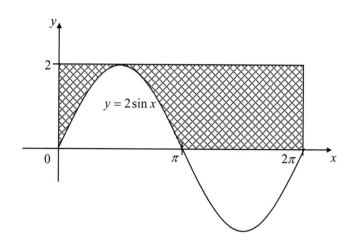

15.   What is the area of the shaded region in the figure above, which is enclosed by the y-axis, the x-axis, $x = 2\pi$, $y = 2$, and the graph of $y = 2\sin x$?

(A) $2\pi - 4$ (B) $2\pi - 2$ (C) $4\pi - 4$ (D) $4 - 2\pi$ (E) $4\pi + 4$

16.   An antiderivative of $\left(x^2 + 1\right)^2$ is

(A) $\dfrac{1}{3}\left(x^2 + 1\right)^3 + C$ (B) $\dfrac{1}{5}x^5 + x + C$ (C) $\dfrac{1}{5}x^5 + \dfrac{2}{3}x^3 + x + C$

(D) $\dfrac{1}{6x}\left(x^2 + 1\right)^3 + C$ (E) $4x\left(x^2 + 1\right) + C$

17.   Which two functions grow at different rates?

(A) $x^2 + 4x$ and x^2 (B) $\ln x$ and $\ln\left(x^2\right)$ (C) $2x^3$ and $5x^3$

(D) e^{x+100} and e^{x-100} (E) x^3 and x^4

18. $\displaystyle\int \sec^2(2x)\,dx =$

(A) $-2\tan(2x) + C$ (B) $\dfrac{1}{2}\tan(2x) + C$ (C) $\tan(2x) + C$

(D) $2\tan(2x) + C$ (E) $\tan^2(2x) + C$

19. What are the *x*-coordinates of the points of inflection on the graph of the function $f(x) = 3x^4 - 4x^3 + 6$?

(A) 0 only   (B) $\frac{2}{3}$ only   (C) 1 only   (D) 0 and $\frac{2}{3}$ (E) 0 and 1

20. $\int_1^3 \left(2 - |x - 2|\right) dx =$

(A) –4   (B) –2   (C) 2   (D) 3   (E) 4

21. If $y = \dfrac{3x+1}{x-1}$, what is $\dfrac{dy}{dx}$?

(A) $\dfrac{-4}{(x-1)^2}$ (B) $\dfrac{-1}{(x-1)^2}$ (C) $\dfrac{1}{(x-1)^2}$

(D) $\dfrac{4}{(x-1)^2}$ (E) $\dfrac{6x+1}{(x-1)^2}$

22. $\dfrac{d}{dx} \int_x^2 \ln(1+t)\, dt =$

(A) $-\dfrac{1}{1+x}$ (B) $\dfrac{1}{3} - \dfrac{1}{1+x}$ (C) $-\ln(1+x)$

(D) $\ln(1+x)$ (E) $\ln 3 - \ln(1+x)$

23. If $f'(x) = -5(x-3)^2(x-2)$, which of the following features does the graph of $f(x)$ have?

(A) a local minimum at $x = 2$ and a local maximum at $x = 3$
(B) a local maximum at $x = 2$ and a local minimum at $x = 3$
(C) a point of inflection at $x = 2$ and a local minimum at $x = 3$
(D) a local minimum at $x = 2$ and a point of inflection at $x = 3$
(E) a local maximum at $x = 2$ and a point of inflection at $x = 3$

24. The area of the region bounded by the graphs of $y = \sin x$, $y = \cos x$, and the x-axis for $0 \le x \le \dfrac{\pi}{2}$ can be represented by

(A) $\displaystyle\int_0^{\pi/2} (\cos x - \sin x)\, dx$

(B) $\displaystyle\int_0^{\pi/2} (\cos x - \sin x)^2\, dx$

(C) $\displaystyle\int_0^{\pi/2} (\cos^2 x - \sin^2 x)\, dx$

(D) $\displaystyle\int_0^{\pi/4} \sin^2 x\, dx + \int_{\pi/4}^{\pi/2} \cos^2 x\, dx$

(E) $\displaystyle\int_0^{\pi/4} \sin x\, dx + \int_{\pi/4}^{\pi/2} \cos x\, dx$

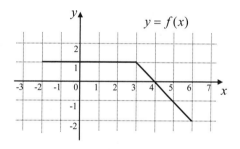

25. The graph of $f(x)$, shown above, consists of two line segments. What is the value of $\displaystyle\int_0^6 f(x)\, dx$?

(A) $7\dfrac{1}{2}$ (B) $5\dfrac{1}{2}$ (C) $3\dfrac{1}{2}$ (D) 3   (E) $1\dfrac{1}{2}$

26. If $\displaystyle\int_1^k \dfrac{1}{\sqrt{x}}\, dx = 4$, then $k =$

(A) $\dfrac{1}{25}$ (B) $\sqrt{3}$ (C) 9   (D) 81   (E) e^8

27. The function $f(x) = e^{x\sqrt{3}} \cos x$ is defined on $0 \le x \le 2\pi$. On what interval(s) is $f(x)$ decreasing?

(A) $\dfrac{\pi}{4} < x < \dfrac{5\pi}{4}$

(B) $\dfrac{\pi}{3} < x < \dfrac{4\pi}{3}$

(C) $\dfrac{2\pi}{3} < x < \dfrac{5\pi}{3}$

(D) $0 < x < \dfrac{\pi}{4}$ and $\dfrac{5\pi}{4} < x < 2\pi$

(E) $0 < x < \dfrac{\pi}{3}$ and $\dfrac{4\pi}{3} < x < 2\pi$

---

28. If $\displaystyle\int_0^b f(x)\,dx = 3a$ and the graph of f is symmetric about the origin, then

$\displaystyle\int_{-b}^b f(x)\,dx =$

(A) 0   (B) $\dfrac{3a}{2}$ (C) $2a$ (D) $3a$ (E) $6a$

Calculus AB
Section I, Part B
Time — 50 minutes
Number of questions — 17

A graphing calculator is required for some questions.

29. If $x + \sin y = \ln y$, then $\dfrac{dy}{dx} =$

    (A) $y + y\cos y$

    (B) $\dfrac{y + \cos y - 1}{y}$

    (C) $\dfrac{1 - y}{y\cos y}$

    (D) $\dfrac{y}{y\cos y - 1}$

    (E) $\dfrac{y}{1 - y\cos y}$

30. If $y = \sin(x - \sin x)$, what is the smallest positive value of x for which the tangent line is parallel to the x-axis?

    (A) 1.677   (B) 2.310   (C) 3.142   (D) 3.973   (E) 6.283

31. If f and g are functions such that $f(g(x)) = x$ for all x in their domains, and if $f(a) = b$ and $f'(a) = c$, then which of the following is true?

    (A) $g'(a) = \dfrac{1}{c}$

    (B) $g'(a) = -\dfrac{1}{c}$

    (C) $g'(b) = \dfrac{1}{c}$

    (D) $g'(b) = -\dfrac{1}{c}$

    (E) $g'(b) = \dfrac{1}{a}$

32. The depth of water in a tidal basin can be represented by the formula $h(t) = -\cos\left(\dfrac{\pi t}{12}\right) + 4$, where t is the time in hours starting at midnight and $h(t)$ is measured in feet. What is the average depth of the water between noon and 3 pm?

    (A) 2.764 ft   (B) 3.100 ft   (C) 4.146 ft   (D) 4.854 ft   (E) 4.900 ft

33. A function $f(x)$, differentiable on $[-7, -4]$, has only one critical number, $x = -5$. What feature must the graph of f have at $(-5, f(-5))$ if $f'(-7) = -1$ and $f'(-4) = \frac{1}{2}$?

    (A) a relative minimum
    (B) a relative maximum
    (C) a point of inflection
    (D) a zero
    (E) none of the above

34. Let f be a function defined for all real numbers. Which of the following statements about f must be true?

    (A) If $\lim\limits_{x \to 2} f(x) = 7$, then $f(2) = 7$.

    (B) If $\lim\limits_{x \to 5} f(x) = -3$, then -3 is in the range of f.

    (C) If $\lim\limits_{x \to 1^-} f(x) = \lim\limits_{x \to 1^+} f(x)$, then $f(1)$ exists.

    (D) If $\lim\limits_{x \to 3^-} f(x) \neq \lim\limits_{x \to 3^+} f(x)$, then $\lim\limits_{x \to 3} f(x)$ does not exist.

    (E) If $\lim\limits_{x \to 4} f(x)$ does not exist, then $f(4)$ does not exist.

35. A particle moves in a straight line, and its velocity at any time t is given by $v(t) = 5 - e^t$. What is the total distance the particle travels from $t = 0$ to $t = 3$?

    (A) 4.086   (B) 5.086   (C) 11.086   (D) 12.180   (E) 19.086

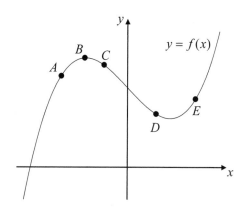

36. At which point on the graph shown above are both the first and second derivatives of $f(x)$ negative?

    (A) A (B) B (C) C (D) D (E) E

37. The total area of the regions bounded by the graphs of $y = \sin(3x)$ and $y = x$ is

    (A) 0   (B) 0.262   (C) 0.316   (D) 0.523   (E) 0.632

38. If $f(x+y) = f(x) \cdot f(y)$ and if $\displaystyle\lim_{h \to 0} \frac{f(h)-1}{h} = 6$, then $f'(x) =$

    (A) 6   (B) $6 + f(x)$ (C) $6 \cdot f(x)$

    (D) $6 + f(h)$ (E) $6 \cdot f(h)$

39. The region in the first quadrant enclosed by the graphs of $y = 2 \ln x$, $y = 2$, and $x = 1$ is rotated about the x-axis. Which of the following integrals represents the volume of the generated solid?

    (A) $\pi \displaystyle\int_1^e (1 - 2\ln x)^2 \, dx$ (B) $\pi \displaystyle\int_1^e (2 - 2\ln x)^2 \, dx$ (C) $\pi \displaystyle\int_1^e (2 - 2\ln x) \, dx$

    (D) $\pi \displaystyle\int_1^e (4 - 2\ln x)^2 \, dx$ (E) $\pi \displaystyle\int_1^e \left(4 - (2\ln x)^2\right) dx$

x	1	2	3	4	5	6
$f(x)$	0.14	0.21	0.28	0.36	0.44	0.54

40. The table above contains values of a continuous function f at several values of x. Estimate $\int_{2}^{5} f(x)\,dx$ using a trapezoidal approximation with three equal subintervals.

   (A) 0.85   (B) 0.965   (C) 1.08   (D) 1.29   (E) 1.93

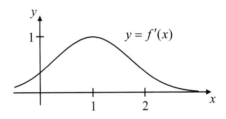

41. The graph of $f'(x)$ is shown above. Which of the following could be the graph of $f(x)$?

(A)

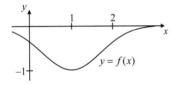

(B)

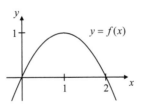

(C)

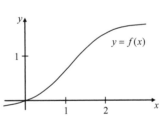

(D)

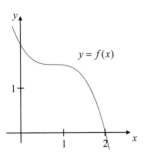

(E)

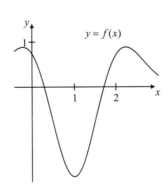

42. The radius of a sphere increases at a constant rate of 2 cm/min. At the time when the volume of the sphere is 40 cm³, what is the rate of increase of the volume in cm³/min? $\left(V = \frac{4}{3}\pi r^3 \right)$

   (A) 2.122   (B) 9.549   (C) 56.562   (D) 113.124   (E) 293.954

43. Suppose g and h are continuous functions and $\int_3^7 g(x)\,dx = 6$, $\int_3^7 h(x)\,dx = 2$, and $\int_3^4 h(x)\,dx = -4$. Which of the following equalities must be true?

   I. $\int_4^7 h(x)\,dx = 6$

   II. $\int_3^7 g(x)\cdot h(x)\,dx = 12$

   III. $\int_3^7 \left(g(x)+2\right)\,dx = 14$

   (A) I only   (B) I and II only   (C) I and III only
   (D) II and III only   (E) I, II, and III

44. If $\frac{dy}{dx} = \frac{\sin x}{x}$ and $y(1) = 4$, then $y(2) =$

   (A) 0.455   (B) 0.659   (C) 4.455   (D) 4.659   (E) 5.289

45. $f(x)$ is a differentiable function with $f(1)=-3$ and $f(5)=4$. Which of the following must be true?

(A) $f(0)=k$ for some k in $(1, 5)$

(B) $f(x)$ is increasing on $(1, 5)$

(C) $f'(x)=\dfrac{7}{4}$ for all x in $(1, 5)$

(D) $f'(k)=0$ for some k in $(1, 5)$

(E) $f'(k)=\dfrac{7}{4}$ for some k in $(1, 5)$

Practice Exam AB-3

Calculus AB
Section II, Part A
Time — 45 minutes
Number of problems — 3

A graphing calculator is required for some problems or parts of problems.

1. Let R be the region bounded by the graphs of $y = \sin\left(x^2\right)$ and $y = 1 - x^2$.

    (a) Find the area of R.

    (b) Find the volume of the solid generated when R is revolved around the line $y = -1$.

    (c) The region R is the base of a solid. For this solid, each cross section perpendicular to the x-axis is an isosceles right triangle with one leg in region R. Find the volume of this solid.

2. Tourists visiting an island resort contracted a mystery illness over a 45 day period. The health authorities recorded the rate of new cases per day and some of the rates are listed in the table below.

t Day	$N(t)$ New cases per day
2	3
6	8
10	15
15	30
25	100
35	50
40	22
45	10

(a) Assuming that the table above shows a sample of values for a continuous function $N(t)$, use a right-hand Riemann sum with 8 subintervals to approximate $\int_0^{45} N(t)\,dt$. Explain the meaning of $\int_0^{45} N(t)\,dt$.

(b) After studying the spread of the disease, the health department authorities decided they could model the number of new cases per day with

$$R(t) = \frac{80000e^{-0.2t}}{\left(1 + 200e^{-0.2t}\right)^2} \text{ for } 0 \le t \le 50 \text{ days. Use } R(t) \text{ to find the average}$$

number of cases over the 45-day period.

(c) The disease is considered eradicated when the number of new cases per day does not exceed 5. Use $R(t)$ to find on what day this will occur.

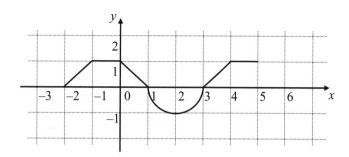

3.   The graph of a function *f*, defined on the closed interval $-2 \leq x \leq 5$, is shown above. It consists of a semicircle and five line segments. Let $g(x) = \int_0^{2x} f(t)\,dt$.

(a)   Find $g(2)$ and $g'(2)$.

(b)   Find the *x*-coordinates of all local maximum points and all local minimum points of $g(x)$ over the interval $-1 < x < 2$. Justify your answer.

(c)   On what intervals is $g(x)$ concave up? Justify your answer.

Calculus AB
Section II, Part B
Time — 45 minutes
Number of problems — 3

No calculator is allowed for these problems.

---

4.   A particle moves along the x-axis with a velocity given by $v(t) = \dfrac{3t}{1+t^2}$ for $t \geq 0$. When $t = 0$, the particle is at the point $(4, 0)$.

   (a)   Determine the maximum velocity for the particle. Justify your answer.

   (b)   Determine the position of the particle at any time t.

   (c)   Find the limit of the velocity as $t \to \infty$.

   (d)   Find the limit of the position as $t \to \infty$.

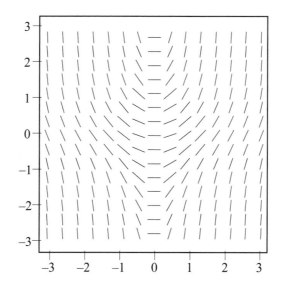

5.   The slope field for the differential equation $\dfrac{dy}{dx} = x\left(y^2 + 1\right)$ is shown above.

(a)   Sketch the solution curve that passes through the point $(0, -3)$ and another solution curve that passes through the point $(0, 1)$.

(b)   Find $\dfrac{d^2 y}{dx^2}$. Use $\dfrac{d^2 y}{dx^2}$ to explain why every point of inflection on the solution curves to this differential equation has a y-coordinate that is less than zero.

(c)   Find the particular solution $y = f(x)$ to the given differential equation with the initial condition $f(0) = 1$.

6.   Consider the ellipse with equation $\dfrac{x^2}{16} + \dfrac{y^2}{9} = 1$ and the hyperbola with equation

$\dfrac{x^2}{9} - \dfrac{y^2}{16} = 1$.

(a)   Find $\dfrac{dy}{dx}$ for any point on the ellipse in terms of x and y.

(b)   Write an equation for the line tangent to the ellipse at the point $\left(\dfrac{12}{5}, \dfrac{12}{5} \right)$.

(c)   Find an expression for the slope at any point on the hyperbola.

(d)   Find a point on the hyperbola such that the tangent line to the hyperbola at that point is perpendicular to the tangent line to the ellipse at $\left(\dfrac{12}{5}, \dfrac{12}{5} \right)$.

Practice Exam BC-1

Calculus BC
Section I, Part A
Time — 55 minutes
Number of questions — 28

No calculator is allowed for these questions.

x	$f(x)$	$f'(x)$
0	1	2
$\dfrac{1}{2}$	2	4
1	3	5
$\dfrac{3}{2}$	$\dfrac{1}{2}$	$-\dfrac{1}{2}$
2	$\dfrac{3}{2}$	-2

Questions 1 and 2 refer to the table above.

1. If f is a differentiable function on the interval $0 < x < 2$, find the derivative of the inverse function $f^{-1}(x)$ at $x = \dfrac{1}{2}$.

   (A) -4 (B) -2 (C) -1 (D) $-\dfrac{1}{8}$ (E) $-\dfrac{1}{16}$

2. Using the table above and the fact that $f'(x)$ is continuous on the interval $0 \le x \le 2$, $\displaystyle\int_0^2 f'(x)\,dx =$

   (A) -4 (B) -2 (C) 0 (D) $\dfrac{1}{2}$ (E) $\dfrac{3}{2}$

3. The glass above is initially empty, then gradually filled with water. Which of the following graphs best represents the volume V of water versus the height h of the water?

(A)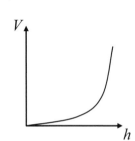

(B)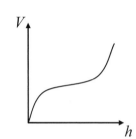

(C)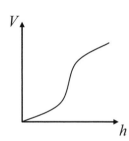

(D)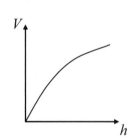

(E)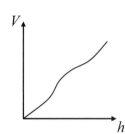

---

4. If $f(x) = \sum_{n=0}^{\infty} \frac{(2x+1)^{n+1}}{n!}$, then $f''\left(-\frac{1}{2}\right) =$

(A) 0   (B) 1   (C) 2   (D) 4   (E) 8

---

5. If $f(x)$ is a continuous and even function and $\int_0^4 f(x)\,dx = -5$ and $\int_4^6 f(t)\,dt = 2$, then the average value of $f(x)$ over the interval from $x = -6$ to $x = 4$ is

(A) −0.2   (B) −0.8   (C) 0.2   (D) 1.2   (E) 2

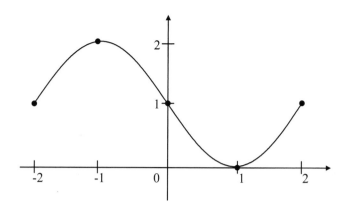

6. Given the graph of $y = f(x)$ shown above, which of the following values is the largest?

    (A) $f(0)$ (B) $f'(0)$ (C) $\lim\limits_{h \to 0} \dfrac{f(h) - 1}{h}$

    (D) $\dfrac{f(1) - f(-1)}{2}$ (E) $\dfrac{f'(1) - f'(-1)}{2}$

7. $\lim\limits_{h \to 0} \left(\dfrac{1}{h} \displaystyle\int_{1}^{1+h} e^{-t^2}\, dt \right) =$

    (A) $-\dfrac{1}{2e}$ (B) $-\dfrac{2}{e}$ (C) 0   (D) $\dfrac{1}{e}$

    (E) the limit does not exist

8. If the differential equation $\dfrac{dy}{dx} = y - 2y^2$ has a solution curve $y = f(x)$ containing point $\left(0, \dfrac{1}{4} \right)$, then $\lim\limits_{x \to \infty} f(x) =$

    (A) 0   (B) $\dfrac{1}{4}$ (C) $\dfrac{1}{2}$ (D) 2

    (E) the limit does not exist

9. If the acceleration vector of a particle in motion on a plane is $\vec{r}''(t) = (t\sin t,\ \sin t)$ and the velocity vector at $t = 0$ is $\vec{r}'(0) = (0, 1)$, then $\vec{r}'(\pi) =$

(A) $(-\pi, 2)$ (B) $(-\pi, 0)$ (C) $(0, 1)$ (D) $(\pi, 1)$ (E) $(\pi, 3)$

10. After the substitution $u = \sqrt{x}$ in $\int \dfrac{\sqrt{x}}{\sqrt{x}+1}\, dx$, the resulting integral is

(A) $\int (1+u)\, du$

(B) $\int \dfrac{1}{u+1}\, du$

(C) $\int \dfrac{u}{u+1}\, du$

(D) $2\int (u+u^2)\, du$

(E) $2\int \dfrac{u^2}{u+1}\, du$

11. If $f(x) = \tan^{-1} x$ then
$$\lim_{x \to \sqrt{3}} \frac{f'(x) - f'\left(\sqrt{3}\right)}{x - \sqrt{3}} =$$

(A) $-\dfrac{\sqrt{3}}{8}$ (B) $-\dfrac{1}{4\sqrt{3}}$ (C) $\dfrac{1}{4}$ (D) $\dfrac{\pi}{6}$ (E) $\dfrac{\pi}{3}$

12. If
$$f(x) = \begin{cases} \dfrac{|x|-2}{x-2}, & x \neq 2 \\ k, & x = 2 \end{cases},$$
then the value of k for which $f(x)$ is continuous for all real values of x is $k =$

(A) –2   (B) –1   (C) 0   (D) 1   (E) 2

13. The radius of convergence of the power series $\displaystyle\sum_{n=1}^{\infty} \frac{(3x+4)^n}{n}$ is

(A) 0   (B) $\dfrac{1}{3}$ (C) $\dfrac{2}{3}$ (D) $\dfrac{3}{4}$ (E) $\dfrac{4}{3}$

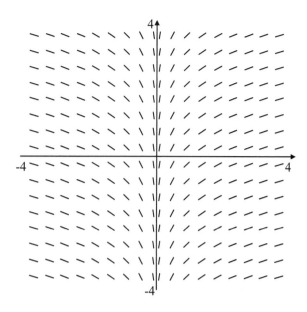

14. The slope field shown above matches which differential equation?

(A) $\dfrac{dy}{dx} = \dfrac{1}{x}$

(B) $\dfrac{dy}{dx} = \dfrac{1}{x^2}$

(C) $\dfrac{dy}{dx} = \dfrac{y}{x}$

(D) $\dfrac{dy}{dx} = \dfrac{\ln x}{x}$

(E) $\dfrac{dy}{dx} = \dfrac{\sin x}{x}$

15. The graph of $f(x) = \dfrac{\sin x}{|x|}$ has

(A) no horizontal asymptotes and no vertical asymptotes
(B) one horizontal asymptote and no vertical asymptotes
(C) one horizontal asymptote and one vertical asymptote
(D) one horizontal asymptote and two vertical asymptotes
(E) two horizontal asymptotes and one vertical asymptote

16. If $f(x) = 4^{3x}$ then $f'(x) =$

(A) $3x\left(4^{3x-1}\right)$

(B) $4^{3x}\left(\ln 4\right)$

(C) $3\left(4^{3x}\right)\left(\ln 4\right)$

(D) $\dfrac{4^{3x}}{\ln 4}$

(E) $\dfrac{4^{3x}}{x \ln 4}$

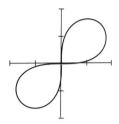

17. The area enclosed by the lemniscate given by the polar equation $r^2 = 4\sin 2\theta$ (whose graph is shown above) is

(A) 2   (B) 4   (C) 2π (D) 8   (E) 4π

18. Which of the following series converges?

(A) $\sum\limits_{n=1}^{\infty} \dfrac{\sqrt{n}}{n+1}$ (B) $\sum\limits_{n=2}^{\infty} \dfrac{1}{\ln n}$ (C) $\sum\limits_{n=1}^{\infty} \sqrt{\dfrac{n}{n+1}}$

(D) $\sum\limits_{n=1}^{\infty} \dfrac{1}{n \ln n}$ (E) $\sum\limits_{n=1}^{\infty} \dfrac{\ln n}{n^2}$

19. If $f(x) = e^x$ then $\dfrac{d}{dx}\left[f(f(x))\right] =$

(A) e^{x^2} (B) e^{e^x} (C) e^{x^e} (D) e^{2e^x} (E) e^{x+e^x}

20. If $\dfrac{dy}{dx} = 1 - \dfrac{x}{y}$ and $y(1) = 1$, then when Euler's method with a step size of 0.5 is used to approximate $y(2)$, the approximation is

(A) 0   (B) 0.375   (C) 0.5   (D) 0.75   (E) 1.5

21. For $x > 0$, $\dfrac{d}{dx} \displaystyle\int_{x}^{2x} \ln t \, dt =$

(A) $-\dfrac{1}{2x}$ (B) $\ln 2$ (C) $\ln 4$ (D) $\ln(2x)$ (E) $\ln(4x)$

22. Suppose the first three terms of the Maclaurin series for e^x are used to approximate $\dfrac{1}{\sqrt{e}}$. If a is the approximate value of $\dfrac{1}{\sqrt{e}}$ obtained and $b = \left| \dfrac{1}{\sqrt{e}} - a \right|$, then

(A) $a = \dfrac{5}{8}$ and $\dfrac{1}{24} \le b < \dfrac{1}{8}$

(B) $a = \dfrac{5}{8}$ and $\dfrac{1}{48} \le b < \dfrac{1}{24}$

(C) $a = \dfrac{5}{8}$ and $b < \dfrac{1}{48}$

(D) $a = \dfrac{3}{4}$ and $\dfrac{1}{24} \le b < \dfrac{1}{8}$

(E) $a = \dfrac{3}{4}$ and $b < \dfrac{1}{48}$

---

23. If the region underneath $y = \dfrac{10}{x^2}$ and above the x-axis for $x \ge 1$ is divided into two regions with equal areas by the line $x = a$, then $a =$

(A) 1   (B) 2   (C) 5   (D) 10   (E) 100

---

24. A series expansion for $f(x) = \dfrac{x}{1+x^2}$ is

(A) $1 - x^2 + x^4 - x^6 + \ldots$

(B) $x + x^3 + x^5 + x^7 + \ldots$

(C) $x - x^3 + x^5 - x^7 + \ldots$

(D) $x^2 - x^3 + x^4 - x^5 + \ldots$

(E) $x^2 - \dfrac{x^4}{3} + \dfrac{x^6}{5} - \dfrac{x^8}{7} + \ldots$

25. $\int_0^2 x\sqrt{4-x^2}\, dx =$

(A) 1 (B) $\dfrac{4}{3}$ (C) $\dfrac{8}{3}$ (D) 2 (E) $\dfrac{16}{3}$

26. The integral expression $\int_1^2 \sqrt{1+\dfrac{4}{x^2}}\, dx$ could represent the arc length from $x=1$ to $x=2$ for the function $f(x) =$

(A) $-\dfrac{4}{x}$ (B) $\dfrac{2}{x}$ (C) $\ln\left(\dfrac{2}{x}\right)$ (D) $\ln\left(x^2\right)$ (E) $\ln\left(x^4\right)$

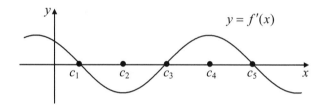

27. The graph of $f'(x)$, the derivative of f, is shown above. $f(x)$ would increase most rapidly at which of the following domain values?

(A) c_1 (B) c_2 (C) c_3 (D) c_4 (E) c_5

28. What is the slope $\dfrac{dy}{dx}$ of the polar curve $r=\dfrac{3}{\theta}$ at $\theta=\dfrac{\pi}{2}$?

(A) $\dfrac{-4}{\pi}$ (B) $\dfrac{-2}{\pi}$ (C) 0 (D) $\dfrac{2}{\pi}$ (E) $\dfrac{\pi}{2}$

Calculus BC
Section I, Part B
Time — 50 minutes
Number of questions — 17

A graphing calculator is required for some questions.

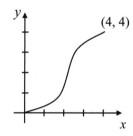

29. For the graph of $y = f(x)$ defined on $0 \le x \le 4$, as shown above, a graph of

$F(x) = \int_{x}^{0} f(t)\, dt$ is best represented by:

(A)

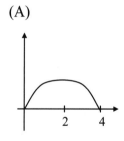

(B)

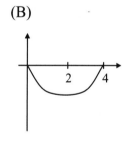

(C)

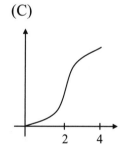

(D)

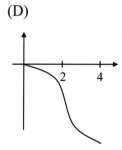

(E)

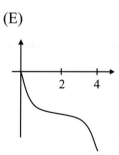

30. If $f(x) = \int_0^x \cos\left(t^2\right) dt$, then a linear approximation for $f(2)$ using the y value on the tangent line to $f(x)$ at $x = \sqrt{\pi}$ is

    (A) −0.891   (B) −0.228   (C) 0.435   (D) 0.663   (E) 1.805

31. If the region bounded by $y = \sin^{-1} x$, $y = \dfrac{\pi}{2}$, and $x = 0$ is rotated about the y-axis, the volume of the solid formed is

    (A) 2.467   (B) 2.605   (C) 2.694   (D) 4.609   (E) 5.636

32. If $f(x)$ is differentiable over the positive real numbers, then which of the following statements must be true?

    I. $f(x)$ is continuous over the positive real numbers
    II. $f(-x)$ is differentiable over the negative real numbers
    III. $f(|x|)$ is differentiable over all real numbers

    (A) I only   (B) II only   (C) I and II
    (D) II and III   (E) I, II, and III

33. The change in price of a popular toy was proportional to the square root of the number of weeks it had been on the market. The toy originally cost \$5.00 and increased by a nickel by the end of the first week. How much did the toy cost (to the nearest penny) at the end of the first 12 weeks?

    (A) \$5.60   (B) \$5.69   (C) \$5.99   (D) \$7.08   (E) \$7.34

34. If $x = \left[f(t)\right]^2$ and $y = f(t)$, then $\dfrac{d^2 y}{dx^2} =$

    (A) $\dfrac{1}{2f(t)}$ (B) $\dfrac{f'(t)}{2f(t)}$ (C) $-\dfrac{1}{4f^2(t)}$

    (D) $-\dfrac{f'(t)}{2f^2(t)}$ (E) $-\dfrac{1}{4f^3(t)}$

35. An equation of a tangent line to the parametric curve $x = e^t$, $y = 2^t$ at $t = 0$ is:

   (A) $y - 1 = \dfrac{1}{2}(x - 1)$ (B) $y - 1 = (\ln 2)(x - 1)$ (C) $y = (\ln 2)x$

   (D) $y = \dfrac{1}{\ln 2}(x - 1)$ (E) $y - 1 = \dfrac{e}{2}(x - 1)$

36. Recall that the diagonal of a cube is $\sqrt{3}$ times its side. The diagonal of a cube is expanding at the rate of 0.5 cm per second. How fast is the volume of the cube changing, in cm^3/sec, when the diagonal is 3 cm?

   (A) 1.5   (B) 2.598   (C) 5.196   (D) 9   (E) 10.392

37. If a particle moves on the curve $x = \sin t$, $y = \sin 2t$, then at time $t = 3$ the speed of the particle is

   (A) 0.930   (B) 0.965   (C) 1.379   (D) 1.645   (E) 2.161

38. The base of a solid is a region bounded by $y = \ln x$, $y = 0$, and $x = e$. Cross sections perpendicular to the base and the y-axis are squares. An integral expression for the volume of this solid is

   (A) $\displaystyle\int_0^1 (e - e^y)^2 \, dy$ (B) $\displaystyle\int_0^e (e - e^y)^2 \, dy$ (C) $\displaystyle\int_1^e (e - \ln y)^2 \, dy$

   (D) $\displaystyle\int_1^e (\ln x)^2 \, dx$ (E) $\displaystyle\int_1^e (e - \ln x)^2 \, dx$

39. The minimum distance from point (5, 6) to the curve $y = x^2 + 1$ is

   (A) 2.358   (B) 2.501   (C) 2.701   (D) 2.913   (E) 3.015

40. If $x + y = \tan^{-1}(xy)$, then $\dfrac{dy}{dx} =$

(A) $\dfrac{1 + x^2 y^2}{x}$ (B) $\dfrac{y}{1 + xy - x}$ (C) $\dfrac{x - 1 - x^2 y^2}{1 + x^2 y^2}$

(D) $\dfrac{1 + x^2 + y^2}{x - x^2 y^2 - 1}$ (E) $\dfrac{y - 1 - x^2 y^2}{1 + x^2 y^2 - x}$

---

41. $\displaystyle\sum_{k=1}^{2n} \sqrt{\dfrac{1 + \dfrac{k}{n}}{n^2}}$ is a Riemann sum for which of the following integrals?

(A) $\displaystyle\int_0^1 \sqrt{1 + x}\, dx$ (B) $\displaystyle\int_0^2 \sqrt{1 + x}\, dx$ (C) $\dfrac{1}{2} \displaystyle\int_0^1 \sqrt{1 + x}\, dx$

(D) $\displaystyle\int_0^1 \sqrt{1 + 2x}\, dx$ (E) $\displaystyle\int_0^2 \sqrt{1 + 2x}\, dx$

---

42. Particle A's velocity function is $v(t) = 4 - t^2$ m/sec over $0 \le t \le 4$ seconds. For the same time interval, particle B's velocity function is $v(t) = t^3 - 4t$ m/sec. The difference in the total meters traveled by the two particles over $0 \le t \le 4$ is

(A) 24   (B) $26\dfrac{2}{3}$ (C) $34\dfrac{2}{3}$ (D) $37\dfrac{1}{3}$ (E) $45\dfrac{1}{3}$

---

43. $\displaystyle\sum_{n=3}^{\infty} \dfrac{e^{\frac{n}{2}}}{\pi^n}$ is

(A) 0.304   (B) 0.525   (C) 4.808   (D) 9.624   (E) divergent

---

44. A line from the point (4, 1) perpendicular to a tangent line to the graph of $f(x) = x^2$ intersects the graph of $y = f(x)$ at $x =$

(A) 1.392   (B) 1.647   (C) 1.939   (D) 4.472   (E) 7.873

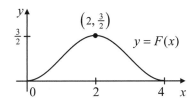

45. The graph of $F(x) = \int_0^x f(t)\, dt$ is shown above, for $0 \le x \le 4$. Which of the following is necessarily true?

I. $\int_0^4 f(t)\, dt = 3$

II. $\int_2^4 f(t)\, dt = \dfrac{3}{2}$

III. $\int_2^0 f(t)\, dt = \int_2^4 f(t)\, dt$

(A) I only　　　　　(B) II only　　　　　(C) III only

(D) I and II　　　　(E) I, II, and III

Practice Exam BC-1

Calculus BC
Section II, Part A
Time — 45 minutes
Number of problems — 3

A graphing calculator is required for some problems or parts of problems.

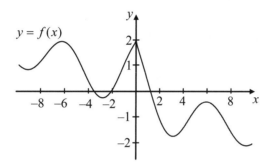

1. The function f is defined by $f(x) = \dfrac{4\cos x - x}{2 + \sqrt[3]{x^2}}$. The graph of $y = f(x)$ is shown above. The functions g and h have derivatives given by $g'(x) = f(x)$ and $h'(x) = \displaystyle\int_0^x f(t)\,dt$. The graph of g contains the point $(0, 1)$.

   (a) Write an equation for the tangent line to the graph of g at $x = 0$.

   (b) Find the absolute maximum value of g on the closed interval $[0, 1]$. Give the reason for your answer.

   (c) Find the x-coordinate of each point of inflection on the graph of $y = h(x)$ for values of x between -10 and 10. Justify your answer.

   (d) Which is larger, $h'(2)$ or $h'(3)$? Justify your answer.

---

2.   A stadium is filling with people from noon until game time at 3:00 p.m. The table below shows the rate of people entering the stadium (measured in people per minute) at particular times.

Time	Noon	1:00	1:30	2:00	2:20	2:40	2:50	2:55	3:00
Rate (people/min)	1	4	18	56	81	74	60	52	44

(a)   Estimate the total number of people attending the game using a trapezoidal approximation for four partitions, starting at noon, 1:30, 2:20, and 2:50.

(b)   The heaviest traffic occurs between 2:00 and 2:50. Estimate the average rate (number per minute) of people entering the stadium for that time period using a right-hand Riemann sum for 3 subintervals.

(c)   One possible model for the rate at which people enter the stadium is given by the function $R(t) = \dfrac{5 \cdot 10^5 e^{-0.05t}}{\left(1 + 1500e^{-0.05t}\right)^2}$ people per minute, where t is the time in minutes since noon. Write and evaluate (rounded to the nearest person) an expression that represents the total number of people at the stadium at 3 p.m., as predicted by this model.

(d)   Using the $R(t)$ model from Part (c), set up and evaluate an expression for the average number of people per minute entering the stadium between 2:00 and 2:50.

3.   A particle moves along a path on a coordinate plane determined by the parametric equations $x(t) = t^2 + 1$ and $y(t) = t^3 - 3t$ over a time interval $0 \le t \le 3$, where the x and y coordinates are measured in centimeters and t is measured in seconds.

(a)   At what time is the particle traveling at a speed of 5 cm/sec?

(b)   How fast is the particle's distance from its initial position changing at time $t = 2$ sec?

(c)   Set up and evaluate an expression for the total distance traveled by the particle from $t = 0$ to $t = 3$ sec.

Calculus BC
Section II, Part B
Time — 45 minutes
Number of problems — 3

No calculator is allowed for these problems.

4.   The function F is defined as $F(x) = \int_1^x \sqrt{t^2 + 3t}\, dt$.

   (a)   Write the second-degree Taylor polynomial for $F(x)$ at $x = 1$.

   (b)   Approximate $\int_1^0 \sqrt{t^2 + 3t}\, dt$ using the Taylor polynomial from Part (a).

   (c)   Approximate $\int_0^2 \sqrt{t^2 + 3t}\, dt$ using the Taylor polynomial from Part (a) and
          properties of integrals.

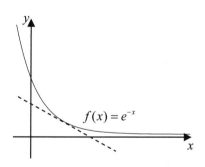

$f(x) = e^{-x}$

5.   Consider the graph of the function $f(x) = e^{-x}$ for $0 \le x < \infty$ above.

   (a)   Set up and evaluate an expression for the area of the region in the first
          quadrant below the graph of $f(x)$.

   (b)   Write an equation for the line tangent to the graph of $f(x)$ at $x = c$ and find
          the area of the right triangle bounded by this tangent line and the x- and
          y-axes.

   (c)   When the triangular area found in Part (b) is removed from the area found in
          Part (a), what is the minimum possible area remaining? Justify your answer.

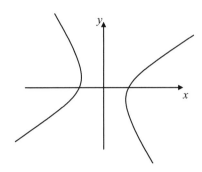

6.   The curve S, graphed above, is defined by the equation $xy + y^2 = x^2 - 5$.

   (a)   Show that S is a solution to the differential equation $\dfrac{dy}{dx} = \dfrac{2x - y}{x + 2y}$.

   (b)   Find the x- and y-coordinates of all points where S has vertical tangents.

   (c)   Sketch the slope field for the differential equation given in Part (a) at the 8 points indicated below.

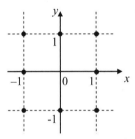

   (d)   Write the equation of the line in the xy-plane that holds all the points with slope 0 in the slope field in Part (c) .

Practice Exam BC-2

Calculus BC
Section I, Part A
Time — 55 minutes
Number of questions — 28

No calculator is allowed for these questions.

---

1. If $f(x) = \dfrac{x^2}{\tan 2x}$, then $f'(x) =$

    (A) $\dfrac{2x}{2\sec^2 2x}$

    (B) $\dfrac{2x \tan 2x - 2x^2 \sec^2 2x}{\tan^2 2x}$

    (C) $\dfrac{2x \tan 2x + 2x^2 \sec^2 2x}{\tan^2 2x}$

    (D) $\dfrac{2x \tan 2x - x^2 \sec^2 2x}{\tan^2 2x}$

    (E) $\dfrac{2x^2 \sec^2 2x - 2x \tan 2x}{\tan^2 2x}$

---

2. What are all the values of t such that the graph of the curve given by the parametric equations $x(t) = t^3 - 6t^2$ and $y(t) = t^2 - 16$ has a vertical tangent?

    (A) $t = 0$ only   (B) $t = 4$ only   (C) $t = 0$ and $t = 4$

    (D) $t = 4$ and $t = -4$ (E) $t = 0$ and $t = 6$

---

3. $\displaystyle\int_2^\infty \frac{\cos\left(\dfrac{2}{x}\right)}{x^2}\,dx =$

   (A) $-\dfrac{1}{2}\sin 1$ (B) $-\dfrac{1}{4}\cos 1$ (C) $\dfrac{1}{2}\sin 1$ (D) $\sin 1$ (E) nonexistent

4. If $\sin 2x + \cos 2y = x - y$, then $\dfrac{dy}{dx} =$

   (A) $\dfrac{1 - 2\cos 2x}{1 - 2\sin 2y}$ (B) $1 - 2\cos 2x + 2\sin 2y$ (C) $\dfrac{1 - 2\cos 2x}{1 + 2\sin 2y}$

   (D) $\dfrac{1 - \cos 2x}{1 - \sin 2y}$ (E) $\dfrac{1 - \cos 2x}{1 + \sin 2y}$

5. $\displaystyle\lim_{w\to 0} \frac{\ln\left(\dfrac{2+w}{2}\right)}{w} =$

   (A) $2\ln 2 - 2$ (B) $\dfrac{1}{2}$ (C) 1 (D) 2 (E) nonexistent

6. If $g(x) = \sin 2x + x^2$ for $-\dfrac{\pi}{4} \le x \le \dfrac{\pi}{4}$, what are all the values of x at which the graph of $y = g(x)$ has a point of inflection?

   (A) $\pm\dfrac{\pi}{12}$ (B) $\pm\dfrac{\pi}{6}$ (C) $\dfrac{\pi}{12}$ only   (D) $\dfrac{\pi}{6}$ only   (E) $\pm\dfrac{\pi}{6}$ and $\pm\dfrac{\pi}{12}$

7. For which of the following functions f does the Mean Value Theorem guarantee that there will be a value c in $(0, 2)$ such that $f'(c) = \dfrac{f(2) - f(0)}{2}$?

   I. $f(x) = 3x^7 - \sqrt{3}x^5 + 17x - \dfrac{\pi}{4}$ II. $f(x) = |x - 1.5|$ III. $f(x) = \dfrac{\pi}{x - 1}$

   (A) I only   (B) II only   (C) III only

   (D) II and III only   (E) I, II, and III

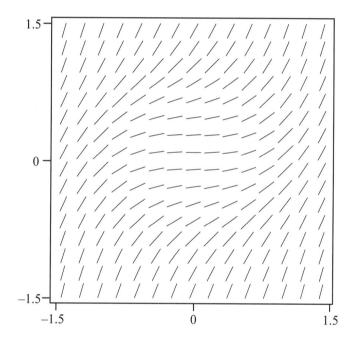

8. The graph above is a slope field for which of the following differential equations?

   (A) $\dfrac{dy}{dx} = -\dfrac{x}{y}$ (B) $\dfrac{dy}{dx} = \arctan x$ (C) $\dfrac{dy}{dx} = x^3$

   (D) $\dfrac{dy}{dx} = x + y$ (E) $\dfrac{dy}{dx} = x^2 + y^2$

9. If $F'(x) = \sec^2 3x$, then $F(x)$ could be

   (A) $\tan 3x$ (B) $3\tan 3x$ (C) $\dfrac{1}{3}\tan 3x$

   (D) $6\sec^2 3x \tan 3x$ (E) $6\sec 3x$

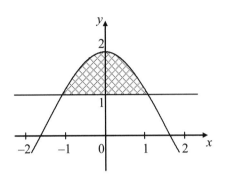

10.   The shaded region in the figure above is bounded by the graphs of $y = 2\cos x$ and $y = 1$. What is its area?

(A) $\dfrac{\pi}{6} - \dfrac{1}{2}$ (B) $\dfrac{\pi}{3} - 1$ (C) $\sqrt{3} - \dfrac{\pi}{3}$ (D) $2\sqrt{3} - \dfrac{2\pi}{3}$ (E) $2\sqrt{3}$

11.   If $f(x)$ is represented by the Maclaurin series

$$-x + \frac{x^3}{3!} - \frac{x^5}{5!} + \ldots + (-1)^n \frac{x^{2n-1}}{(2n-1)!} + \ldots,$$ which of the following is an equation for the

line tangent to the graph of f at the point where $x = \dfrac{2\pi}{3}$?

(A) $y - \dfrac{\sqrt{3}}{2} = -\dfrac{1}{2}\left(x - \dfrac{2\pi}{3}\right)$

(B) $y - \dfrac{1}{2} = -\dfrac{\sqrt{3}}{2}\left(x - \dfrac{2\pi}{3}\right)$

(C) $y + \dfrac{1}{2} = -\dfrac{\sqrt{3}}{2}\left(x - \dfrac{2\pi}{3}\right)$

(D) $y + \dfrac{1}{2} = -\dfrac{1}{2}\left(x - \dfrac{2\pi}{3}\right)$

(E) $y + \dfrac{\sqrt{3}}{2} = \dfrac{1}{2}\left(x - \dfrac{2\pi}{3}\right)$

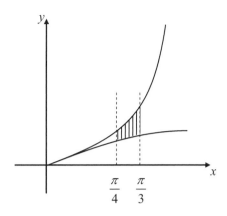

12. The figure above shows the graphs of $y = \tan x$ and $y = \sin x$. Between $x = \dfrac{\pi}{4}$ and $x = \dfrac{\pi}{3}$, vertical line segments are drawn between the two graphs. What is the average length of all such line segments?

(A) $\dfrac{12}{\pi}\left(\ln\sqrt{2} + \dfrac{1-\sqrt{2}}{2} \right)$ (B) $\dfrac{12}{\pi}\left(\ln\sqrt{2} + \dfrac{\sqrt{2}-1}{2} \right)$ (C) $\dfrac{12}{\pi}\left(2 + \dfrac{1-\sqrt{2}}{2} \right)$

(D) $\dfrac{12}{\pi}\left(2 + \dfrac{\sqrt{2}-1}{2} \right)$ (E) $\dfrac{12}{\pi}\left(\dfrac{\sqrt{3}+\sqrt{2}}{2} - 1 \right)$

13. The derivatives of the functions f, g, and h are given below. Which of the functions f, g, and h are decreasing on $[1, 2]$?

$f'(x) = -x^4$ $g'(x) = \sin(\pi x)$ $h'(x) = 3 - 2x$

(A) f only   (B) g only   (C) h only

(D) f and g only   (E) f, g, and h

14. To what number does the infinite series $1 - \pi^2 + \dfrac{\pi^4}{2!} - \dfrac{\pi^6}{3!} + \ldots + \dfrac{(-1)^n \pi^{2n}}{n!} + \ldots$ converge?

(A) $\cos\left(\pi^2\right)$ (B) $e^{-\pi^2}$ (C) e^{π} (D) $e^{2\pi}$ (E) e^{π^2}

15. Which one of the following could be the graph of the solution to the differential equation $\frac{dy}{dx} = -0.5x$ with $y(0) = 0$?

(A)

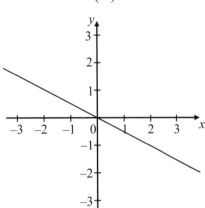

(B)

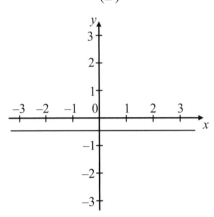

(C)

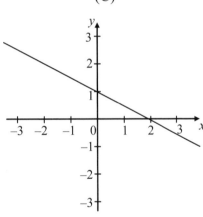

(D)

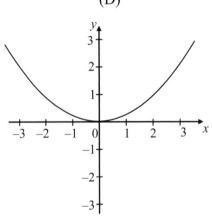

(E)

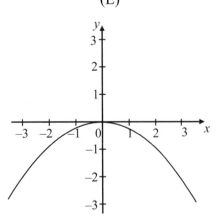

16. If $g(x) = \lim\limits_{h \to 0} \dfrac{\tan(x+h) - \tan x}{h}$, then $g'\left(\dfrac{\pi}{3}\right) =$

(A) $8\sqrt{3}$ (B) $2\sqrt{3}$ (C) $\sqrt{3}$ (D) 2   (E) 1

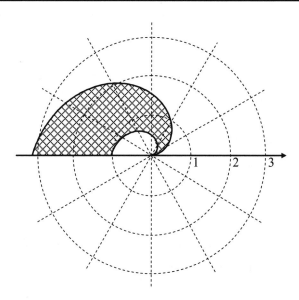

17. The figure above shows the polar graphs of $r_1(\theta) = \dfrac{\theta}{3}$ and $r_2(\theta) = \theta$. What is the area of the shaded region shown, bounded by the graphs of $r_1(\theta)$, $r_2(\theta)$, and the ray $\theta = \pi$?

(A) $\dfrac{\pi^3}{54}$ (B) $\dfrac{\pi^2}{6}$ (C) $\dfrac{4\pi^3}{27}$ (D) $\dfrac{\pi^3}{6}$ (E) $\dfrac{8\pi^3}{27}$

x	0	0.5	1
$f'(x)$	-2	$-\dfrac{3}{2}$	$-\dfrac{3}{4}$

18. The table above gives selected values of the derivative of a function f. If $f(0) = 3$, what is the approximation of $f(1)$ using Euler's method with step size 0.5?

(A) $-\dfrac{7}{4}$ (B) $-\dfrac{1}{2}$ (C) $\dfrac{7}{8}$ (D) $\dfrac{5}{4}$ (E) 2

19. If $\dfrac{dy}{dx} = \dfrac{2x+1}{3y^2}$ and $y = 2$ when $x = 1$, what is y when $x = 3$?

(A) $\dfrac{5}{2}$ (B) $\sqrt[3]{7}$ (C) $\sqrt[3]{12}$ (D) $\sqrt[3]{16}$ (E) $\sqrt[3]{18}$

---

20. If $f(x) = \begin{cases} \ln 4 - \dfrac{x}{4}, & \text{if } x \le 4 \\ \ln x - \dfrac{x^2}{16}, & \text{if } x > 4 \end{cases}$, which of the following statements are true?

I. f is continuous for all real numbers
II. f is differentiable for all real numbers
III. $\displaystyle\int_3^5 f(x)\,dx$ exists

(A) I only   (B) III only   (C) I and II only
(D) I and III only   (E) I, II, and III

---

21. $\displaystyle\int \dfrac{dx}{5x - x^2} =$

(A) $\ln\left|5x - x^2\right| + C$ (B) $\ln\left|5x\right| - \ln\left(x^2\right) + C$ (C) $\ln\left|\dfrac{x}{5-x}\right| + C$

(D) $\dfrac{1}{5}\ln\left|\dfrac{x}{5-x}\right| + C$ (E) $\dfrac{1}{5}\ln\left|x \cdot (5-x)\right| + C$

---

22. If $g(x) = 6 + \displaystyle\int_{-2}^{x} \cos(w^5)\,dw$, then $g'(x) =$

(A) 0 (B) $\sin\left(x^5\right)$ (C) $\cos\left(x^5\right)$

(D) $6 + 5x^4 \sin\left(x^5\right)$ (E) $6 + \sin\left(x^5\right) - \sin(-32)$

23. $\int 2\arctan x\,dx =$

(A) $\dfrac{2}{1+x^2}+C$

(B) $2x\arctan x - \ln\left(1+x^2\right)+C$

(C) $2x\arctan x + \ln\left(1+x^2\right)+C$

(D) $2\ln\left|\operatorname{arcsec} x\right|+C$

(E) $2x\arctan x + 2\sqrt{1-x^2}+C$

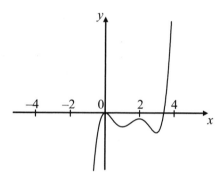

24.   The graph of $y = f(x)$ is shown above. If $g(x) = \displaystyle\int_0^x f(t)\,dt - \int_2^x f(t)\,dt$, then how many points of inflection are there on the graph of $y = g(x)$?

(A) 0   (B) 1   (C) 2   (D) 3   (E) 4

25.   What is $\displaystyle\lim_{x\to\infty}\dfrac{\arctan x + 2x}{x+e^{-x}}$?

(A) 0   (B) 1   (C) $\dfrac{\pi}{2}$ (D) 2   (E) nonexistent

26. If the third-degree Taylor polynomial for $\sin x$ at $x = \pi$ is used to approximate $\sin 3$, what is the approximation?

(A) $-(3-\pi) + \dfrac{(3-\pi)^3}{6}$ (B) $-(3-\pi) + \dfrac{(3-\pi)^3}{3}$ (C) $(3-\pi) - \dfrac{(3-\pi)^3}{6}$

(D) $(3-\pi) + \dfrac{(3-\pi)^3}{3}$ (E) $-(3-\pi) + \dfrac{(3-\pi)^2}{2} - \dfrac{(3-\pi)^3}{6}$

27. A particle moves in the xy-plane so that its position at any time t is given by $x(t) = 2\sin 4t$ and $y(t) = -3\cos 4t$. Which of the following expressions gives the length of the acceleration vector of the particle at $t = \dfrac{\pi}{4}$?

(A) -48 (B) $\sqrt{48}$ (C) 8 (D) 48 (E) $\sqrt{32^2 + 48^2}$

28. If $\dfrac{dy}{dt} = 0.3y\left(1 - \dfrac{y}{0.06}\right)$ and $y(0) = 0.1$, which of the following intervals is the largest such interval where $t > 0$ and $y(t)$ is decreasing?

(A) $0 < t < 0.06$ (B) $0 < t < 0.03$ (C) $0.03 < t < 0.06$

(D) $t > 0.03$ (E) all positive values of t

Calculus BC
Section I, Part B
Time — 50 minutes
Number of questions — 17

A graphing calculator is required for some questions.

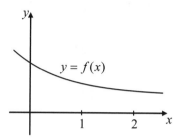

29. Let $g(x) = \int_1^x f(t)\,dt$, where f is the function whose graph is shown above. Which of the following could be a table of values for g?

(A)		(B)		(C)		(D)		(E)	
x	$g(x)$	x	$g(x)$	x	$g(x)$	x	$g(x)$	x	$g(x)$
0	3	0	−2	0	−3	0	−2	0	3
1	2	1	−1	1	0	1	0	1	0
2	1	2	−0.5	2	2	2	3	2	2

30. If $f'(x) < 0$ for all real numbers, and $g(x) = f\left(x^3 + e^{-x^2}\right)$, then g has a local maximum at $x =$

(A) −0.806   (B) 0   (C) 0.266   (D) 0.508   (E) 0.513

31. If $y = x^3 - e^x$ for all values of x in the closed interval $[0, 2]$, for what value of x is the instantaneous rate of change of y with respect to x the same as the average rate of change on the interval $[0, 2]$?

(A) 0.183   (B) 0.842   (C) 1.007   (D) 1.149   (E) 1.460

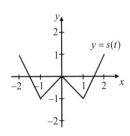

32. The graph of $y = s(t)$ is shown above. Which of the following could be the graph of a function whose derivative is the same as the derivative of s?

(A)

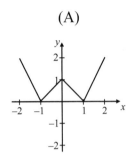

(B)

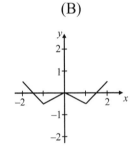

(C)

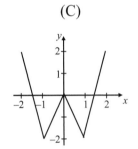

(D)

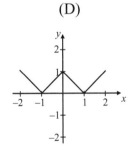

(E)

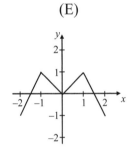

33. If $\lim\limits_{x \to 0} \dfrac{\sin 3x + ax + bx^3}{x^3} = 0$, then what are the values of a and b?

(A) $a = -1, b = 0$

(B) $a = -3, b = 0$

(C) $a = 3, b = 0$

(D) $a = -1, b = \dfrac{9}{2}$

(E) $a = -3, b = \dfrac{9}{2}$

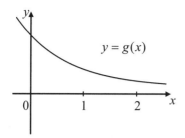

34. The graph of $y = g(x)$ is shown above. Which of the following approximations is greater than $\int_1^2 g(x)\,dx$?

    I.   Left-hand Riemann sum
    II.   Trapezoidal sum
    III. Right-hand Riemann sum

(A) I only   (B) II only   (C) I and II only

(D) II and III only   (E) I, II, and III

---

35. What is the radius of convergence of the series $\displaystyle\sum_{n=1}^{\infty} \frac{n}{3^n}(6x-5)^n$?

(A) $\dfrac{1}{3}$ (B) $\dfrac{1}{2}$ (C) $\dfrac{5}{6}$ (D) $\dfrac{6}{5}$ (E) $\dfrac{4}{3}$

---

t (hours)	0	1	2	3	4	5	6	7	8
R (gals/hr)	120	130	116	118	110	108	109	120	125

36. During an eight-hour cruise, a ship consumes fuel at the rates R given in the table above. What is the approximation of the total fuel consumption in gallons over the eight hours, using a midpoint Riemann sum with 4 subintervals of equal width?

(A) 457.5   (B) 476   (C) 910   (D) 915   (E) 952

---

37. What are all the values of p for which $\displaystyle\int_2^{\infty} \frac{dx}{x(\ln x)^p}$ converges?

(A) $p < 1$ (B) $p > 1$ (C) $p \le 1$ (D) $p \ge 1$ (E) $-1 < p < 1$

38. If $m'(x) = \cos\left(1-x^2\right)$ and $m(2) = 1$, what is $m(3)$?

(A) −0.990   (B) 0.010   (C) 0.104   (D) 1.104   (E) 2.031

---

39. The path of a racetrack follows the curve defined by the parametric equations $x(t) = 2\cos t + \cos^2 t$ and $y(t) = 2\sin t + \sin t \cos t$. Which of the following represents the distance around the track?

(A) $\displaystyle \int_0^{2\pi} \left(-2\sin t - 2\cos t \sin t\right)^2 + \left(2\cos t - \sin^2 t + \cos^2 t\right)^2 \, dt$

(B) $\displaystyle \int_0^{2\pi} \sqrt{\left(-2\sin t - 2\cos t \sin t\right)^2 + \left(2\cos t - \sin^2 t + \cos^2 t\right)^2} \, dt$

(C) $\displaystyle \int_0^{2\pi} \sqrt{1 + \left(\frac{2\sin t + \sin t \cos t}{2\cos t + \cos^2 t}\right)^2} \, dt$

(D) $\displaystyle \int_0^{2\pi} \sqrt{1 + \left(\frac{2\cos t - \sin^2 t + \cos^2 t}{-2\sin t - 2\cos t \sin t}\right)^2} \, dt$

(E) $\displaystyle \int_0^{2\pi} 1 + \left(\frac{2\cos t - \sin^2 t + \cos^2 t}{-2\sin t - 2\cos t \sin t}\right)^2 \, dt$

---

x	$F(x)$	$G(x)$	$G'(x)$
3	−2	7	0
5	4	15	$\sqrt{5}$

40. The functions F and G are continuous for all real numbers, and $F'(x) = G(x)$. F, G, and G' have the values indicated in the table above. What is $\displaystyle \int_3^5 G(x)\,dx$?

(A) $\sqrt{5}$ (B) $2\sqrt{5}$ (C) 6   (D) 8   (E) 12

x	0	1	2	3	4
f(x)	1.7	0	0	1.3	2

41. The function *f* is defined by $f(x) = \int_2^x g(t)\,dt$, where *g* is a continuous function for all real numbers in the closed interval [0, 4]. A table of values for *f* is shown above. Which of the following could be the graph of $y = g(x)$?

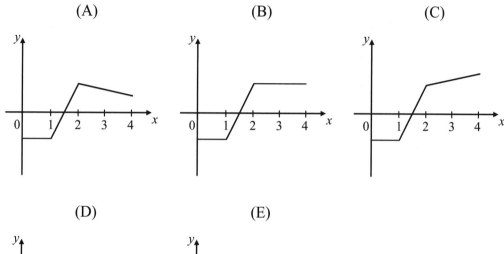

(A)   (B)   (C)

(D)   (E)

42. The base of a solid is the region in the first quadrant bounded by the axes and the graph of $y = 3^{-x^2} - \dfrac{1}{3}$. Each cross section of the solid perpendicular to the *x*-axis is a rectangle with a base in the *xy*-plane and a height that is 2 units more than the length of the base. What is the volume of the solid?

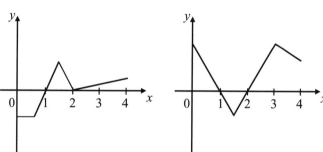

(A) .395   (B) .403   (C) .634   (D) .992   (E) 2.202

43. If $f''(x) = 2^{-\frac{x^2}{5}} \cdot \cos x + \frac{x}{6} - 0.1$, how many points of inflection are there on the graph of $y = f(x)$ between $x = -6$ and $x = 6$?

(A) 1   (B) 2   (C) 3   (D) 4   (E) 5

x	$f(x)$	$f'(x)$
-1	2	$-\dfrac{1}{2}$
2	3	-7

44. If f and its derivative f' have the values given in the table above and $h(x) = f^{-1}(x)$, what is $h'(2)$?

(A) -7 (B) -2 (C) $-\dfrac{1}{2}$ (D) $-\dfrac{1}{7}$ (E) $\dfrac{1}{2}$

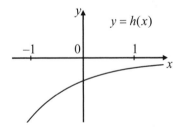

$y = h(x)$

45. The graph of $y = h(x)$ is shown above. Which of the following could be a table of values for the derivative of h, $h'(x)$?

	(A)		(B)		(C)		(D)		(E)
x	$h'(x)$	x	$h'(x)$	x	$h'(x)$	x	$h'(x)$	x	$h'(x)$
-1	$\dfrac{1}{2}$	-1	$-\dfrac{1}{3}$	-1	$\dfrac{1}{2}$	-1	$\dfrac{1}{4}$	-1	-1
0	$\dfrac{1}{2}$	0	$-\dfrac{1}{2}$	0	$\dfrac{1}{3}$	0	$\dfrac{1}{3}$	0	$-\dfrac{1}{2}$
1	$\dfrac{1}{2}$	1	-1	1	$\dfrac{1}{4}$	1	$\dfrac{1}{2}$	1	$-\dfrac{1}{3}$

Practice Exam BC-2

Calculus BC
Section II, Part A
Time — 45 minutes
Number of problems — 3

A graphing calculator is required for some problems or parts of problems.

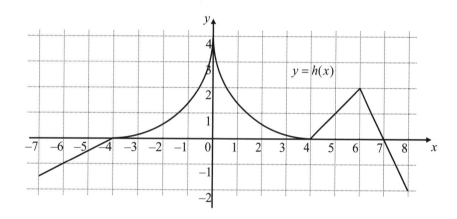

1. The graph of $y = h(x)$, shown above, consists of three line segments and two quarter-circles. The function g is defined by $g(x) = \int_4^x h(t)\,dt$ for all x in the closed interval $[-7, 8]$.

    (a) Write an equation for the line tangent to the graph of $y = g(x)$ at $x = 6$.

    (b) Which is larger, $g(-4)$ or $g(-3)$? Give a reason for your answer.

    (c) What is the absolute maximum of g on the closed interval $[-7, 8]$? Justify your answer.

    (d) Find the x-coordinate of each point of inflection on the graph of $y = g(x)$. Give a reason for your answer.

2.   Consider the differential equation $\dfrac{dy}{dx} = 2\cos\left(e^x\right)$ with the initial condition

$y(0) = -0.5$.

(a)   Use Euler's method with three equal steps to find an approximation of $y(1)$.
        Show the calculations you use to find the approximation.

(b)   Sketch the solution to the differential equation with the given initial condition
        on the slope field shown below.

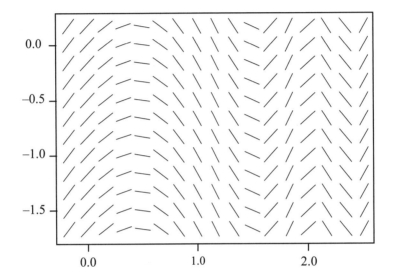

(c)   Use the Fundamental Theorem of Calculus to write an expression for $y(x)$
        and evaluate that expression at $x = 1$.

(d)   Use $\dfrac{d^2y}{dx^2} = -2e^x \sin\left(e^x\right)$ to explain why the approximation of $y(1)$ found in
        Part (a) is greater than the result found in Part (c).

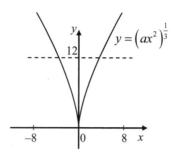

3.   Consider the graph of $y = \left(ax^2\right)^{\frac{1}{3}}$ shown in the picture above.

   (a)   Find the value of a for which the average value of y on the closed interval $-8 \leq x \leq 8$ is equal to 12.

   (b)   Using the value of a determined in Part (a), set up and evaluate the volume of a solid formed by rotating the region bounded by the curve $y = \left(ax^2\right)^{\frac{1}{3}}$ and the line $y = 12$ about that line.

Calculus BC
Section II, Part B
Time — 45 minutes
Number of problems — 3

No calculator is allowed for these problems.

---

4.   Let R be the region in the first quadrant enclosed by the graphs of $f(x) = x^3$ and $g(x) = kx$, where k is a positive real number. The value of k is increasing at a constant rate of 9 units per second.

(a)   Find the area of R in terms of k.

(b)   How fast is the area of R changing at the instant when $k = 4$?

(c)   The region R is revolved around the x-axis. Find the volume of the generated solid in terms of k.

(d)   How fast is the volume of the solid described in Part (c) changing at the instant when $k = 4$?

---

5.   Consider the function f defined by the power series $f(x) = \sum_{n=1}^{\infty} \frac{nx^{n-1}}{3^n}$ for all x in the interval of convergence.

(a)   Find the radius of convergence.

(b)   Let g be the function such that $g'(x) = f(x)$ and $g(1) = 3$. Find the first three non-zero terms and the general term for the series representation of $g(x)$.

(c)   If possible, evaluate $g(3)$. If this is not possible, explain why not.

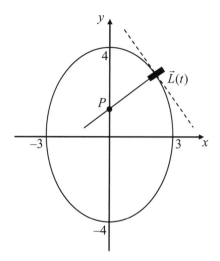

6.   A small laser moves counterclockwise along the ellipse $\dfrac{x^2}{9} + \dfrac{y^2}{16} = 1$. The position

vector of the laser at any time t is given by $\vec{L}(t) = \left(3\cos\left(\dfrac{\pi t}{12}\right),\, 4\sin\left(\dfrac{\pi t}{12}\right) \right)$. From

every position $\vec{L}(t)$ on its path, the laser emits a beam in the direction perpendicular to the tangent line to the ellipse at the laser's current position, towards the y-axis, as shown in the figure above.

(a)   Write an expression for the velocity vector of the laser in terms of t. Write an expression for the slope of the tangent line to the ellipse at the point $\vec{L}(t)$.

(b)   Write an equation for the laser beam line that goes through the point $\vec{L}(t)$. (The laser beam line is perpendicular to the tangent line to the ellipse at $\vec{L}(t)$.)

(c)   The laser beam, emitted at time t, intersects the y-axis at the point P with coordinates $\big(0,\, h(t)\big)$ (unless the beam is vertical). Write an expression for $h(t)$ in terms of t (where $h(t)$ is defined). Is the point P moving up or down at $t = 2$? Justify your answer.

(d) $h(t)$, the vertical position of the point P, is undefined when the laser beam is vertical. What values should be assigned to $h(6)$ and $h(18)$ so that $h(t)$ became continuous on the interval $0 \le t \le 24$? With this adjustment, what is the total distance traveled by the point P over the time interval $0 \le t \le 24$?

Answers and Solutions

AB-1

SECTION I: MULTIPLE CHOICE

1.	E	11.	E	21.	D	31.	D	41.	D
2.	E	12.	B	22.	D	32.	E	42.	C
3.	E	13.	C	23.	E	33.	E	43.	C
4.	D	14.	B	24.	C	34.	A	44.	E
5.	B	15.	A	25.	D	35.	B	45.	B
6.	B	16.	C	26.	C	36.	A		
7.	B	17.	B	27.	D	37.	D		
8.	B	18.	A	28.	E	38.	C		
9.	A	19.	C	29.	A	39.	E		
10.	A	20.	C	30.	C	40.	C		

Notes:

2. $3x^{\frac{1}{3}}\Big|_1^8 = 3(2-1)$.

4. $6x - 2y - 2xy' + 2yy' = 0 \Rightarrow y'\big|_{x=1,y=-2} = \dfrac{6 - 2 \cdot (-2)}{2 - 2 \cdot (-2)}$.

6. $u = \sin 2x;\ du = 2\cos(2x)\,dx$. You get $\int_{\sqrt{3}/2}^1 u\,du$.

9. You can cancel $(x-5)$ from the numerator and denominator, but it may be faster to use l'Hôpital's Rule: $\displaystyle\lim_{x\to 5} \frac{2x^2 - 50}{x^2 - 15x + 50} = \lim_{x\to 5} \frac{4x}{2x - 15} = \frac{4 \cdot 5}{10 - 15}$.

11. Use u-substitution with $u = 4 - x^2$, $du = -2x\,dx$.

12. $-72 + \displaystyle\int_0^t \tau^2\,d\tau = 0 \Rightarrow \frac{t^3}{3} = 72$.

13. $\dfrac{d^2y}{dx^2} = 6x + 6 = 0$.

15. $f(-3.1) = f(-3) + f'(-3)(-0.1) = 4 - 2 \cdot (0.1) = 3.8$.

16. $\dfrac{1}{4} \int_{1}^{5} (x-1)^{\frac{1}{2}} dx = \dfrac{1}{4} \cdot \dfrac{2}{3}(x-1)^{\frac{3}{2}} \Big|_{1}^{5} = \dfrac{1}{4} \cdot \dfrac{2}{3}(8-0)$.

17. f' changes from decreasing to increasing.

18. f' changes sign from $+$ to $-$.

19. $f(-2) = f(2) + \int_{2}^{-2} f'(x)\,dx = 1 - \dfrac{1}{2}$.

20. $2\big((-3)^2 + (-1)^2 + 1^2\big)$.

21. The slope does not depend on x; it is positive everywhere.

22. $\int_{0}^{2}(4-t^2)\,dt + \int_{2}^{3}(t^2-4)\,dt = 2\left(8 - \dfrac{8}{3}\right) + (9-12)$.

23. $\pi \int_{-1}^{1}\left(1^2 - (x^2)^2\right)dx = \pi\left(x - \dfrac{x^5}{5}\right)\Big|_{-1}^{1} = 2\pi\left(x - \dfrac{x^5}{5}\right)\Big|_{0}^{1}$.

24. $y' = ky \;\Rightarrow\; y = y(0)e^{kx}$. Here $y(0) = 6$, $k = -4$.

25. $\dfrac{d}{dx}\left[\int_{0}^{u}\cos(t^3)\,dt\right] = \cos(u^3) \cdot \dfrac{du}{dx}$, where $u = x^2$.

26. $\int_{1}^{e}\dfrac{\ln x}{x}\,dx = \int_{0}^{1} u\,du$, where $u = \ln x$.

27. $\dfrac{d}{dx}\left[x^3 + 12x(3-x)\right] = 3x^2 - 24x + 36 = 0 \;\Rightarrow\; x^2 - 8x + 12 = 0 \;\Rightarrow\; x = 2 \;\Rightarrow\; y = 1$.

28. $x^3 = 27 \;\Rightarrow\; x = 3$; $\dfrac{d}{dt}(6x^2)\Big|_{x=3} = 12x\dfrac{dx}{dt}\Big|_{x=3} = 12 \cdot 3 \cdot 2 = 72$.

29. $f'(x) = \tan^{-1}(x^3 - x) = 2$? Never: $\tan^{-1}(x) < \dfrac{\pi}{2}$.

31. $v(t)$ is positive and increasing at $t = 1$.

32. $f'(x)$ must go from negative to positive; $2^x = 2\pi$.

36. $\dfrac{g(5) - g(0)}{5} = \dfrac{5-2}{5}$.

37. $60 + \displaystyle\int_{10}^{20} g(t)\, dt \approx 60 + 22$.

39. I: MVT; II: IVT; III: IVT.

40. $\dfrac{4}{3}\pi r^3 = 36\pi \implies r = 3$; $6 \cdot 2 = \dfrac{dV}{dt} = \dfrac{4}{3}\pi \cdot 3r^2 \dfrac{dr}{dt} \implies \dfrac{dr}{dt} = \dfrac{12}{36\pi}$.

42. $\dfrac{d}{dx} g(x)\Big|_{x=1} = \dfrac{1}{2} e^{-\frac{x^2}{4}}\Big|_{x=1} \approx 0.389$.

43. `Y₁=cos(X³)-sin(X³)`. Calculate Zero. Ans→A. `fnInt(Y₁,X,0,A)`.

45. `375-fnInt(ln(1+X)+cos(X+√(X)),X,0,60)`.

SECTION II: FREE RESPONSE

1.

(a) $\dfrac{2x}{3} = \sqrt{\cos^3(x)} \Rightarrow \blacksquare\ P = (\tilde{x}, \tilde{y}) \approx (0.830687,\ 0.553791)$.

The slope of the tangent line $= \dfrac{d}{dx}\sqrt{\cos^3(x)}\bigg|_{x=\tilde{x}} \blacksquare \approx -0.910$.

(b) Area of $R = \displaystyle\int_0^{\tilde{x}}\left(\sqrt{\cos^3(x)} - \dfrac{2x}{3}\right)dx \ \blacksquare \approx 0.469$.

(c) Area of $S = \displaystyle\int_0^{\tilde{x}}\dfrac{2x}{3}\,dx + \int_{\tilde{x}}^{\frac{\pi}{2}}\sqrt{\cos^3(x)}\,dx \ \blacksquare \approx 0.230004 + 0.174653 \approx 0.405$.

(d) $\displaystyle\int_0^{\tilde{x}}\left(\pi\cos^3 x - \dfrac{4\pi x^2}{9}\right)dx = \pi\int_0^{\tilde{x}}\left(\cos^3 x - \dfrac{4x^2}{9}\right)dx \ \blacksquare \approx 0.519\pi$.

2.

(a) Average acceleration $= \dfrac{v(10)-v(0)}{10} = \dfrac{\ln(101)}{10} \ \blacksquare \approx 0.462$ inches/min^2.

$a(10) = v'(10) = \dfrac{2t}{1+t^2}\bigg|_{t=10} = \dfrac{20}{101} \ \blacksquare \approx 0.198$ inches/min^2.

(b) Distance traveled $= \displaystyle\int_0^{10}|v(t)|\,dt \ \blacksquare \approx 29.09346 \approx 29.093$ inches.

(c) The average velocity is $\dfrac{1}{10}\displaystyle\int_0^{10}v(t)\,dt \ \blacksquare \approx 2.909346$.

$v(t) = \ln(1+t^2) = 2.909346 \Rightarrow \ \blacksquare\ t \approx 4.165$ minutes.

(d) The hare's velocity at time t is $2(t-9.5)+b$ inches per minute for $t \geq 9.5$ minutes. The distance traveled between time $t = 9.5$ and $t = 10$ minutes is

$\displaystyle\int_{9.5}^{10}\big(2(t-9.5)+b\big)\,dt = \int_0^{0.5}\big(2t+b\big)\,dt = 0.25 + 0.5b$.

$0.25 + 0.5b \approx 29.09346 \Rightarrow \blacksquare\ b \approx 57.687$ inches/min.

3.

(a) On the interval $0 \le t \le 8$, $r'(t) = \dfrac{\pi}{4}\cos(\pi t/4)e^{\sin(\pi t/4)}$ is equal to 0 and changes

from positive to negative only once, when $\dfrac{\pi}{4}t = \dfrac{\pi}{2} \Rightarrow t = 2$ hours.

(b) The total amount of methane $= 20 + \displaystyle\int_0^8 r(t)\,dt$ ▪ $\approx 20 + 10.1285 \approx 30.129$ liters.

(c) The average rate of methane accumulation $= \dfrac{\displaystyle\int_0^8 r(t)\,dt + \int_8^{24}\big(r(t)-1.5\big)\,dt}{24} =$

$\dfrac{3 \cdot \displaystyle\int_0^8 r(t)\,dt - 16 \cdot (1.5)}{24} = \dfrac{3 \cdot 10.1285 - 24}{24}$ ▪ ≈ 0.266 liters/hour.

(d) The absolute maximum amount of methane in the cave over the time interval $0 \le t \le 24$ hours may occur when $r(t)-1.5$ changes sign from positive to negative $(t > 8)$. This occurs at times $t_1 = 11.4684$ and $t_2 = t_1 + 8$ hours. The amount of methane at t_1 is $30.1285 + \displaystyle\int_8^{t_1}\big(r(t)-1.5\big)\,dt$ ▪ ≈ 32.174 liters. The amount of methane at t_2 is smaller because $\displaystyle\int_{t_1}^{t_2}\big(r(t)-1.5\big)\,dt \approx -1.87 < 0$. The answer is 32.174 liters.

4.

(a) $g'(3) = f(3) = 2$ and $g''(3) = f'(3) = \dfrac{2}{5}$, the slope of the tangent line.

(b) g has a relative maximum where f changes sign from positive to negative. This occurs at $x = -6$.

(c) g has a point of inflection where f changes from increasing to decreasing or vice-versa. This occurs at $x = -3$ and at $x = 4$.

(d) $C(x) = \dfrac{\displaystyle\int_0^x f(t)\,dt}{x} = \dfrac{g(x)}{x}$. $C(3) = \dfrac{1}{3} \cdot \dfrac{11}{5}$. $C'(x) = \dfrac{x\,f(x) - g(x)}{x^2} \Rightarrow$

$C'(3) = \dfrac{3 \cdot 2 - (11/5)}{3^2}$.

5.

(a) $2[f(-2)+f(0)+f(2)]=2(3-1-4)=-4$.

(b) $\dfrac{f(0)-f(-3)}{3}=\dfrac{-1-5}{3}=\dfrac{-6}{3}$; $\dfrac{f(3)-f(0)}{3}=\dfrac{-7-(-1)}{3}=\dfrac{-6}{3}$.

(c) f changes sign between $x=0$ and $x=1$ and again between $x=1$ and $x=2$. By the Intermediate Value Theorem for continuous functions (and f is continuous by its differentiability), there must be two points a and b between 0 and 2 such that $f(a)=f(b)=0$. By the Mean Value Theorem, there must be at least one value c between a and b (and, therefore, between 0 and 2) such that $f'(c)=0$.

(d)   From Part (b) and the Mean Value Theorem, there is a value r between -3 and 0 such that $f'(r)=-2$ and there is a value s between 0 and 3 such that $f'(s)=-2$. Using the Mean Value Theorem again, there must be a value d between r and s such that $f''(d)=0$. (One could also argue that f decreases, increases, and decreases on the interval, therefore there are two points where $f'(x)=0$ and between them there must be at least one point where $f''(d)=0$.)

6.

(a)

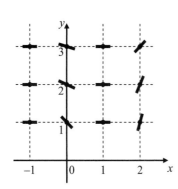

(b) $\dfrac{d^2y}{dx^2} = \dfrac{(2x)y - y'(x^2-1)}{y^2}$. For $x = 2$, $y = 1$, $y' = \dfrac{2^2-1}{1} = 3$ and so

$f''(2) = \dfrac{(2 \cdot 2)(1) - 3(3)}{1^2} = -5$.

(c) $\displaystyle\int y\,dy = \int (x^2-1)\,dx \Rightarrow \dfrac{y^2}{2} = \dfrac{x^3}{3} - x + C$. Substituting $y = 1$ and $x = 2$ to solve

for C: $\dfrac{1}{2} = \dfrac{8}{3} - 2 + C \Rightarrow C = -\dfrac{1}{6} \Rightarrow$

$\dfrac{y^2}{2} = \dfrac{x^3}{3} - x - \dfrac{1}{6} \Rightarrow y = \pm\sqrt{\dfrac{2x^3}{3} - 2x - \dfrac{1}{3}}$. $y(2) > 0 \Rightarrow y = \sqrt{\dfrac{2x^3}{3} - 2x - \dfrac{1}{3}}$.

Answers and Solutions

AB-2

SECTION I: MULTIPLE CHOICE

1.	E	11.	A	21.	C	31.	C	41.	C
2.	B	12.	A	22.	C	32.	B	42.	D
3.	B	13.	D	23.	C	33.	E	43.	A
4.	B	14.	C	24.	B	34.	B	44.	D
5.	A	15.	E	25.	E	35.	A	45.	D
6.	C	16.	E	26.	E	36.	C		
7.	E	17.	E	27.	E	37.	C		
8.	D	18.	A	28.	A	38.	D		
9.	D	19.	D	29.	B	39.	B		
10.	B	20.	D	30.	A	40.	D		

Notes:

2. $\int \dfrac{1}{\sqrt{2x+3}}\, dx = \sqrt{2x+3}$.

3. Continuity does not necessarily imply differentiability.

5. The slope should be 0 when $x = 0$ or $y = 0$.

8. f'' changes sign at $x = -2$ and $x = -\dfrac{6}{5}$.

11. Divide the numerator and the denominator by $\sqrt{x^2} = |x|$.

18. $\dfrac{1}{4}\displaystyle\int_0^4 e^{3x}\, dx$.

22. Let $u = x^2 + 1 \implies du = 2x\, dx$.

26. $\displaystyle\lim_{x \to 2} \dfrac{e^{2x} - e^4}{x - 2} = \dfrac{d}{dx} e^{2x}\bigg|_{x=2}$.

27. $\dfrac{1}{\cos^2 u} = \sec^2 u$ with $u = \sqrt{x}$.

28. Sketch the graph of $f(x)$ to see that A is the only realistic choice.

30. From the quotient rule we must have $f(x) = f'(x) \Rightarrow f(x) = e^x$.

39. $x'(t) = \dfrac{x(3) - x(0)}{3}$.

40. $f(4) = f(1) + \displaystyle\int_1^4 f'(x)\,dx$.

41. $f(c) = 0; f'(c) < 0; f''(c) > 0$.

43. $f(x) = \sin x$, $\Delta x = \dfrac{\pi - 0}{n}$.

SECTION II: FREE RESPONSE

1.

(a)   Intersection is at ▮ $x = A \approx 1.479$. Area $= \int_0^A \left[\left(4e^{-x} \right) - \left(\tan \dfrac{x}{2} \right) \right] dx$ ▮ ≈ 2.483.

(b)   Volume $= \pi \int_0^A \left[\left(4e^{-x} \right)^2 - \left(\tan \dfrac{x}{2} \right)^2 \right] dx$ ▮ ≈ 22.743.

(c)   Radius of semicircle $= \dfrac{4e^{-x} - \tan \left(\dfrac{x}{2} \right)}{2}$.

Volume $= \dfrac{\pi}{8} \int_0^A \left[\left(4e^{-x} \right) - \left(\tan \dfrac{x}{2} \right) \right]^2 dx$ ▮ ≈ 2.373.

2.

(a) f is decreasing on (0, 2) because f' is negative.

(b) f has a relative maximum at $x = 0$ because f' goes from positive to negative at $x = 0$.

(c) f is concave up on $(-3, -1)$ and on $(1, 4)$ because f' is increasing on these intervals.

(d)

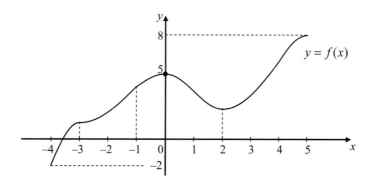

3.

(a) $\dfrac{8(221+82+22)}{24} \approx 108.333$ grams per minute.

(b) $R'(12) \approx \dfrac{R(12)-R(8)}{4} = \dfrac{82-130}{4} = -12$ grams/min^2 or

$R'(12) \approx \dfrac{R(16)-R(12)}{4} = \dfrac{39-82}{4} = -10.75$ grams/min^2 or

$R'(12) \approx \dfrac{R(16)-R(8)}{8} = \dfrac{39-130}{8} = -11.375$ grams/min^2.

(c) $\dfrac{1}{24} \displaystyle\int_0^{24} 320(0.882)^t \, dt \approx 100.972$ grams/min.

(d) $G'(12) \approx -8.905$ grams/min^2. At time t = 12 minutes, the decay rate is decreasing at the rate of 8.905 grams per minute per minute.

4.

(a) $v(t) = y'(t) = 2t - \dfrac{4}{t+1}$.

(b) $v(t) = 2t - \dfrac{4}{t+1}$. The critical numbers for $y(t)$ are $t = -2$ and $t = 1$ ($v(t) = 0$) and $t = -1$ ($v(t)$ does not exist). Only one of them is positive, $t = 1$. $v(t) < 0$ when $0 \le t < 1$ and $v(t) > 0$ for $t > 1$. Therefore, by the First Derivative Test, $y(1)$ is a local minimum, and, since there are no other critical points for $t \ge 0$, $y(1)$ is the global minimum. The lowest point is $y(1) = -4\ln 2$.

(c)   The speed of the particle is increasing when $a(t)$ and $v(t)$ have the same sign. $a(t) = v'(t) = 2 + \dfrac{4}{(t+1)^2} > 0$ for all t. $v(t) = 2t - \dfrac{4}{t+1} > 0$ for $t > 1$. The speed of the particle is increasing for $t > 1$.

(d) $y(0) = -1;\ y(1) = -4\ln 2;\ y(2) = 3 - 4\ln 3$. The total distance traveled is $|y(1) - y(0)| + |y(2) - y(1)| = |-4\ln 2 + 1| + |3 - 4\ln 3 + 4\ln 2|$.

5.

(a)

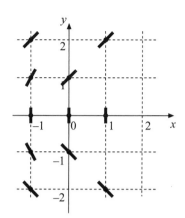

(b)   The slope field has positive slopes for all coordinates above the *x*-axis, which implies no decreasing behavior for the solution $y = f(x)$ above the *x*-axis.

(c) $y\, dy = (1 + x^2)\, dx \Rightarrow \dfrac{y^2}{2} = x + \dfrac{x^3}{3} + C$.

$y(3) = -4 \Rightarrow \dfrac{(-4)^2}{2} = 3 + \dfrac{3^3}{3} + C \Rightarrow C = -4 \Rightarrow$

$\dfrac{y^2}{2} = x + \dfrac{x^3}{3} - 4 \Rightarrow y = -\sqrt{2x + \dfrac{2x^3}{3} - 8}$ (we choose the minus sign because $y(3) = -4$).

6.

(a) $A = \pi r^2$. When $A = 100\pi \, \text{m}^2$, $r = 10$ m.

$\dfrac{dA}{dt} = 2\pi r \dfrac{dr}{dt} = 2\pi \cdot 10 \cdot 3 = 60\pi \ \text{m}^2/\text{min}.$

(b) $V = A \cdot h$. $\dfrac{dV}{dt} = 0$. When $t = T$, $h = \dfrac{1000}{\pi}$ m.

$0 = \dfrac{dV}{dt} = A\dfrac{dh}{dt} + h\dfrac{dA}{dt} \Rightarrow 100\pi \dfrac{dh}{dt} + \dfrac{1000}{\pi} 60\pi = 0 \Rightarrow \dfrac{dh}{dt} = -\dfrac{600}{\pi}$ m/min.

(c) $\dfrac{dA}{dh} = \dfrac{\dfrac{dA}{dt}}{\dfrac{dh}{dt}} = \dfrac{60\pi \ \text{m}^2/\text{min}}{-\dfrac{600}{\pi} \ \text{m/min}} = -\dfrac{\pi^2}{10}\,\text{m}^2/\text{m}.$

Answers and Solutions

AB-3

SECTION I: MULTIPLE CHOICE

1. A	11. A	21. A	31. C	41. C
2. A	12. C	22. C	32. E	42. D
3. B	13. D	23. E	33. A	43. C
4. A	14. D	24. E	34. D	44. D
5. A	15. C	25. E	35. D	45. E
6. B	16. C	26. C	36. C	
7. B	17. E	27. B	37. D	
8. E	18. B	28. A	38. C	
9. D	19. D	29. E	39. E	
10. B	20. D	30. B	40. B	

Notes:

1. $f(x) = x^2$.

2. $\int_0^1 (x - x^2) - (x^3 - x)\, dx = \int_0^1 2x - x^2 - x^3\, dx = \left(x^2 - \frac{x^3}{3} - \frac{x^4}{4} \right)\Big|_0^1 = 1 - \frac{1}{3} - \frac{1}{4} = \frac{5}{12}$.

3. $-1 < f'(1) < 0$.

5. $v(t) = 3t^2 + 18t;\ a(t) = 6t + 18;\ a = 0 \Rightarrow t = -3 \Rightarrow v = -27$.

6. For $-1 \le x < 0,\ [x] = -1$.

7. $y = \int [\sin(3t - 3) + 4]\, dt = -\frac{1}{3}\cos(3x - 3) + 4x + C.\ y(1) = 7 \Rightarrow C = \frac{10}{3}$.

8. $\tan^{-1}(\sqrt{3}) = \frac{\pi}{3};\ \frac{d}{dx}\tan^{-1}(x)\Big|_{x=\sqrt{3}} = \frac{1}{1 + x^2}\Big|_{x=\sqrt{3}} = \frac{1}{4}$.

9. I from IVT; III from MVT.

10. $f'(2) = 4 \cdot 2^3 + 2a \cdot 2 = 0$.

11. $u = 2 + \cos x;\ du = -\sin x\, dx \;\Rightarrow\; \int_{-\pi}^{0} \dfrac{\sin x}{2 + \cos x}\, dx = -\int_{1}^{3} \dfrac{du}{u} = -\ln 3$.

12. $\dfrac{dy}{dx}$ should be 0 when $x = 1$ or $y = 2$.

14. $-2xe^{-2x} + e^{-2x} = 0 \;\Rightarrow\; x = \dfrac{1}{2},\ y = \dfrac{1}{2}e^{-1}$.

15.   Area $= 2 \cdot 2\pi - \int_{0}^{\pi} 2\sin x\, dx = 4\pi + 2\cos x\big|_{0}^{\pi}$.

17. $\ln\left(x^2\right) = 2\ln x;\ e^{x+100} = e^{200} \cdot e^{x-100}$.

19. $f'(x) = 12x^3 - 12x^2;\ f''(x) = 36x^2 - 24x = 0 \;\Rightarrow\; x = 0$ or $x = \dfrac{2}{3}$.

20. $\int_{1}^{3}\left(2 - |x - 2|\right)dx = \int_{1}^{2}\left(2 + x - 2\right)dx + \int_{2}^{3}\left(2 - x + 2\right)dx$.

21. $y' = \dfrac{(x-1)(3) - (3x+1)}{(x-1)^2}$.

23. $f'(x)$ changes from positive to negative at $x = 2$. $f''(x) = -5(x-3)(3x-7)$ changes sign at $x = 3$ (and $x = \dfrac{7}{3}$ — another inflection point).

24.

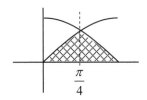

26. $2\sqrt{k} - 2 = 4$.

27. $-\sin x + \sqrt{3}\cos x = 0 \;\Rightarrow\; \tan x = \sqrt{3} \;\Rightarrow\; x = \dfrac{\pi}{3}, \dfrac{4\pi}{3}$.

29. $1 + y' \cos y = y' \dfrac{1}{y}$.

30. $y' = \cos(x - \sin x)(1 - \cos x) = 0$. Or just plot `nDeriv(sin(x-sin(x)),...)`.

32. $h_{avg} = \dfrac{1}{3} \displaystyle\int_{12}^{15} h(t)\, dt$.

33. f' changes from negative to positive at $x = -5$.

35. $\displaystyle\int_0^3 |5 - e^t|\, dt$.

37. Area $= 2 \displaystyle\int_0^A [\sin(3x) - x]\, dx$, where $\sin(3A) = A$.

38. $\dfrac{f(x+h) - f(x)}{h} = \dfrac{f(x)f(h) - f(x)}{h} = f(x)\dfrac{f(h) - 1}{h}$.

40. $\dfrac{1}{2} 0.21 + 0.28 + 0.36 + \dfrac{1}{2} 0.44$.

41. $f(x)$ should be always increasing (and it changes from concave up to concave down at $x = 1$).

42. $\dfrac{dV}{dt} = 4\pi r^2 \dfrac{dr}{dt} = 4\pi r^2 \cdot 2$, where $V = \dfrac{4}{3}\pi r^3 = 40$.

44. $y(2) = 4 + \displaystyle\int_1^2 \dfrac{\sin x}{x}\, dx$.

SECTION II: FREE RESPONSE

1.

    (a)   The curves intersect at ▨ $a \approx -0.7148$ and $b \approx 0.7148$.

            Area $= \int_a^b \left[\left(1 - x^2\right) - \sin\left(x^2\right) \right] dx$ ▨ ≈ 0.947.

    (b)   Volume $= \int_a^b \left[\left(1 - x^2 - (-1)\right)^2 - \left(\sin\left(x^2\right) - (-1)\right)^2 \right] dx$ ▨ ≈ 8.923 or 8.924.

    (c)   Volume $= \int_a^b \frac{1}{2} \left[1 - x^2 - \sin\left(x^2\right) \right]^2 dx$ ▨ ≈ 0.378.

2.

    (a) $\int_0^{45} N(t)\, dt \approx 3 \cdot 2 + 8 \cdot 4 + 15 \cdot 4 + 30 \cdot 5 + 100 \cdot 10 + 50 \cdot 10 + 22 \cdot 5 + 10 \cdot 5 = 1908$.

            $\int_0^{45} N(t)\, dt$ represents the total number of new cases of the illness over the 45-day period.

    (b)   Average number of new cases per day $= \frac{1}{45} \int_0^{45} R(t)\, dt$ ▨ ≈ 43.153.

    (c)   Solve $R(t) = 5 \Rightarrow$ ▨ $t \approx 48.274$. The model shows that the disease will be eradicated on the 49-th day.

3.

(a) $g(2) = \int_0^4 f(t)\,dt = 1 - \dfrac{\pi}{2}$. For $x > -1$, $g'(x) = 2 \cdot f(2x)$, so

$g'(2) = 2 \cdot f(4) = 2$.

(b) For $x > -1$, $g'(x) = 2 \cdot f(2x)$. Since $g'(x)$ changes from positive to

negative at $x = \dfrac{1}{2}$, $g(x)$ has a local maximum at $x = \dfrac{1}{2}$. Since $g'(x)$

changes from negative to positive at $x = \dfrac{3}{2}$, $g(x)$ has a local minimum at

$x = \dfrac{3}{2}$.

(c) $g(x)$ is concave up when $g'(x) = 2 \cdot f(2x)$ is increasing, namely the

intervals $-1 < x < -\dfrac{1}{2}$ and $1 < x < 2$.

4.

(a) $v'(t) = \dfrac{(1+t^2)3 - 3t(2t)}{(1+t^2)^2} = \dfrac{3 - 3t^2}{(1+t^2)^2}$. Since $v'(t)$ changes from positive to

negative at $t = 1$, and $t = 1$ is the only critical number, $v(1) = \dfrac{3}{2}$ is the absolute

maximum for velocity.

(b) $s(t) = \int \dfrac{3t}{1+t^2}\,dt = \dfrac{3}{2}\ln(1+t^2) + C$. Since $s(0) = 4$,

$C = 4 \Rightarrow s(t) = \dfrac{3}{2}\ln(1+t^2) + 4$.

(c) $\lim\limits_{t \to \infty} \dfrac{3t}{1+t^2} = 0$, since in this rational function the degree of the numerator is

smaller than the degree of the denominator.

(d) $\lim\limits_{t \to \infty} \dfrac{3}{2}\ln(1+t^2) + 4 = \infty$

5.

(a)

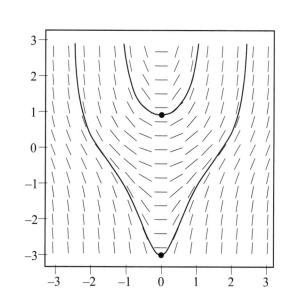

(b) $\dfrac{dy}{dx} = x\left(y^2 + 1\right) \Rightarrow$

$\dfrac{d^2 y}{dx^2} = \left(y^2 + 1\right)(1) + x\left(2 y y'\right) = \left(y^2 + 1\right) + 2xy\left(x\left(y^2 + 1\right)\right) = \left(y^2 + 1\right)\left(1 + 2x^2 y\right)$.

At a point of inflection, we must have $\dfrac{d^2 y}{dx^2} = 0 \Rightarrow \left(y^2 + 1\right)\left(1 + 2x^2 y\right) = 0$.

Since $y^2 + 1 > 0$ for all values of y, $1 + 2x^2 y = 0 \Rightarrow y = -\dfrac{1}{2x^2} < 0$.

(c) $\dfrac{dy}{dx} = x\left(y^2 + 1\right) \Rightarrow \displaystyle\int \dfrac{dy}{y^2 + 1} = \int x\, dx \Rightarrow \tan^{-1} y = \dfrac{1}{2} x^2 + C$. Since $y(0) = 1$,

$C = \tan^{-1}(1) = \dfrac{\pi}{4} \Rightarrow y = \tan\left(\dfrac{1}{2} x^2 + \dfrac{\pi}{4}\right)$.

6.

(a) $\dfrac{x^2}{16}+\dfrac{y^2}{9}=1 \;\Rightarrow\; \dfrac{2x}{16}+\dfrac{2yy'}{9}=0 \;\Rightarrow\; y'=-\dfrac{9x}{16y}\,.$

(b) $\left.\dfrac{dy}{dx}\right|_{x=\frac{12}{5},\,y=\frac{12}{5}}=-\dfrac{9}{16}\,.$ An equation for the tangent line is $y-\dfrac{12}{5}=-\dfrac{9}{16}\left(x-\dfrac{12}{5}\right).$

(c) $\dfrac{x^2}{9}-\dfrac{y^2}{16}=1 \;\Rightarrow\; \dfrac{2x}{9}-\dfrac{2yy'}{16}=0 \;\Rightarrow\; y'=\dfrac{16x}{9y}\,.$

(d)   The slope of the tangent line to the hyperbola must be negative reciprocal to the slope found in Part (b). $\dfrac{16x}{9y}=-\dfrac{1}{-\dfrac{9}{16}}=\dfrac{16}{9} \;\Rightarrow\; y=x\,.$

$\dfrac{x^2}{9}-\dfrac{y^2}{16}=1 \;\Rightarrow\; \dfrac{x^2}{9}-\dfrac{x^2}{16}=\dfrac{7}{9\cdot16}x^2=1 \;\Rightarrow\; x=y=\pm\dfrac{12}{\sqrt{7}}\,.$ The point is $\left(\dfrac{12}{\sqrt{7}},\dfrac{12}{\sqrt{7}}\right)$ or $\left(-\dfrac{12}{\sqrt{7}},-\dfrac{12}{\sqrt{7}}\right).$

Answers and Solutions

BC-1

1.	B	11.	A	21.	E	31.	A	41.	B
2.	D	12.	D	22.	C	32.	C	42.	A
3.	A	13.	B	23.	B	33.	D	43.	A
4.	E	14.	A	24.	C	34.	E	44.	A
5.	B	15.	B	25.	C	35.	B	45.	C
6.	A	16.	C	26.	D	36.	B		
7.	D	17.	B	27.	D	37.	E		
8.	C	18.	E	28.	D	38.	A		
9.	E	19.	E	29.	D	39.	C		
10.	E	20.	D	30.	C	40.	E		

Notes:

1. $f^{-1}\left(\dfrac{1}{2}\right) = \dfrac{3}{2}$; $f'\left(\dfrac{3}{2}\right) = -\dfrac{1}{2}$; $\left(f^{-1}\right)'\left(\dfrac{1}{2}\right) = \dfrac{1}{\left[f'\left(f^{-1}\left(\dfrac{1}{2}\right)\right)\right]} = \dfrac{1}{\left[-\dfrac{1}{2}\right]} = -2$.

2. $\displaystyle\int_0^2 f'(x)\,dx = f(2) - f(0)$.

3. We are looking at $V(h)$, not $V(t)$! The rate of change of V increases as h increases.

4. Brute force approach:
    $f(x) = (2x+1)e^{2x+1} \Rightarrow f'(x) = 2e^{2x+1} + 2(2x+1)e^{2x+1} \Rightarrow$
    $f''(x) = 4e^{2x+1} + 4e^{2x+1} + 4(2x+1)e^{2x+1} \Rightarrow f''\left(-\dfrac{1}{2}\right) = 8$.

    Clever approach: $f(x) = \displaystyle\sum_{n=0}^{\infty} \dfrac{(2x+1)^{n+1}}{n!} = \sum_{n=0}^{\infty}\left[2^{n+1}(n+1)\dfrac{\left(x+\dfrac{1}{2}\right)^{n+1}}{(n+1)!}\right]$; this is the

    Taylor series for f at $x = -\dfrac{1}{2} \Rightarrow f''\left(-\dfrac{1}{2}\right) =$ the coefficient in the second-degree

    term $= 2^{n+1}(n+1)$, where $n+1 = 2$.

7.   This is $\dfrac{d}{dh}\displaystyle\int_1^{1+h} e^{-t^2}\,dt = e^{-t^2}\Big|_{t=1}$.

8.   This is a logistics equation with $y(0)=\dfrac{1}{4}$.

$\dfrac{dy}{dx}>0$ for $y<\dfrac{1}{2}$, so f increases towards the

asymptote $y=\dfrac{1}{2}$.

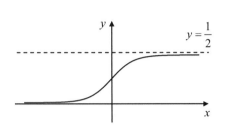

9. $r'(t)=\big(\ldots,\,-\cos t+C\big);\ r'(0)=(\ldots,\,1)\ \Rightarrow\ C=2\ \Rightarrow\ r'(\pi)=(\ldots,\,3)$.

Or: $r'(\pi)=r'(0)+\left(\displaystyle\int_0^\pi t\sin t\,dt,\ \int_0^\pi \sin t\,dt\right)=(0,\,1)+(\ldots,\,2)$.

10. $du=\dfrac{1}{2\sqrt{x}}\,dx\ \Rightarrow\ dx=2u\,du$.

11. $\big(\tan^{-1}(x)\big)'=\dfrac{1}{1+x^2};\ \big(\tan^{-1}(x)\big)''=-\dfrac{2x}{\left(1+x^2\right)^2}$.

13. $-1<3x+4<1\ \Rightarrow\ -\dfrac{1}{3}<x+\dfrac{4}{3}<\dfrac{1}{3}$.

17. $2\displaystyle\int_0^{\frac{\pi}{2}}\dfrac{4\sin 2\theta}{2}\,d\theta=-2\cos 2\theta\Big|_0^{\frac{\pi}{2}}=4$.

20. $y(1.5)=y(1)+0.5\left(1-\dfrac{1}{1}\right)=1;\ y(2)=y(1.5)+0.5\left(1-\dfrac{1.5}{y(1.5)}\right)=1-\dfrac{1}{4}=\dfrac{3}{4}$.

21. $\dfrac{d}{dx}\displaystyle\int_x^{2x}\ln t\,dt=\dfrac{d}{dx}\int_1^{2x}\ln t\,dt-\dfrac{d}{dx}\int_1^{x}\ln t\,dt=2\ln(2x)-\ln x$.

22. $\dfrac{1}{\sqrt{e}}=e^{-\frac{1}{2}}=1-\dfrac{1}{2}+\dfrac{1}{8}-\ldots=\dfrac{5}{8}-\ldots$ The error is less than the absolute value of the

next term $=\left|\left(-\dfrac{1}{2}\right)^3\dfrac{1}{3!}\right|=\dfrac{1}{48}$.

23. $\displaystyle\int_a^\infty\dfrac{10}{x^2}\,dx=\dfrac{1}{2}\int_1^\infty\dfrac{10}{x^2}\,dx=5\ \Rightarrow\ a=2$.

24. $x\left(1-x^2+x^4-x^6+\ldots\right)$.

25. $u=4-x^2$.

28. $\dfrac{dy}{dx}=\dfrac{r'\sin\theta+r\cos\theta}{r'\cos\theta-r\sin\theta}$. $\theta=\dfrac{\pi}{2}\Rightarrow\dfrac{dy}{dx}=\dfrac{r'\sin\theta}{-r\sin\theta}=-\dfrac{r'}{r}=-\dfrac{\dfrac{-3}{\theta^2}}{\dfrac{3}{\theta}}=\dfrac{1}{\theta}$.

29. Decreasing, first slower, then faster.

30. $f(2)\approx f\left(\sqrt{\pi}\right)+\left(2-\sqrt{\pi}\right)f'\left(\sqrt{\pi}\right)=\displaystyle\int_0^{\sqrt{\pi}}\cos\left(t^2\right)dt+\left(2-\sqrt{\pi}\right)\cos\pi$.

31. $\pi\displaystyle\int_0^{\frac{\pi}{2}}\sin^2 y\,dy$.

33. $\dfrac{dP}{dt}=k\sqrt{t}\Rightarrow P-P_0=\dfrac{2}{3}kt^{\frac{3}{2}}$. $\dfrac{2}{3}k\left(1^{\frac{3}{2}}\right)=0.05\Rightarrow\dfrac{2}{3}k\left(12^{\frac{3}{2}}\right)\approx 2.08$.

     $5 + $2.08 = 7.08.

34. $y=x^{\frac{1}{2}}\Rightarrow\dfrac{d^2y}{dx^2}=\dfrac{1}{2}\left(-\dfrac{1}{2}\right)x^{-\frac{3}{2}}=-\dfrac{1}{4}\left(f(t)\right)^{-3}$.

35. $\dfrac{dx}{dt}=e^t;\ \dfrac{dy}{dt}=\left(\ln 2\right)2^t\Rightarrow\left.\dfrac{dy}{dx}\right|_{t=0}=\ln 2$.

37. $\sqrt{\left(\cos 3\right)^2+\left(2\cos 6\right)^2}$.

39. $d=\sqrt{\left(x-5\right)^2+\left(x^2+1-6\right)^2}$. Graph and find the minimum.

40. $1+y'=\dfrac{1}{1+\left(xy\right)^2}\left(xy'+y\right)$.

42. $\left| \int_0^4 \left| 4 - t^2 \right| dt - \int_0^4 \left| t^3 - 4t \right| dt \right|$.

43. This is a geometric series.

44. $\dfrac{y-1}{x-4} = \dfrac{x^2-1}{x-4} = -\dfrac{1}{2x} \Rightarrow x \approx 1.392$.

SECTION II: FREE RESPONSE

1.

(a) $g'(0) = f(0) = \dfrac{4\cos 0 - 0}{2 + \sqrt[3]{0^2}} = 2$. The equation of the tangent line at $x = 0$ is

$y - 1 = 2x$.

(b) g is continuous on $[0, 1]$ since its derivative exists for all x.

$g'(x) = \dfrac{4\cos x - x}{2 + \sqrt[3]{x^2}} > 0$ for all x in the closed interval $[0, 1]$. Therefore, g is

increasing on $[0, 1]$ and its absolute maximum occurs at the right endpoint,

$x = 1$. By the Fundamental Theorem, $g(x) = 1 + \displaystyle\int_0^x \dfrac{4\cos t - t}{2 + \sqrt[3]{t^2}}\,dt$, and so

$g(1) = 1 + \displaystyle\int_0^1 \dfrac{4\cos t - t}{2 + \sqrt[3]{t^2}}\,dt \approx 2.146$.

(c) The graph of h has a point of inflection where $h''(x) = f(x)$ changes sign.
This happens at $x \approx -3.595$, $x \approx -2.133$ and $x \approx 1.252$.

(d) By the Fundamental Theorem, $h''(x) = f(x)$, and $f(x) < 0$ for all x between 2
and 3. Therefore, $h'(x)$ is decreasing on $[2, 3]$, and $h'(2) > h'(3)$.

2.

(a) Number of people $= \dfrac{1 + 18}{2} 90 + \dfrac{18 + 81}{2} 50 + \dfrac{81 + 60}{2} 30 + \dfrac{60 + 44}{2} 10$.

(b) Average rate $= \dfrac{81 \cdot 20 + 74 \cdot 20 + 60 \cdot 10}{50} = 74$.

(c) Number of people $= \displaystyle\int_0^{180} R(t)\,dt \approx 5620.890 \approx 5621$.

(d) Average rate $= \dfrac{1}{170 - 120} \displaystyle\int_{120}^{170} R(t)\,dt \approx 73.895$.

3.

(a) $x'(t) = 2t$; $y'(t) = 3t^2 - 3$. Speed $= \sqrt{4t^2 + (3t^2 - 3)^2} = 5 \Rightarrow$ ▮ $t \approx 1.5236$ sec.

(b) Distance $r = \sqrt{(t^2 + 1 - 1)^2 + (t^3 - 3t)^2} = \sqrt{t^4 + (t^3 - 3t)^2}$.

$\left. \dfrac{dr}{dt} \right|_{t=2}$ ▮ ≈ 7.603 cm/sec.

(c) Distance traveled $= \displaystyle\int_0^3 \sqrt{4t^2 + (3t^2 - 3)^2}\, dt$ ▮ ≈ 24.385 cm.

4.

(a) $F(1) = 0$; $F'(1) = \sqrt{1^2 + 3 \cdot 1} = 2$;

$F''(x) = \dfrac{d}{dx}\sqrt{x^2 + 3x} = \dfrac{2x + 3}{2\sqrt{x^2 + 3x}} \Rightarrow F''(1) = \dfrac{5}{4}$.

$P(x) = 2(x - 1) + \dfrac{5}{4}\dfrac{(x - 1)^2}{2}$.

(b) $\displaystyle\int_1^0 \sqrt{t^2 + 3t}\, dt \approx P(0) = -2 + \dfrac{5}{8}$.

(c) $\displaystyle\int_0^2 \sqrt{t^2 + 3t}\, dt = \int_1^2 \sqrt{t^2 + 3t}\, dt - \int_1^0 \sqrt{t^2 + 3t}\, dt \approx P(2) - P(0) =$

$\left(2 + \dfrac{5}{8}\right) - \left(-2 + \dfrac{5}{8}\right) = 4$.

5.

(a)   Area $= \int_0^\infty e^{-x}\, dx = \lim_{b\to\infty} \int_0^b e^{-x}\, dx = \lim_{b\to\infty}\left[-e^{-x}\Big|_0^b\right] = \lim_{b\to\infty}\left[-\left(e^{-b}-1\right)\right] = 1$.

(b) $f'(c) = -e^{-c} \Rightarrow y - e^{-c} = -e^{-c}(x-c)$. Intercepts:
$y = (c+1)e^{-c}$ and $x = c+1$.

The area of triangle $A(c) = \dfrac{1}{2}e^{-c}(c+1)^2$.

(c) $\dfrac{dA}{dc} = -\dfrac{1}{2}e^{-c}(c+1)^2 + \dfrac{1}{2}e^{-c}2(c+1) = \dfrac{1}{2}e^{-c}(c+1)\left[2-(c+1)\right]$.

$\dfrac{dA}{dc} > 0$ when $0 < (c+1) < 2$ and $\dfrac{dA}{dc} < 0$ when $2 < (c+1) \Rightarrow A$ reaches a relative maximum at $c+1 = 2 \Rightarrow c = 1$. Since $c = 1$ is the only critical number for $c > 0$, this is the absolute maximum. The maximum area of triangle is $2e^{-1} = \dfrac{2}{e}$. The minimum remaining area is $1 - \dfrac{2}{e}$.

6.

(a)

$$\left.\begin{array}{l} \dfrac{d}{dx}\left(xy+y^2\right)=y+x\dfrac{dy}{dx}+2y\dfrac{dy}{dx} \\[3mm] \dfrac{d}{dx}\left(x^2-5\right)=2x \end{array}\right\} \Rightarrow 2x=y+\left(x+2y\right)\dfrac{dy}{dx} \Rightarrow \dfrac{dy}{dx}=\dfrac{2x-y}{x+2y}.$$

(b) $x+2y=0 \Rightarrow y=-\dfrac{x}{2}$. Substituting into the equation we get:

$$x\left(-\dfrac{x}{2}\right)+\dfrac{x^2}{4}=x^2-5 \Rightarrow -2x^2+x^2=4x^2-20 \Rightarrow x^2=4 \Rightarrow x=\pm 2.$$

$\left(2,-1\right)$ and $\left(-2,1\right)$.

(c)

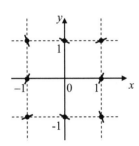

(d) $2x-y=0 \Rightarrow y=2x$.

Answers and Solutions

BC-2

SECTION I: MULTIPLE CHOICE

1.	B	11.	E	21.	D	31.	D	41.	A
2.	B	12.	A	22.	C	32.	A	42.	D
3.	C	13.	D	23.	B	33.	E	43.	C
4.	A	14.	B	24.	A	34.	C	44.	B
5.	B	15.	E	25.	D	35.	B	45.	C
6.	C	16.	A	26.	A	36.	E		
7.	A	17.	C	27.	D	37.	B		
8.	E	18.	D	28.	E	38.	D		
9.	C	19.	E	29.	C	39.	B		
10.	D	20.	E	30.	E	40.	C		

Notes:

2. We must have $\dfrac{dx}{dt} = 3t^2 - 12t = 0$ while $\dfrac{dy}{dt} = 2t \neq 0$. When both are 0, we must investigate $\dfrac{dy}{dx}$ using limits. Here, $\lim\limits_{t \to 0} \dfrac{dy}{dx} = -\dfrac{1}{6}$.

3. $u = \dfrac{2}{x} \Rightarrow u(2) = 1,\ \lim\limits_{x \to \infty} u(x) = 0,\ du = -\dfrac{2}{x^2}\, dx \Rightarrow \displaystyle\int_2^\infty \dfrac{\cos\left(\dfrac{2}{x}\right)}{x^2}\, dx = -\dfrac{1}{2} \int_1^0 \cos u\, du$.

4. $\sin 2x + \cos 2y = x - y \Rightarrow 2\cos 2x - 2\sin 2y \cdot y' = 1 - y'$.

5. Using l'Hôpital's Rule, $\lim\limits_{w \to 0} \dfrac{\ln\left(\dfrac{2+w}{2}\right)}{w} = \lim\limits_{w \to 0} \dfrac{\dfrac{2}{2+w} \cdot \dfrac{1}{2}}{1}$.

6. $g''(x) = -4\sin 2x + 2 = 4\left(\dfrac{1}{2} - \sin 2x\right)$. On the given interval, this changes sign only at $\dfrac{\pi}{12}$.

383

7. $f(x) = |x - 1.5|$ has a corner at $x = 1.5$ and $f(x) = \dfrac{\pi}{x-1}$ has an asymptote at $x = 1$.

8. $\dfrac{dy}{dx}$ depends upon both x and y and is greater than or equal to 0 throughout. (Also has symmetry about both axes.)

10.   Area $= 2\displaystyle\int_0^{\frac{\pi}{3}} (2\cos x - 1)\, dx = 2(2\sin x - x)\Big|_0^{\frac{\pi}{3}} = 2\left(\sqrt{3} - \dfrac{\pi}{3}\right).$

11. $f(x) = -\sin x \ \Rightarrow \ f\left(\dfrac{2\pi}{3}\right) = -\dfrac{\sqrt{3}}{2},\ f'\left(\dfrac{2\pi}{3}\right) = -\cos\left(\dfrac{2\pi}{3}\right) = \dfrac{1}{2}.$

12. $\dfrac{\displaystyle\int_{\frac{\pi}{4}}^{\frac{\pi}{3}} (\tan x - \sin x)\, dx}{\dfrac{\pi}{3} - \dfrac{\pi}{4}} = \dfrac{12}{\pi}\left(-\ln|\cos x| + \cos x\right)\Big|_{\frac{\pi}{4}}^{\frac{\pi}{3}} = \dfrac{12}{\pi}\left(\ln\sqrt{2} + \dfrac{1 - \sqrt{2}}{2}\right).$

13. $h'(1.1) > 0.$

14.   This is $e^x = 1 + x + \dfrac{x^2}{2} + \ldots + \dfrac{x^n}{n!} + \ldots$ for $x = -\pi^2.$

15.   Solve to get $y = -\dfrac{x^2}{4}$ or notice that $\dfrac{dy}{dx}$ must be positive for $x < 0$ and negative for $x > 0$.

16. $g(x) = \displaystyle\lim_{h \to 0} \dfrac{\tan(x+h) - \tan x}{h} = \dfrac{d}{dx}\tan x = \sec^2 x.$

$g'(x) = \dfrac{d}{dx}\sec^2 x = 2\sec x \cdot \tan x \cdot \sec x \ \Rightarrow \ g'\left(\dfrac{\pi}{3}\right) = 2 \cdot 2 \cdot \sqrt{3} \cdot 2.$

17.   Area $= \displaystyle\int_0^{\pi}\left(\dfrac{r_2^2}{2} - \dfrac{r_1^2}{2}\right) d\theta = \dfrac{4}{9} \cdot \dfrac{\theta^3}{3}\Big|_0^{\pi}.$

18. $E(0.5) = f(0) + 0.5 f'(0) = 3 + 0.5(-2) = 2.$

$E(1) = E(0.5) + 0.5 f'(0.5) = 2 + 0.5\left(-\dfrac{3}{2}\right).$

19. $3y^2 dy = (2x+1)dx \Rightarrow y^3 = x^2 + x + C$. $y(1) = 2 \Rightarrow C = 6 \Rightarrow [y(3)]^3 = 18$.

20. The domain is all real numbers. $f(4) = \ln 4 - 1 = \lim_{x \to 4^-} f(x) = \lim_{x \to 4^+} f(x)$.

$\lim_{x \to 4^-} f'(x) = -\frac{1}{4}$; $\lim_{x \to 4^+} f'(x) = \lim_{x \to 4^+} \left(\frac{1}{x} - \frac{x}{8} \right) = -\frac{1}{4}$. An integral of any continuous function on a bounded closed interval exists.

21. $\int \dfrac{dx}{5x - x^2} = \int \dfrac{dx}{x(5-x)} = \dfrac{1}{5} \int \dfrac{1}{x} + \dfrac{1}{5-x} \, dx$.

23. By parts: $\int 2 \arctan x \, dx = 2x \arctan x - \int \dfrac{2x}{1+x^2} \, dx = 2x \arctan x - \ln\left(1 + x^2\right) + C$.

24. $g(x) = \int_0^2 f(t) \, dt$ — a constant.

25. $\lim\limits_{x \to \infty} \dfrac{2x}{x+1} \leq \lim\limits_{x \to \infty} \dfrac{\arctan x + 2x}{x + e^{-x}} \leq \lim\limits_{x \to \infty} \dfrac{\dfrac{\pi}{2} + 2x}{x}$. Or, using l'Hôpital's Rule,

$\lim\limits_{x \to \infty} \dfrac{\arctan x + 2x}{x + e^{-x}} = \lim\limits_{x \to \infty} \dfrac{\dfrac{1}{1+x^2} + 2}{1 - e^{-x}}$.

26. $\sin \pi = 0$; $\sin'(\pi) = \cos \pi = -1$; $\sin''(\pi) = -\sin \pi = 0$; $\sin'''(\pi) = -\cos \pi = 1$.

$\sin(3) = 0 + (-1)(3 - \pi) + 0 + \dfrac{1}{3!}(3 - \pi)^3$.

27. $\vec{v}(t) = (8 \cos 4t, \, -12 \sin 4t) \Rightarrow \vec{a}(t) = (-32 \sin 4t, \, -48 \cos 4t)$.

$\sin\left(4 \cdot \dfrac{\pi}{4}\right) = 0$, $\cos\left(4 \cdot \dfrac{\pi}{4}\right) = -1 \Rightarrow \left| a\left(\dfrac{\pi}{4}\right) \right| = \sqrt{0^2 + 48^2} = 48$.

28. This is a logistic curve with the asymptote $y = 0.06$ as $t \to \infty$. $y(0)$ is above the asymptote, so $\dfrac{dy}{dt}$ always remains negative.

29. $g(0) < 0$; $g(1) = 0$; $g(2) - g(1) < g(1) - g(0)$.

30. f is decreasing, so g has a local maximum where $x^3 + e^{-x^2}$ has a local minimum.

31. $3x^2 - e^x = \dfrac{\left(8 - e^2\right) - \left(0 - 1\right)}{2 - 0} \Rightarrow$ ▯ $x \approx 1.149$.

33.   Using l'Hôpital's Rule, $\displaystyle\lim_{x \to 0} \dfrac{3\cos 3x + a + 3bx^2}{3x^2} = 0 \Rightarrow a = -3\cos 0 = -3$. Using
       l'Hôpital's Rule again, twice,
       $$\lim_{x \to 0} \dfrac{-9\sin 3x + 6bx}{6x} = \lim_{x \to 0} \dfrac{-27\cos 3x + 6b}{6} = 0 \Rightarrow b = \dfrac{27}{6}\cos 0 = \dfrac{9}{2}$$

35. $\displaystyle\lim_{n \to \infty} \left| \dfrac{a_{n+1}}{a_n} \right| = \lim_{n \to \infty} \left| \dfrac{n+1}{n} \dfrac{3^n}{3^{n+1}} \dfrac{(6x-5)^{n+1}}{(6x-5)^n} \right| = \left| \dfrac{6x-5}{3} \right| \Rightarrow \left| x - \dfrac{5}{6} \right| < \dfrac{1}{2}$.

36. $2(130+118+108+120)$.

37. $u = \ln x \Rightarrow du = \dfrac{1}{x}dx \Rightarrow \displaystyle\int_2^\infty \dfrac{dx}{x(\ln x)^p} = \int_{\ln 2}^\infty u^{-p}\,du$ — converges for p > 1.

38. $m(3) = m(2) + \displaystyle\int_2^3 \cos\left(1 - x^2\right)dx$.

39. $\displaystyle\int_0^{2\pi} \sqrt{\left(\dfrac{dx}{dt}\right)^2 + \left(\dfrac{dy}{dt}\right)^2}\,dt$.

40. $\displaystyle\int_3^5 G(x)\,dx = F(5) - F(3)$.

41. $f(1) = 0 \Rightarrow \displaystyle\int_2^1 g(t)\,dt = 0$ and $f(3) - f(2) > f(4) - f(3) \Rightarrow \displaystyle\int_2^3 g(t)\,dt > \int_3^4 g(t)\,dt$.

42. $\displaystyle\int_0^1 \left(3^{-x^2} - \dfrac{1}{3}\right)\left(3^{-x^2} - \dfrac{1}{3} + 2\right)dx$.

43.   Plot $y = f''(x)$ in the window $-6 \le x \le 6; -1 \le y \le 1$ and count the changes of sign.

44. $f(-1) = 2;\ h'(2) = \dfrac{1}{f'(-1)} = \dfrac{1}{-\dfrac{1}{2}}$.

45. $h'(x)$ is positive and decreasing everywhere.

SECTION II: FREE RESPONSE

1.

(a) $g(6) = \int_4^6 h(t)\,dt = 2$ and $g'6) = h(6) = 2$. An equation for the tangent line is $y - 2 = 2(x - 6)$.

(b) $g(-3)$ is larger. $g'(x) = h(x)$ and $h(x) \geq 0$ on [–4, –3]. Therefore, g is increasing on that interval, and $g(-3) > g(-4)$.

(c) The candidates are $x = -7$, an endpoint maximum (since $g'(x) = h(x) < 0$ to the right of -7), and $x = 7$, where $g'(x)$ changes sign from positive to negative. $g(-7) = \int_4^{-7} h(t)\,dt < 0$ because there is more area under the quarter circles from 4 to -4 (which contributes negatively to the integral) than area above the linear piece from -4 to -7 (which contributes positively to the integral). (The former is more than 4 sq. units; the latter is less than 3 sq. units.) Therefore, $g(7) = \int_4^7 h(t)\,dt = 2 + 1 = 3$ is the absolute maximum.

(d) Points of inflection on the graph of $y = g(x)$ occur where $g''(x)$ changes sign. $g''(x) = h'(x)$, and $h'(x)$ changes sign from positive to negative at $x = 0$ and $x = 6$, and from negative to positive at $x = 4$. Therefore, the points of inflection occur at $x = 0$, $x = 4$, and $x = 6$.

2.

(a) $y\left(\dfrac{1}{3}\right) \approx -0.5 + 2\cos\left(e^0\right)\cdot\dfrac{1}{3}$ ▮ $\approx -.1397985$

 $y\left(\dfrac{2}{3}\right) \approx -.1397985 + 2\cos\left(e^{1/3}\right)\cdot\dfrac{1}{3}$ ▮ $\approx -.0236056$.

 $y(1) \approx -.0236056 + 2\cos\left(e^{2/3}\right)\cdot\dfrac{1}{3}$ ▮ $\approx -.269$.

(b)

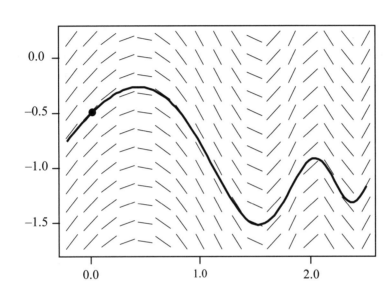

(c) $y(x) = -0.5 + \displaystyle\int_0^x 2\cos\left(e^t\right)dt$; $y(1) = -0.5 + \displaystyle\int_0^1 2\cos\left(e^t\right)dt$ ▮ $\approx -.747$.

(d) The Euler's method uses line segments, which have constant slope, to approximate the solution curve. $\dfrac{d^2y}{dx^2} = -2e^x\sin\left(e^x\right) < 0$ for all x between 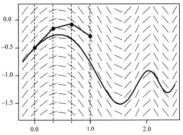 0 and 1, so the actual slope of the solution curve is decreasing. Therefore, the points obtained by the Euler's Method are above the curve and the approximation in Part (a) is greater than the answer in Part (c).

3.

(a) Average =

$$\frac{1}{16}\int_{-8}^{8}\left(ax^{2}\right)^{\frac{1}{3}}dx=\frac{a^{\frac{1}{3}}}{16}\int_{-8}^{8}x^{\frac{2}{3}}dx=\frac{a^{\frac{1}{3}}}{16}\cdot\frac{3}{5}x^{\frac{5}{3}}\Bigg|_{-8}^{8}=\frac{a^{\frac{1}{3}}}{16}\cdot\frac{3}{5}\cdot64=12\ \Rightarrow\ a=125.$$

(b) $5\tilde{x}^{\frac{2}{3}}=12\Rightarrow\blacksquare\ \tilde{x}\approx3.718064$.

Volume = $\pi\int_{-\tilde{x}}^{\tilde{x}}\left(12-\left(125x^{2}\right)^{\frac{1}{3}}\right)^{2}dx=2\pi\int_{0}^{\tilde{x}}\left(12-5x^{\frac{2}{3}}\right)^{2}dx\ \blacksquare\approx2\pi\cdot122.377$.

4.

(a) $x^{3}=kx\ \Rightarrow\ kx-x^{3}=0\ \Rightarrow\ x\left(k-x^{2}\right)=0\ \Rightarrow$ the graphs of f and g intersect at $x=0$ and $x=\sqrt{k}$. The area is

$$\int_{0}^{\sqrt{k}}\left(kx-x^{3}\right)dx=\left(\frac{kx^{2}}{2}-\frac{x^{4}}{4}\right)\Bigg|_{0}^{\sqrt{k}}=\frac{k^{2}}{2}-\frac{k^{2}}{4}=\frac{k^{2}}{4}.$$

(b) $A=\dfrac{k^{2}}{4}\ \Rightarrow\ \dfrac{dA}{dt}=\dfrac{k}{2}\dfrac{dk}{dt}$. Since $\dfrac{dk}{dt}=9$ units/sec , at $k=4$, we have

$$\frac{dA}{dt}\Bigg|_{k=4}=\frac{4}{2}\cdot9=18\ \text{units}^{2}/\text{sec} .$$

(c) $V=\pi\int_{0}^{\sqrt{k}}\left(kx\right)^{2}-\left(x^{3}\right)^{2}dx=\pi\left(\dfrac{k^{2}x^{3}}{3}-\dfrac{x^{7}}{7}\right)\Bigg|_{0}^{\sqrt{k}}=$

$$\pi\left(\frac{k^{2}k^{3/2}}{3}-\frac{k^{7/2}}{7}\right)=\frac{\pi}{3}k^{7/2}-\frac{\pi}{7}k^{7/2}=\frac{4}{21}\pi k^{\frac{7}{2}} .$$

(d) $V=\dfrac{4}{21}\pi k^{\frac{7}{2}}\ \Rightarrow\ \dfrac{dV}{dt}=\dfrac{4}{21}\cdot\dfrac{7}{2}\pi k^{\frac{5}{2}}\dfrac{dk}{dt}=\dfrac{2}{3}\pi k^{\frac{5}{2}}\dfrac{dk}{dt}$. At $k=4$, we have

$$\frac{dV}{dt}\Bigg|_{k=4}=\frac{2}{3}\pi\cdot32\cdot9=192\pi\ \text{units}^{3}/\text{sec} .$$

5.

(a) Using the ratio test,

$$\lim_{n\to\infty}\left|\frac{a_{n+1}}{a_n}\right| = \lim_{n\to\infty}\left|\frac{\dfrac{(n+1)x^n}{3^{n+1}}}{\dfrac{nx^{n-1}}{3^n}}\right| = \lim_{n\to\infty}\left|\frac{(n+1)x^n}{3^{n+1}}\cdot\frac{3^n}{nx^{n-1}}\right| = \lim_{n\to\infty}\left|\frac{(n+1)}{n}\cdot\frac{x}{3}\right| = \left|\frac{x}{3}\right| < 1 \Leftrightarrow$$

$|x| < 3$. The radius of convergence is 3.

(b) Since $g'(x) = f(x)$, $g(x) = \int f(x)dx = \int \sum_{n=1}^{\infty}\frac{nx^{n-1}}{3^n}dx = \sum_{n=1}^{\infty}\frac{x^n}{3^n} + C$.

$$g(1) = 3 \Rightarrow \sum_{n=1}^{\infty}\frac{1}{3^n} + C = \frac{\dfrac{1}{3}}{1-\dfrac{1}{3}} + C = \frac{1}{2} + C = 3 \Rightarrow C = \frac{5}{2} \Rightarrow$$

$$g(x) = \frac{5}{2} + \frac{x}{3} + \frac{x^2}{9} + \dots + \frac{x^n}{3^n} + \dots$$

(c) g is a geometric series with a common ratio of $\dfrac{x}{3}$. If $x = 3$, the ratio is 1 and the series diverges. So $g(3)$ cannot be evaluated.

6.

(a) $\dfrac{dx}{dt} = -3\sin\left(\dfrac{\pi t}{12}\right)\cdot\dfrac{\pi}{12} = -\dfrac{\pi}{4}\sin\left(\dfrac{\pi t}{12}\right)$. $\dfrac{dy}{dt} = 4\cos\left(\dfrac{\pi t}{12}\right)\cdot\dfrac{\pi}{12} = \dfrac{\pi}{3}\cos\left(\dfrac{\pi t}{12}\right)$. The

velocity vector of the laser is $\vec{v}(t) = \left(-\dfrac{\pi}{4}\sin\left(\dfrac{\pi t}{12}\right),\ \dfrac{\pi}{3}\cos\left(\dfrac{\pi t}{12}\right)\right)$. Slope of the

tangent line $= \dfrac{dy}{dx} = \dfrac{dy}{dt}\Big/\dfrac{dx}{dt} = \dfrac{\dfrac{\pi}{3}\cos\left(\dfrac{\pi t}{12}\right)}{-\dfrac{\pi}{4}\sin\left(\dfrac{\pi t}{12}\right)} = -\dfrac{4}{3}\cot\left(\dfrac{\pi t}{12}\right)$.

(b) Slope of the line perpendicular to the tangent is $-\dfrac{1}{-\dfrac{4}{3}\cot\left(\dfrac{\pi t}{12}\right)} = \dfrac{3}{4}\tan\left(\dfrac{\pi t}{12}\right)$.

The equation for the laser beam line at $\vec{L}(t)$ is

$y - 4\sin\left(\dfrac{\pi t}{12}\right) = \dfrac{3}{4}\tan\left(\dfrac{\pi t}{12}\right)\left(x - 3\cos\left(\dfrac{\pi t}{12}\right)\right)$, when $\cos\dfrac{\pi t}{12} \neq 0$, and $x = 0$,

when $\cos\dfrac{\pi t}{12} = 0$.

(c) The point P, when defined, is the y-intercept of the laser beam line.

$h(t) - 4\sin\left(\dfrac{\pi t}{12}\right) = \dfrac{3}{4}\tan\left(\dfrac{\pi t}{12}\right)\left(0 - 3\cos\left(\dfrac{\pi t}{12}\right)\right) = -\dfrac{9}{4}\sin\left(\dfrac{\pi t}{12}\right) \Rightarrow$

$h(t) = 4\sin\left(\dfrac{\pi t}{12}\right) - \dfrac{9}{4}\sin\left(\dfrac{\pi t}{12}\right) = \dfrac{7}{4}\sin\left(\dfrac{\pi t}{12}\right)$. $h'(2) = \dfrac{7}{4}\cos\left(\dfrac{\pi t}{12}\right)\cdot\dfrac{\pi}{12}\Big|_{t=2} > 0 \Rightarrow$

at $t = 2$, P is moving up.

(d) For $h(t)$ to be continuous on [0, 24], we must set $h(6) = \lim\limits_{t\to 6} h(t)$ and

$h(18) = \lim\limits_{t\to 18} h(t)$. Since $\dfrac{7}{4}\sin\left(\dfrac{\pi t}{12}\right)$ is continuous for all t, we must set

$h(6) = \dfrac{7}{4}\sin\left(\dfrac{\pi\cdot 6}{12}\right) = \dfrac{7}{4}$ and $h(18) = \dfrac{7}{4}\sin\left(\dfrac{\pi\cdot 18}{12}\right) = -\dfrac{7}{4}$. P covers the interval

$-\dfrac{7}{4} \leq y \leq \dfrac{7}{4}$ twice from $t = 0$ to $t = 24$, so the total distance traveled by P over

that time interval is $2\left[\dfrac{7}{4} - \left(-\dfrac{7}{4}\right)\right] = 7$.

Index

Other Math and Computer Science
Titles from
Skylight Publishing

Be Prepared for the AP Computer Science Exam in Java, 3rd Edition
ISBN 978-0-9727055-6-1

800 Questions in Calculus
ISBN 978-0-9727055-4-7

Solutions to 800 Questions in Calculus
Part 978-0-9727055-C-D

Calculus Calculator Labs Student Pack
Part 978-0-9727055-S-L

Calculus Calculator Labs Teacher Pack
Part 978-0-9727055-T-L

Mathematics for the Digital Age and Programming in Python
ISBN 978-0-9727055-8-5

Java Methods A&AB: Object-Oriented Programming and Data Structures
ISBN 978-0-9727055-7-8

175 Multiple-Choice Questions in Java
ISBN 978-0-9727055-1-6

100 Multiple-Choice Questions in C++
ISBN 978-0-9654853-0-2

http://www.skylit.com
sales@skylit.com
Toll free: 888-476-1940
Fax: 978-475-1431

Skylight Publishing, 9 Bartlet Street, Suite 70, Andover, MA 01810